Lecture Notes in Computer Science 9544

Commenced Publication in 1973
Founding and Former Series Editors:
Gerhard Goos, Juris Hartmanis, and Jan van Leeuwen

More information about this series at http://www.springer.com/series/7409

Guilin Qi · Kouji Kozaki
Jeff Z. Pan · Siwei Yu (Eds.)

Semantic Technology

5th Joint International Conference, JIST 2015
Yichang, China, November 11–13, 2015
Revised Selected Papers

 Springer

Editors

Guilin Qi
Southeast University
Nanjing
China

Jeff Z. Pan
The University of Aberdeen
Aberdeen
UK

Kouji Kozaki
Osaka University
Ibaraki
Japan

Siwei Yu
Zhongnan Hospital of Wuhan University
Wuhan
China

ISSN 0302-9743   ISSN 1611-3349   (electronic)
Lecture Notes in Computer Science
ISBN 978-3-319-31675-8   ISBN 978-3-319-31676-5   (eBook)
DOI 10.1007/978-3-319-31676-5

Library of Congress Control Number: 2016933481

LNCS Sublibrary: SL3 – Information Systems and Applications, incl. Internet/Web, and HCI

Printed on acid-free paper

his Springer imprint is published by Springer Nature
The registered company is Springer International Publishing AG Switzerland

Preface

This volume contains the papers presented at JIST 2015: The 5th Joint International Semantic Technology Conference held during November 11–13, 2015, in Yichang, China. JIST 2015 was co-hosted by Wuhan University of Science and Technology and China Three Gorges University, Computer Technology Center (NECTEC). JIST is a regional federation of semantic technology-related conferences. It attracts many participants from mainly the Asia region and often Europe and the USA. The mission of JIST is to bring together researchers in semantic technology research and other areas of semantic-related technologies to present their innovative research results or novel applications of semantic technologies.

The main topics of the conference include semantic technologies such as the Semantic Web, social web, ontology, reasoning, linked data, knowledge graph, among others. The theme of the JIST 2015 conference was "Big Data and Social Media." JIST 2015 consisted of main technical tracks including two keynotes and two invited talks, a regular technical paper track (full and short papers), an in-use track, a poster and demo session, a workshop, and a tutorial. There were a total of 43 submissions for the regular paper track and six papers for the in-use track from 15 countries, which included Asian, European, African, North American, and South American countries. Most papers were reviewed by three reviewers and the results were rigorously discussed by the program chairs. In all, 14 full papers (33 %) and seven short paper were accepted in the regular track. While two full papers (33 %) and one short paper were accepted in the in-use track.

The paper topics are divided into eight categories: ontology and reasoning, linked data, learning and discovery, RDF and query, knowledge graph, knowledge integration, query and recommendation, and applications of semantic technologies.

We would like to thank the JIST Steering Committee, Organizing Committee, and Program Committee for their significant contributions. We would also like to especially thank the co-hosts and sponsors for their support in making JIST 2015 a successful and memorable event. Finally, we would like to express our appreciation to all speakers and participants of JIST 2015. This book is an outcome of their contributions.

November 2015

Guilin Qi
Kouji Kozaki
Jeff Z. Pan
Siwei Yu

Organization

Program Committee

Guilin Qi	Southeast University, China
Haofen Wang	East China University of Science and Technology, China
Seungwoo Lee	KISTI, South Korea
Ekawit Nantajeewarawat	Sirindhorn International Institute of Technology, Thammasat University, Thailand
Guohui Xiao	KRDB Research Centre, Free University of Bozen-Bolzano, Italy
Jason Jung	Chung-Ang University, South Korea
Nopphadol Chalortham	Silpakorn University, Thailand
Marco Ronchetti	Università di Trento, Italy
Gong Cheng	Nanjing University, China
Yasunori Yamamoto	Database Center for Life Science, The University of Tokyo, Japan
Panos Alexopoulos	Expert System Iberia, Madrid, Spain
Paolo Bouquet	University of Trento, Italy
Jeff Z. Pan	University of Aberdeen, UK
Takahira Yamaguchi	Keio University, Japan
Yuting Zhao	University of Aberdeen, UK
Dejing Dou	University of Oregon, USA
Honghan Wu	Nanjing University of Information Science and Technology, China
Peng Wang	Southeast University, China
Naoki Fukuta	Shizuoka University, Japan
Paola Di Maio	ISTCS.org/IIT Mandi, India
Giorgos Stoilos	National Technical University of Athens (NTUA), Greece
Seiji Koide	National Institute of Informatics, Tokyo, Japan
Anni-Yasmin Turhan	University of Oxford, UK
Ralf Möller	Universität zu Lübeck, Germany
Riichiro Mizoguchi	Japan Advanced Institute of Science and Technology, Japan
Wei Hu	Nanjing University, China
Amal Zouaq	Royal Military College of Canada, Canada
Eero Hyvönen	Aalto University, Finland
Sungkook Han	Wonkwang University, South Korea
Volker Haarslev	Concordia University, Canada
Pyung Kim	Jeonju National University of Education, South Korea

Krzysztof Wecel	Poznan University of Economics, Poland
Vahid Jalali	Indiana University, USA
Takahiro Kawamura	Toshiba Corp., Japan
Jae-Hong Eom	Seoul National University, South Korea
Yuan-Fang Li	Monash University, Australia
Ryutaro Ichise	National Institute of Informatics, Tokyo, Japan
Kerry Taylor	Australian Bureau of Statistics and Australian National University, Australia
Jun Zhao	Lancaster University, UK
Boontawee Suntisrivaraporn	Sirindhorn International Institute of Technology, Thailand
Ikki Ohmukai	National Institute of Informatics, Tokyo, Japan
Artemis Parvizi	Oxford University Press, UK
Yue Ma	LRI-CNRS, University of Paris Sud, France
Boris Villazn-Terrazas	Expert System Iberia, Spain
Umberto Straccia	ISTI-CNR, Italy
Shinichi Nagano	Toshiba Corporation, Japan
Hanmin Jung	KISTI, South Korea
C. Chantrapornchai	Kasetsart University, Thailand
Stefan Dietze	L3S Research Center, Germany
Gang Wu	College of Information Science and Engineering, Northeastern University, China
Shenghui Wang	OCLC Research, The Netherlands
Masahiro Hamasaki	National Institute of Advanced Industrial Science and Technology (AIST), Japan
Ulrich Reimer	University of Applied Sciences St. Gallen, Switzerland
Leo Obrst	MITRE, USA
Koiti Hasida	AIST, Japan
Yoshinobu Kitamura	I.S.I.R., Osaka University, Japan
Hideaki Takeda	National Institute of Informatics, Japan
Alessandro Faraotti	IBM, Italy
Zhiqiang Gao	Southeast University, China
Xiang Zhang	Southeast University, China
Holger Wache	University of Applied Science Northwestern, Switzerland
Jinguang Gu	Wuhan University of Science and Technology, China
Zhicun Wang	Beijing Normal University, China
Yanjun Ma	Baidu, China
Yongbin Liu	Tsinghua University, China
Yi Zeng	Institute of Automation, Chinese Academy of Sciences, China
Yinglong Ma	INCEPU, China
Marut Buranarach	Rajburi Orchids Cluster, Thailand
Hua Luan	Beijing Normal University, China

Zhixing Li	Chongqing University of Posts and Telecommunications, China
Gaya Nadarajan	Seoul National University, South Korea
Hong Gee Kim	Seoul National University, South Korea
Huajun Chen	Zhejiang University, China
Kouji Kozaki	I.S.I.R., Osaka University, Japan
Xiaowang Zhang	Tianjin University, China
Zhichun Wang	Beijing Normal University, China
Kun Zhang	Sogou, China
Tong Ruan	ECUST, China
Zhisheng Huang	Vrije University Amsterdam, The Netherlands
Donghong Ji	Wuhan University, China
Nansu Zong	Seoul National University, South Korea
Thomas Juettemann	European Bioinformatics Institute, UK

Organizing Committee

Conference General Chairs

| Wendy Hall | University of Southampton, UK |
| Juanzi Li | Tsinghua University, China |

Program Committee Chairs

| Guilin Qi | Southeast University, China |
| Kouji Kozaki | Osaka University, Japan |

Publicity Chairs

| Jeff Z. Pan | University of Aberdeen, UK |
| Siwei Yu | Wuhan University, China |

Workshop Chairs

| Hong-Gee Kim | Seoul National University, South Korea |
| Guohui Xiao | Free University of Bozen-Bolzano, Italy |

Sponsorship Chair

| Huajun Chen | Zhejiang University, China |

Autumn School Chairs

| Dongyan Zhao | Peking University, China |
| Haofen Wang | East China University of Science and Technology, China |

Poster and Demo Chairs

Zhichun Wang   Beijing Normal University, China
Boontawee Suntisrivaraporn   Sirindhorn International Institute of Technology,
                                  Thailand

Tutorial Chairs

Yuan-Fang Li   Monash University, Australia
Hanmin Jung   Korea Institute of Science and Technology
                                  Information, South Korea

In-Use Chairs

Kun Zhang   Sogou, China
Yanjun Ma   Baidu, China

Competition Chairs

Kang Liu   Institute of Automation, Chinese Academy of Sciences,
                                  China
Jianfeng Du   Guangdong University of Foreign Studies, China

Proceedings and Publication Chair

Yue Ma   LRI UMR CNRS, University of Paris-Saclay, France

Local Chairs

Jinguang Gu   Wuhan University of Science and Technology, China
Tingyao Jiang   China Three Gorges University, China

Additional Reviewers

Tianxing Wu   Elem Güzel
Zhangquan Zhou   Sarah Komla-Ebri
Sabin Kafle

Contents

In-Use

Linked Open Data and Data Warehouses

Modeling and Querying Spatial Data Warehouses on the Semantic Web

Nurefşan Gür[1]([✉]), Katja Hose[1], Torben Bach Pedersen[1],
and Esteban Zimányi[2]

[1] Aalborg University, Aalborg, Denmark
{nurefsan,khose,tbp}@cs.aau.dk
[2] Université Libre de Bruxelles, Brussels, Belgium
ezimanyi@ulb.ac.be

Abstract. The Semantic Web (SW) has drawn the attention of data
enthusiasts, and also inspired the exploitation and design of multidimensional data warehouses, in an unconventional way. Traditional data warehouses (DW) operate over static data. However multidimensional (MD)
data modeling approach can be dynamically extended by defining both
the schema and instances of MD data as RDF graphs. The importance
and applicability of MD data warehouses over RDF is widely studied
yet none of the works support a spatially enhanced MD model on the
SW. Spatial support in DWs is a desirable feature for enhanced analysis, since adding encoded spatial information of the data allows to query
with spatial functions. In this paper we propose to empower the spatial
dimension of data warehouses by adding spatial data types and topological relationships to the existing QB4OLAP vocabulary, which already
supports the representation of the constructs of the MD models in RDF.
With QB4SOLAP, spatial constructs of the MD models can be also published in RDF, which allows to implement spatial and metric analysis
on spatial members along with OLAP operations. In our contribution,
we describe a set of spatial OLAP (SOLAP) operations, demonstrate a
spatially extended metamodel as, QB4SOLAP, and apply it on a use case
scenario. Finally, we show how these SOLAP queries can be expressed
in SPARQL.

1 Introduction

The evolution of the Semantic Web (SW) and its tools allow to employ complex analysis over multidimensional (MD) data models via On-Line Analytical Processing (OLAP) style queries. OLAP emerges when executing complex
queries over data warehouses (DW) to support decision making. DWs store large
volumes of data which are designed with MD modeling approach and usually
perceived as *data cubes*. Cells of the cube represent the observation *facts* for
analysis with a set of attributes called *measures* (*e.g.* a sales fact cube with
measures of product quantity and prices). Facts are linked to *dimensions* which
give contextual information (*e.g.* sales date, product, and location). Dimensions
are perspectives which are used to analyze data, organized into *hierarchies* and

© Springer International Publishing Switzerland 2016
G. Qi et al. (Eds.): JIST 2015, LNCS 9544, pp. 3–22, 2016.
DOI: 10.1007/978-3-319-31676-5_1

levels that allow users to analyze and aggregate measures at different levels of detail. Levels have a set of *attributes* that describe the characteristics of the level members.

In traditional DWs, the "location" dimension is widely used as a conventional dimension which is represented in an alphanumeric manner with only nominal reference to the place names. This neither allow manipulating location-based data nor deriving topological relations among the hierarchy levels of the location dimension. This issue yields a demand for truly spatial DWs for better analysis purposes. Including encoded geometric information of the location data significantly improves the analysis process (*i.e.* proximity analysis of the locations) with comprehensive perspectives by revealing dynamic spatial hierarchy levels and new spatial members. The scope of this work is first focuses on enhancing the spatial characteristics of the cube members on the SW, and then describing and utilizing SOLAP operators for advanced analysis and decision making.

In our approach we consider enabling SOLAP capabilities directly over Resource Description Framework (RDF) data on the SW. Importance and applicability of performing OLAP operations directly over RDF data is studied in [9,12]. To perform SOLAP over the SW consistently, an explicit and precise vocabulary is needed for the modeling process. The key concepts of spatial cube members need to be defined in advance to realize SOLAP operations since they employ spatial measures with spatial aggregate functions (*e.g.* union, buffer, and, convex-hull) and topological relations among spatial dimension and hierarchy level members (*e.g.* within, intersects, and, overlaps). Current state of the art RDF and OLAP technologies is limited to support conventional dimension schema and analysis along it's levels. Spatial dimension schema and SOLAP require an advanced specialized data model. As a first effort to overcome the limitations of modeling and querying spatial data warehouses on the Semantic Web we give our contributions in the following.

Contributions. We propose an extended metamodel solution that enables representation and RDF implementation of spatial DWs. We base our metamodel on the most recent QB4OLAP vocabulary and present an extension to support the spatial functions and spatial elements of the MD cubes. We discuss the notion of a SOLAP operator and observe it with examples, then we give the semantics of each SOLAP operator formally and finally, show how to implement them in SPARQL by using sub-queries and nested set of operators.

In the remainder of the paper, we first present the state of the art, in Sect. 2. As a prerequisite for our contribution in Sect. 3, we give the preliminary concepts and explain the structure of a SOLAP operator. Then, in Sect. 4 we define the semantics of MD data cube elements in RDF, present QB4SOLAP and formalize the SOLAP operators over MD data cube elements. We present a QB4SOLAP use case in Sect. 5 and then, we show how to write the defined SOLAP queries over this use case in SPARQL in Sect. 6. Finally, in Sect. 7, we conclude and remark to the future work directions.

2 State of the Art

DW and OLAP technologies have been proven a successful approach for analysis purposes on large volumes of data [1]. Aligning DW/OLAP technologies with RDF data makes external data sources available and brings up dynamic scenarios for analysis. The following studies are found concerning DW/OLAP with the SW.

DW/OLAP and Semantic Web: The potential of OLAP to analyze SW data is recognized in several approaches, thus MD modeling from ontologies is studied in the works of [8,17]. However these approaches do not support standard querying of RDF data in SPARQL but require a MD or a relational database query engine, which limits the access to frequently updated RDF data. Kämpgen *et al.* propose an extended model [12] on top of RDF Data Cube Vocabulary (QB) [6] for interacting with statistical linked data via OLAP operations in SPARQL, but it has the limitations of the QB and thus cannot support full OLAP dimension structure with aggregate functions. It also has only limited support for complete MD data model members (*e.g.* hierarchies and levels). Etcheverry *et al.* introduce QB4OLAP [9] as an extended vocabulary of QB with a full MD metamodel, which supports OLAP operations directly over RDF data with SPARQL queries. However, none of these vocabularies and approaches support spatial DWs, unlike our proposal.

Spatial DW and OLAP: The constraint representation of spatial data has been focus in many fields from databases to AI [18]. Extending OLAP with spatial features has attracted the attention of data warehousing communities as well. Several conceptual models are proposed for representing spatial data in data warehouses. Stefanovic *et al.* [11] investigates on constructing and materializing the spatial cubes in their proposed model. The MultiDim conceptual model is introduced by Malinowski and Zimányi [16] which copes with spatial features and extended in [20], to include complex geometric features (*i.e.* continuous fields), with a set of operations and MD calculus supporting spatial data types. Gómez *et al.* [10] propose an algebra and a very general framework for OLAP cube analysis on discrete and continuous spatial data. Even though spatial data warehousing is widely studied, it has not implemented yet on the Semantic Web.

Geospatial Semantic Web: The Open Geospatial Consortium – OGC pursue an important line of work for geospatial SW with GeoSPARQL [3] as a vocabulary to represent and query spatial data in RDF with an extension to SPARQL. Kyzirakos *et al.* presents a comprehensive survey in data models and query languages for linked geospatial data in [14], and propose a semantic geospatial data store - Strabon with an extensive query language – stSPARQL in [13], which is yet limited to a specific environment. LinkedGeoData is a significant contribution on interactively transforming OpenStreetMap[1] data to RDF data [19]. GeoKnow [15] is a more recent project with focus on linking geospatial data from heterogeneous sources.

---

[1] http://www.openstreetmap.org.

The studies shows that, SW and RDF technologies evolve to give better functionality and standards for spatial data representation and querying. It is also argued above that spatial data is very much needed for DW/OLAP applications. However modeling and querying of spatial DWs on the SW is not addressed in any of the above papers. There are recent efforts on creating an Extract-Load-Transform (ETL) framework from semantic data warehouses [7] and publishing/-converting open spatial data as Linked Open Data [2], which motivates modeling and querying spatial data warehouses on the Semantic Web. Spatial data requires specific treatment techniques, particular encoding, special functions and different manipulation methods, which should be considered during the design and modeling process. Current state of the art geospatial Semantic Web focuses on techniques for publishing, linking and querying spatial data however does not elaborate on analytical spatial queries for MD data. In order to address these issues we propose a generic and extensible metamodel based on the best practices of MD data publishing in RDF. Then we show how to create spatial analytical queries with SOLAP on MD data models. We base ourselves on existing works by extending the most recent version of the QB4OLAP vocabulary with spatial concepts. Furthermore, we introduce the new concept of SOLAP operators that navigate along spatial dynamic hierarchy levels and implement these analytical spatial queries in SPARQL.

3   Spatial and OLAP Operations

In this section we give define the spatial and spatial OLAP (SOLAP) operations.

3.1   Spatial Operations

In order to understand spatial operations, it is important to understand what is a *spatial object*. A spatial object is the data or information that identifies a real-world entity of geographic features, boundaries, places etc. Spatial objects can be represented in object/vector or image/raster mode. Database applications that can store spatial objects need to specify the spatial characteristics, encoded as specific information such as *geometry* data type which is the most common and supports planar or Euclidean (flat-earth) data. *Point, Line, and, Polygon* are the basic instantiable types of the geometry data type.

Geometries are associated with a *spatial reference system* (SRS) which describes the coordinate space in which the geometry is defined. There are several SRSs and each of them are identified with a *spatial reference system identifier* (SRID). The World Geodetic System (WGS) is the most well-known SRS and the latest version is called WGS84, which is also used in our use case.

Spatial data types have a set of operators that can function among applications. We grouped these operations into classes. Our classification is based on the common functionality of the operators. These classes are defined as follows:

Spatial Aggregation. The operators in the spatial aggregation, $\mathcal{S}_{agg}$ class aggregate two or more spatial objects. The result of these operators returns a new composite spatial object. Union, Intersection, Buffer, ConvexHull, and, MBR - Minimum Bounding Rectangle are example operators of this class.

Topological Relation. The operators in the topological relation, $\mathcal{T}_{rel}$ class are commonly contained in the RCC8[2] and DE-9DIM[3] models. Topological relations are standardized by OGC as Boolean operators which specify how two spatial objects are related to each other with a set of spatial predicates for example: Intersects, Disjoint, Equals, Overlaps, Contains, Within, Touches, Covers, CoveredBy, and, Crosses.

Numeric Operation. The operators in the class of numeric operation, $\mathcal{N}_{op}$ take one or more spatial objects and return a numeric value. Perimeter, Area, # of Interior Rings, Distance, Haversine Distance, Nearest Neighbor (NN), and # of Geometries are some of the example operators of this class.

3.2 SOLAP Operations

OLAP operations emerge when executing complex queries over multidimensional (MD) data models. OLAP operations let us interpret data from different perspectives at different levels of detail. *Spatially extended multidimensional models* incorporate spatial data during the analysis and decision making process by revealing new patterns, which are difficult to discover otherwise. In connection with our definition of MD models in the first paragraph of Sect. 1, hereafter we enhance and describe the spatially extended MD data cube elements.

A spatially extended MD model contains both conventional and spatial dimensions. A *spatial dimension* is a dimension which includes at least one *spatial hierarchy*. Dimensions usually have more than one level which are represented through hierarchies and there is always a unique top level *All* with just one member. A hierarchy is a *spatial hierarchy* if it has at least one *spatial level* in which the application should store the spatial characteristics of the data, which is captured by it is geometry and can be recorded in the *spatial attributes* of the level. A *spatial fact* is a fact that relates several dimensions in which, two or more are spatial. For example, consider a "Sales" *spatial fact* cube, which has "Customer" and "Supplier" (company) as *spatial dimensions* with a *spatial hierarchy* as "Geography" that expands into (hierarchic) *spatial levels*; "City → State → Country → Continent → *All*" from the customer and supplier's location. All these *spatial levels* record the spatial characteristics *i.e.* with a *spatial attribute* of a city (center) as "point" coordinates. Measures in the cube express additional and essential information for each MD data cell which is not exhibited through the dimensions and level attributes. Typically, spatially extended MD models have *spatial measures* which are represented by a geometry *i.e.* point, polygon, etc.

---

[2] RCC8 – Region Connection Calculus describes regions in Euclidean space, or in a topological space by their possible relations to each other.
[3] DE-9DIM – Dimensionally Extended Nine-Intersection Model is a topological model that describes spatial relations of two geometries in two-dimensions.

Spatial OLAP operates on spatially extended MD models. SOLAP enhances the analytical capabilities of OLAP with the spatial information of the cube members. The term SOLAP used in [4] and their similar works as a visual platform, which is designed to analyze huge volumes of geo-referenced data by providing visualization of pivot tables, graphical displays and interactive maps. We define the term SOLAP concisely as a platform (and query language) independent high-level concept, which is applicable on any spatial multidimensional data. We explain and exemplify in the following how SOLAP operators are interpreted.

Each operator in SOLAP should include at least one *spatial condition* by using the aforementioned operators from the spatial operation classes defined in Sect. 3.1. Spatial operations in SOLAP create a dynamic interpretation of the cube members as a *dynamic spatial hierarchy or level*. These interpretations allow new perspectives to analyze the spatial MD data which cannot be accessed in a traditional MD model. For instance, the classical OLAP operator *roll-up* aggregates measures along a hierarchy to obtain data at a coarser granularity. In the spatial dimension schema Fig. 1, the (classical) roll-up relation, *Customer to City* is shown with black straight arrows. On the other hand, in SOLAP, a new "dynamic spatial hierarchy" is created on the fly to roll-up among spatial levels by a spatial condition (closest distance), which is given as *Customer to Closest-City (of the Supplier)*, shown with curved arrows in gray. The details of this operator in comparison with OLAP and SOLAP are given in the following example.

Example: Roll-up. The user wants to sum the total amount of the sales to customers up to the city level with the roll-up operator. The instance data for *Sales fact* is given in Table 1 and shown on the map in Fig. 2.

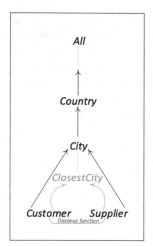

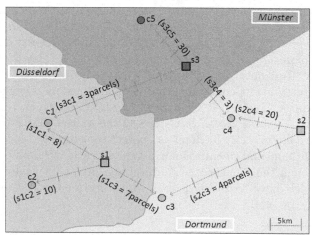

Fig. 1. S-Dim. Schema　　**Fig. 2.** Example map of sales (instance) data

Table 1. Sample (instance) data for sales

City	Customer	Supplier sales			Total sales
		s1	s2	s3	
Düsseldorf	c1	8pcs.	–	3pcs.	11pcs.
	c2	10pcs.	–	–	10pcs.
Dortmund	c3	7pcs.	4pcs.	–	11pcs.
	c4	–	20pcs.	3pcs.	23pcs.
Münster	c5	–	–	30pcs.	30pcs.

Table 2. Roll-up

City	Sales
Düsseldorf	21pcs.
Dortmund	34pcs.
Münster	30pcs.

Table 4. Customer to Supplier Distance (km.s)

Sup. City		Düsseldorf	Dortmund	Münster
Cust. City	Sup. Cust.	s1	s2	s3
Düsseldorf	c1	15 km.s	45km.s	30 km.s
	c2	15 km.s	60 km.s	60 km.s
Dortmund	c3	*15 km.s*	30 km.s	45 km.s
	c4	45 km.s	15 km.s	*15 km.s*
Münster	c5	60 km.s	45 km.s	15 km.s

Table 3. S-Roll-up

City	Sales
Düsseldorf	25pcs.
Dortmund	20pcs.
Münster	33pcs.

The amount of the sales are shown in parentheses along with the quantities of the sold parcels (from supplier to customer). The arrows on the map, between the supplier and customer locations are used to represent the distance. The summarized data for sale instances (Table 1) does not originally contain the records of the supplier – customer distance (as given in Table 4) which can lead to increase in the storage space. If there are no sales to customers from the corresponding suppliers, a dash (–) is used in Table 1. The syntax of the traditional roll-up operator is **ROLLUP(Sales, (Customer → City), SUM(SalesAmount))** which aggregates the "*total sales to customers up to city level*" (results in Table 2). Alternatively, the user may like to view the "*total sales to customers by city of the closest suppliers*", in which some customers can be closer to their suppliers from other cities, as emphasized in Table 4. This query is possible with traditional OLAP, if only Table 4 is recorded in the base data which requires extra storage space. For a better support and flexibility we define a spatial roll-up operator that aggregates the total sales along the dynamic spatial hierarchy, which is created based on a spatial condition (in this example, distance between customer and supplier locations). The syntax for s-roll-up is; **S-ROLLUP(Sales,[CLOSEST(Customer,Supplier)] → City',SUM(SalesAmount))**. The spatial condition transforms the customer→city hierarchy as a dynamic spatial hierarchy, which depends on the proximity of the suppliers that is calculated during runtime. The user has the flexibility to make analyses with different spatial operations in SOLAP. Spatial extensions of common OLAP operators (roll-up, drill-down, slice, and dice) are formally defined in Sect. 4.3.

4   Semantics of Spatial MD Data and OLAP Operations

In this section, we first present our approach on how to support spatial MD data in RDF by using the QB4SOLAP vocabulary. Afterwards, we define the general semantics of each SOLAP operator to be implemented in SPARQL as a proof of concept to the QB4SOLAP metamodel. The concepts introduced in this metamodel are an extension to the most recent QB4OLAP vocabulary [9].

Figure 3 shows the proposed and extended QB4OLAP vocabulary for the *cube schema* RDF triples. Capitalized terms in the figure represent RDF classes and non-capitalized terms represent RDF properties. Classes in external vocabularies are depicted in light gray background and font. RDF Cube (QB), QB4OLAP, QB4SOLAP classes are shown with white, light gray, dark gray backgrounds, respectively. Original QB terms are prefixed with qb:. QB4OLAP and QB4SOLAP classes and properties are prefixed with qb4o: and qb4so:. In order to represent spatial classes and properties an external prefix from OGC, geo: is used in the metamodel. Since QB4OLAP and QB4SOLAP are RDF-based multidimensional model schemas, we first define formally what an RDF triple is, and then discuss the basics of describing MD data using QB4OLAP and spatially enhanced MD data in QB4SOLAP.

An *RDF triple t* consists of three components; s is the subject, p is the predicate, and o is the object, which is defined as: *triple (s,p,o)* $\in t = (\mathcal{I} \cup \mathcal{B}) \times \mathcal{I} \times (\mathcal{I} \cup \mathcal{B} \cup \mathcal{L}i)$ where the set of *IRIs* is $\mathcal{I}$, the set of *blank nodes* is $\mathcal{B}$, and the set of *literals* is $\mathcal{L}i$. Given an MD element x of the cube schema $\mathcal{CS}$, $\mathcal{CS}(x) \in (\mathcal{I} \cup \mathcal{B} \cup \mathcal{L}i)$ returns a set of triples $\mathcal{T}$, with IRIs, blank nodes and literals. The notation of the triples in the following definitions is given as (x rdf:type ex:SomeProperty). If the concepts are defined with a set of triples, after the first triple, we use a semicolon ";" to link predicates (p) and objects (o) with the subject (s) concept x. The blank nodes $\mathcal{B}$ are expressed as _: and nesting unlabeled blank nodes are abbreviated in square brackets "[]".

4.1   Defining MD Data in QB4OLAP

In order to explain the spatially enhanced MD models we first describe MD elements described in Sect. 1 with the RDF formalization in QB4OLAP vocabulary.

Cube Schema. A data structure definition (DSD) specifies the schema of a data set (*i.e.*, a cube, which is an instance of the class qb:DataSet). The DSD can be shared among different data sets. The DSD of a data set represents dimensions, levels, measures, and attributes with component properties. The DSD is defined through a conceptual MD cube schema $\mathcal{CS}$, which has a set of dimension types $\mathcal{D}$, a set of measures $\mathcal{M}$ and, with a fact type $\mathcal{F}$ as $\mathcal{CS} = (\mathcal{D}, \mathcal{M}, \mathcal{F})$. For example, a cube schema $\mathcal{CS}$ can be used to define a physical structure of a company's sales data to be represented as a MD data cube. We define the cube schema elements in the following definitions.

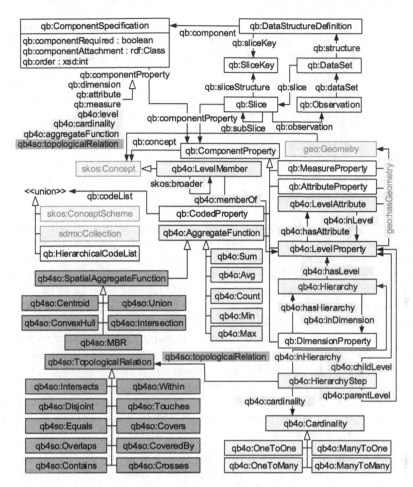

Fig. 3. QB4SOLAP vocabulary meta-model

Attributes. An attribute $a \in \mathcal{A} = \{a_1, a_2, \ldots, a_n\}$ has a domain $\langle a : dom \rangle$ in the cube schema $\mathcal{CS}$ with a set of triples $t_a \in \mathcal{T}$ where t_a is encoded as (a `rdf:type qb:AttributeProperty; rdfs:domain xsd:Schema`)

The domain of the attribute is given with the property `rdfs:domain`[4] from the corresponding schema and `rdfs:range` defines what values the property can take *i.e.*; *integer*, *decimal*, etc. from the given, *xsd:Schema* elements[5]. Attributes are the finest granular elements of the cube, which exists in levels to describe the characteristics of level members *e.g.*, customer level attributes could be as; name, id, address, etc.

Levels. A level $l \in \mathcal{L} = \{l_1, l_2, \ldots, l_n\}$ consists of a set of attributes $\mathcal{A}_l$, which is defined by a schema $l(a_1 : dom_1, \ldots, a_n : dom_n)$, where l is the level and each attribute a is defined over the domain dom. For each level $l \in \mathcal{L}$

---

[4] RDF Schema http://www.w3.org/TR/rdf-schema/.

[5] XML Schema http://www.w3.org/TR/xmlschema11-1/.

in the cube schema $\mathcal{CS}$, there is a set of triples $t_l \in \mathcal{T}$ which is encoded as (l rdf:type qb4o:LevelProperty; qb4o:hasAttribute a). Relevant levels for customer data include; customer level, city level, country level, etc.

Hierarchies. A hierarchy $h \in \mathcal{H} = \{h_1, h_2, \ldots, h_n\}$ in the cube schema $\mathcal{CS}$, is defined with a set of triples $t_h \in \mathcal{T}$, and encoded as (h rdf:type qb4o:HierarchyProperty; qb4o:hasLevel l; qb4o:inDimension $\mathcal{D}$).

Each hierarchy $h \in \mathcal{H}$ is defined as $h = (\mathcal{L}_h, \mathcal{R}_h)$; with a set of $\mathcal{L}_h$ (hierarchy) levels, which is a subset of the set $\mathcal{L}_d$ levels of the dimension $\mathcal{D}$ where $\mathcal{L}_h \subseteq \mathcal{L}_d \in \mathcal{D}$. $\mathcal{L}_d$ contains the initial base level of the dimension in addition to hierarchy levels L_h. For example, customer–location hierarchy can be defined by the levels; customer, city, country, etc. where customer is the base level and contained only in $\mathcal{L}_d$.

Due to the nature of the hierarchies, a hierarchy entails a roll-up relation $\mathcal{R}_h$ between its levels, $\mathcal{R}_h = (\mathcal{L}_c, \mathcal{L}_p, card)$ where $\mathcal{L}_c$ and $\mathcal{L}_p$ are respectively child and parent levels, where the lower level is called child and higher level is called parent. Cardinality $card \in \{1-1, 1-n, n-1, n-n\}$ describes the minimum and maximum number of members in one level that can be related to a member in another level, e.g., $\mathcal{R}_h = (city, country, many-to-one)$ shows that the roll-up relation between the child level city to parent level country is many-to-one, which means that each country can have many cities. In order to represent cardinalities between the child and the parent levels, blank nodes are created as hierarchy steps, $\_:h_{hs} \in \mathcal{B}$. Hierarchy steps relate the levels of the hierarchy from a bottom (child) level to an upper (parent) level, which is defined with a set of triples $t_{hs} \in \mathcal{T}$ and encoded as ($\_:h_{hs}$ rdf:type qb4o:HierarchyStep; qb4o:childLevel lh_c; qb4o:parentLevel lh_p; qb4o:cardinality $card$) where $lh_c \in \mathcal{L}_c$, $lh_p \in \mathcal{L}_p$ and $card \in \{1-1, 1-n, n-1, n-n\}$.

Dimensions. An n-dimensional cube schema has a set of dimensions $\mathcal{D} = \{d_1, d_2, \ldots, d_n\}$. And each $d \in \mathcal{D}$ is defined as a tuple $d = (\mathcal{L}, \mathcal{H})$; with a set of $\mathcal{L}_d$ levels, organized into $\mathcal{H}_d$ hierarchies. Dimensions, inherently have all the levels from the hierarchies they have, and an initial base level. For each dimension $d \in \mathcal{D}$, in the cube schema $\mathcal{CS}$, there is a set of triples $t_d \in \mathcal{T}$, which is encoded as (d rdf:type qb:DimensionProperty; qb4o:hasHierarchy h). For example, customerDim is a dimension with a location and a customer type hierarchy where location expands to levels of customer's location (e.g., city, country, etc.) and customer type expands to levels of customer's type (e.g.,profession, branch, etc.).

Measures. A measure $m \in \mathcal{M} = \{m_1, m_2, \ldots, m_n\}$, is a property, which is associated to the facts. Measures are given in the cube schema $\mathcal{CS}$ with a set of triples $t_m \in \mathcal{T}$, which is encoded as ($m$rdf:type qb:MeasureProperty; rdfs:subPropertyOf sdmx-measure:obsValue; rdfs:domain xsd:Schema).

Measures are defined with a sub-property from the Statistical Data and Metadata Exchange - ($sdmx$) definitions, sdmx-measure:obsValue which is the value of a particular variable for a particular observation[6]. Similarly to

---

[6] http://sdmx.org/.

the attributes rdfs:domain specifies the schema of the measure property and, rdfs:range defines what values the property can take *i.e.*; *integer, decimal, etc.* in the instances. For example, quantity and price are measures of a fact (e.g., sales) where the instance values can be given respectively, in the form: "13" ^^xsd:positiveInteger and "42.40"^^xsd:decimal. Measures are associated with observations (facts) and related to dimension levels in the DSD as explained in the following.

Facts. A fact $f \in \mathcal{F} = \{f_1, f_2, \ldots, f_n\}$ is related to values of dimensions and measures. The relation is described in *components* in the schema level of the facts cube definition, by a set of triples $t_f \in \mathcal{T}$, which is encoded as ($\mathcal{F}$ rdf:type qb:DataStructureDefinition; qb:component[qb4o:level l; qb4o:cardinality $card$]; qb:component[qb:measure m; qb4o:aggregate Function BIF]). Cardinality, $card \in \{1 - 1, 1 - n, n - 1, n - n\}$ represents the cardinality of the relationship between facts and level members. The specification of the aggregate functions for measures is required in the definition of the cube schema. Standard way of representing typical aggregate functions is defined by QB4OLAP namely built-in functions such as; $BIF \in \{Sum, Avg, Count, Min, Max\}$. For example, a fact schema $\mathcal{F}$ can be sales of a company which has associated dimensions and measures defined as components respectively e.g. product and price.

Finally, the facts $\mathcal{F} = \{f_1, f_2, \ldots, f_n\}$ are given on the instance level where each fact f has a unique IRI $\mathcal{I}$, which are observations. This is encoded as (f rdf:type qb:Observation). An example of a fact instance f with it's relation to measure values and dimension levels is a "sale" transacted to customer "John" (value of the dimension level), for a product "chocalate" (value of another dimension level), which has a unit price of "29.99" (value of a measure) euros, and quantity of "20" (value of another measure) boxes. Cardinality of the dimension level customer and fact member is many–to–one where several sales can be transacted to the same customer (i.e. John). Specification of the aggregate function for measure unit price is "average" while quantity can be specified as "sum".

We gave the cube schema $\mathcal{CS} = (\mathcal{D}, \mathcal{M}, \mathcal{F})$ members above where dimensions $d \in \mathcal{D}$ are defined as a tuple of dimension levels $\mathcal{L}_d$ and hierarchies $\mathcal{H}$, $d = (\mathcal{L}_d, \mathcal{H})$, and a hierarchy $h \in \mathcal{H}$ is defined with hierarchy levels $\mathcal{L}_h$ such that $h = (\mathcal{L}_h)$, where $\mathcal{L}_h \subseteq \mathcal{L}_d$, and a level l contains attributes $\mathcal{A}_l$ as $l = (\mathcal{A}_l)$.

4.2 Defining Spatially Enhanced MD Data in QB4SOLAP

QB4SOLAP adds a new concept to the metamodel, which is geo:Geometry class from OGC schemas[7]. We define the QB4SOLAP extension to the cube schema in the description of the following spatial MD data elements, which are explained in Sect. 3.2.

Spatial Attributes. Each attribute is defined over a domain (Sect. 4.1). Every attribute with geometry domain ($a : dom_g \in \mathcal{A}$) is a member of geo:Geometry

---

[7] OGC Schemas http://schemas.opengis.net/.

class and they are called spatial attributes a_s, which are defined in the cube schema $\mathcal{CS}$ by a set of triples $t_{ag} \in \mathcal{T}$ and encoded as (a_s rdf:type qb:AttributeProperty; rdfs:domain geo:Geometry). The type (point, polygon, line, *etc.*) of the each spatial attribute is assigned with rdfs:range predicate in the instances. For example, a spatial attribute can be the "capital city" of a country which is represented through a point geometry.

Spatial Levels. Each spatial level $l_s \in \mathcal{L}$ is defined with a set of triples $t_{l_s} \in \mathcal{T}$ in the cube schema $\mathcal{CS}$, and encoded as (l_s rdf:type qb4o:LevelProperty; qb4o:hasAttribute a, a_s; geo:hasGeometry geo:Geometry). Spatial levels must be a member of geo:Geometry class and might have spatial attributes. For example country level is a spatial level which has a polygon geometry and might also record geometry of the capital city in the level attributes as a point type.

Spatial Hierarchies. Each hierarchy $h_s \in \mathcal{H}$ is spatial, if it relates two or more spatial levels l_s. Spatial hierarchy step defines the relation between the spatial levels with a roll-up relation as in conventional hierarchy steps (Sect. 4.1). QB4SOLAP introduces topological relations, $\mathcal{T}_{rel}$ (Sect. 3.1) besides cardinalities in the roll-up relation which is encoded as $\mathcal{R} = (\mathcal{L}_c, \mathcal{L}_p, card, \mathcal{T}_{rel})$ for the spatial hierarchy steps.

Let $t_{shs} \in \mathcal{T}$ a set of triples to represent a hierarchy step for spatial levels in hierarchies, which is given with a blank node :_$shh_{hs} \in \mathcal{B}$ and encoded as (:_shh_{hs} rdf:type qb4o:HierarchyStep; qb4o:childLevel slh_{ci}; qb4o:parentLevel slh_{pi}; qb4o:cardinality $card$; qb4so:hasTopologicalRelation $\mathcal{T}_{rel}$) where $slh_{ci} \in \mathcal{L}_c$, $slh_{pi} \in \mathcal{L}_p$. For example, a spatial hierarchy is "geography" which should have spatial levels (e.g. customer, city, country, and continent) with the roll-up relation $\mathcal{R}_h = (city, country, many - to - one, within)$, which also specifies that child level city is "within" the parent level country, in addition to the hierarchy steps from Sect. 4.1.

Spatial Dimensions. Dimensions are identified as spatial if only they have at least one spatial hierarchy. More than one dimension can share the same spatial hierarchy and the spatial levels, which belongs to that hierarchy. QB4SOLAP uses the same schema definitions of the dimensions as in Sect. 4.1. For example, a spatial dimension is customer dimension, which has a spatial hierarchy geography.

Spatial Measures. Each spatial measure $m_s \in \mathcal{M}$ is defined in the cube schema $\mathcal{CS}$ by a set of triples $t_{ms} \in \mathcal{T}$ and encoded as (m_s rdf:type qb:MeasurePro perty; rdfs:subPropertyOf sdmx-measure:obsValue; rdfs:domain geo: Geometry). The class of the numeric value is given with the property rdfs:domain and rdfs:range assigns the values from geo:Geometry class, *i.e.*, point, *polygon*, *etc.* at the instance level.

Spatial measures are represented by a geometry thus they use a different schema than conventional (numeric) measures. The schemas for spatial measures have common geometry serialization standards[8] that are used in OGC schemas. For example a spatial measure is coordinates of an accident location, which is given as a point geometry type and associated to an observation fact of accidents.

Spatial Facts. Spatial facts $\mathcal{F}_s$ relates several dimensions of which two or more are spatial. If there are several spatial dimension levels (l_s), related to the fact, topological relations $\mathcal{T}_{rel}$ (Sect. 3.1) between the spatial members of the fact instance may be required which is not necessarily imposed for all the spatial fact cubes. Ideally a spatial fact cube has spatial measures (m_s), as its members which makes it possible to aggregate along spatial measures with the spatial aggregation functions $\mathcal{S}_{agg}$ (Sect. 3.1). Representation of a complete spatial fact cube at the schema level in RDF is given by a set of triples $t_{fs} \in \mathcal{T}$, and encoded as ($\mathcal{F}_s$ a qb:DataStructureDefinition; qb:component [qb4o:levell_s; rdfs:subPropertyOf sdmx-dimension:refArea; qb4o:cardinality *card*; qb4so:TopologicalRelation $\mathcal{T}_{rel}$]; qb: component[qb:measure m_s,sdmx-measure:obsValue;qb4o:aggregateFunction *BIF'*]). QB4SOLAP extends the built-in functions of QB4OLAP with spatial aggregation functions as $BIF' = BIF \cup \mathcal{S}_{agg}$ which is added with a class qb4so:SpatialAggregateFunction to the metamodel in Fig. 3. An example of a spatial fact instance f_s with it's relation to measure values and dimension levels is a traffic "accident" incident occured on a highway "E–45" (value of the highway spatial dimension level) with coordinate points of the location "57.013, 9.939" (value of the location spatial measure). Cardinality of the dimension level highway and fact member is many–to–one where several accidents might take place in the same highway. Specification of the spatial aggregate function for spatial measure location (coordinate points) can be specified as "convex hull" area of the accident locations.

4.3 SOLAP Operators

The proposed vocabulary QB4SOLAP allows publishing spatially enhanced multidimensional RDF data which allows us to query with SOLAP operations. Subqueries and aggregation functions in SPARQL 1.1[9] make it easily possible to operate with OLAP queries on multidimensional RDF data. Moreover, spatially enhanced RDF stores, provide functions to an extent for querying with topological relations and spatial numeric operations. In the following, we define common OLAP operators with spatial conditions in order to formalize spatial OLAP query classes. Spatial conditions can be selected from a range of operation classes that can be applied on spatial data types (Sect. 3.1). Let $\mathcal{S}$ be any spatial operation where $\mathcal{S} = (\mathcal{S}_{agg} \cup \mathcal{T}_{rel} \cup \mathcal{N}_{op})$ to represent a spatial condition in a SOLAP operation. The following OLAP operators are given with a spatial extension to the well-known OLAP operators defined over cubes based in Cube Algebra operators [5].

---

[8] The Well Known Text (WKT) serialization aligns the geometry types with ISO 19125 Simple Features [ISO 19125-1], and the GML serialization aligns the geometry types with [ISO 19107] Spatial Schema.

[9] http://www.w3.org/TR/sparql11-query/.

S–Roll–up. Given a cube $\mathcal{C}$, a dimension $d \in \mathcal{C}$, and an upper dimension level $l_u \in d$, such that $l\langle a : dom_g \rangle \rightarrow^* l_u$, where $l\langle a : dom_g \rangle$ represents the level in dimension d with attributes (a_s) whose domain is a geometry type. Let $\mathcal{R}_s$ be the spatial roll-up relation which comprises $\mathcal{S}$ and traditional roll-up relation $\mathcal{R}$ such that $\mathcal{R}_s = \mathcal{S}(d, l\langle a : dom_g \rangle) \cup \mathcal{R}(\mathcal{C}, d, l_u) \rightarrow \mathcal{C}'$.

Initially, in the semantics of *S–Roll–up* above, spatial constraint $\mathcal{S}$ is applied over a dimension d on the spatial attributes a_s along levels l. As a result of the roll–up relation $\mathcal{R}$, the measures are aggregated up to level l_u along d which returns a new cube $\mathcal{C}'$. Note that applying $\mathcal{S}$, on spatial level attributes a_s of dimension $\mathcal{D}$, operates on the hierarchy step $l \rightarrow l_u$ with a dynamic spatial hierarchy (Ref. Sect. 3.2). For example, the query "total sales to customers by city of the closest suppliers" implies a S-Roll-up operator.

S–Drill–down. Analogously, *S–Drill–down* is an inverse operation of *S–Roll–up*, which disaggregates previously summarized data down to a child level. For example, the query "average sales of the employees from the biggest city in its country" implies a S-Drill-down operator by disaggregating data from (parent) country level to (child) city level by imposing also a spatial condition (area from $\mathcal{N}_{op}$ to choose the biggest city).

S–Slice. Given a cube $\mathcal{C}$ with n dimensions $\mathcal{D} = \{d_1, d_2, \ldots, d_n\} \in \mathcal{C}$, let $\mathcal{S}'$ be the traditional slice operator which removes a dimension d from the cube $\mathcal{C}$. And let $\mathcal{S}_s$ be the spatial slice operator, which comprises $\mathcal{S}$, the spatial function to fix a single value in the level $\mathcal{L} = \{l_1, l_2, \ldots, l_n\} \in d$ defined as follows; $\mathcal{S}_s = \mathcal{S}'(\mathcal{C}, d) \cup \mathcal{S}(d, l\langle a : dom_g \rangle) \rightarrow \mathcal{C}'$.

Note that the spatial function is applied on the spatial attributes of the selected level, measures are aggregated along dimension d up to level *All*. The result returns a new cube $\mathcal{C}'$ with $n - 1$ dimensions $\mathcal{D}' = \{d_1, d_2, \ldots, d_{n-1}\} \in \mathcal{C}'$. For example, the query "total sales to the customers located in the city within a 10 km buffer area from a given point" implies a S-Slice operator, which dynamically defines the city level by (fixing) a specified buffer area around a given custom point in the city.

S–Dice. Dice operation is analogous to relational algebra - R *selection*; $\sigma_\phi(R)$, instead the argument is a cube $\mathcal{C}$; $\sigma_\phi(\mathcal{C})$. In SOLAP dice is not a select operation rather a nested "select" and a "spatial filter" operation. S-Dice $\mathcal{D}_s$ keeps the cells of a cube $\mathcal{C}$ that satisfy a spatial Boolean $\mathcal{S}(\phi)$ condition over spatial dimension levels, attributes and measures which is defined as; $\mathcal{D}_s = (\mathcal{C}, \mathcal{S}(\phi)) \rightarrow \mathcal{C}'$ where $\mathcal{S}(\phi) = \mathcal{S}(\sigma_{a\phi b}(\mathcal{C})) \vee \mathcal{S}(\sigma_{a\phi v}(\mathcal{C}))$ and a, b are spatial levels (l_s), geometry attributes $(a : dom_g)$ or measures (m, m_s) while v is a constant value and the result returns a sub-cube $\mathcal{C}' \subset \mathcal{C}$. For example, the query "total sales to the customers which are located less than 5 Km from their city" implies a S-Dice operator.

In this paper, we focus on direct querying of single data cubes. The integration of several cubes through *S–Drill–across* or set-oriented operations such as *Union, Intersection, and Difference*[5] is out of scope and remained as future work. The actual use of these query classes in SPARQL with the instance data is given in Sect. 6.

5   Use Case Scenario: GeoNorthwind Data Warehouse

Figure 4 consists of the conceptual schema of the GeoNorthwind DW use case. GeoNorthwind DW has synthetic data about companies and their sales, however it is well suited for representing MD data modeling concepts due to its rich dimensions and hierarchies. It is a good proof of concept use case to show how to implement spatial data cube concepts on the SW. We show next how to express the conceptual schema of GeoNorthwind in QB4SOLAP.

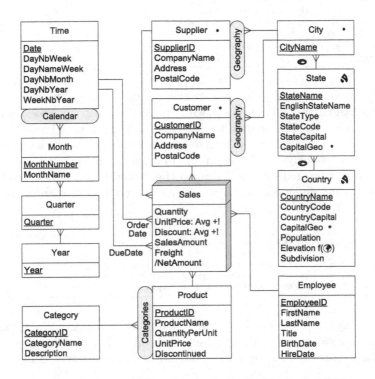

Fig. 4. Conceptual MD schema of the GeoNorthwind DW

In the use case, measures are given in the Sales cube. All measures are conventional. The members of the GeoNorthwind DW are given with gnw: prefix. The underlying syntax for RDF representation is given in Turtle[10] syntax in the boxes. An example of a measure in the cube schema is given in the following as defined in Sect. 4.1.

```
gnw:quantity a rdf:Property, qb:MeasureProperty;
rdfs:subPropertyOf sdmx-measure:obsValue;rdfs:range xsd:integer.
```

---

[10] http://www.w3.org/TR/turtle/.

In the following, a spatial attribute of a spatial level `gnw:state` is given along with the level and attribute properties. Spatial level has a geometry as `gnw:statePolygon` independently having a spatial attribute `gnw:capitalGeo`. Each spatial attribute in the schema is defined separately by using common RDF and standard spatial schemas[11] to represent their domain and data type as described in Sect. 4.2.

```
gnw:state a qb4o:LevelProperty; qb4o:hasAttribute gnw:stateName,
   gnw:stateType, gnw:stateCapital, gnw:capitalGeo;
   geo:hasGeometry gnw:statePolygon.
gnw:captialGeo a qb:AttributeProperty;
   rdfs:domain geo:Geometry; rdfs:range geo:Point, geo:wktLiteral, virtrdf:Geometry.
```

In the next listing, an example of a spatial dimension from the use case data is `gnw:customerDim`, which is given with its spatial hiearchy `gnw:geography` (Sects. 4.1 and 4.2). The spatial hierarchy is organized into levels (*i.e.* city, state, country *etc.*) where `qb4o:hasLevel` predicate indicates the levels that compose the hierarchy. Each hierarchy in dimensions is represented with `qb4o:inDimension` predicate, referring to the dimension(s) it belongs to. The levels given in the dimension hierarchy are all spatial, and the sample representation of a spatial level is given above.

```
gnw:customerDim a rdf:Property, qb:DimensionProperty;
   qb4o:hasHierarchy gnw:geography.
gnw:geography a qb4o:HierarchyProperty; qb4o:hasLevel gnw:city,
   gnw:state, gnw:region, gnw:country, gnw:continent;
   qb4o:inDimension gnw:customerDim, gnw:supplierDim.
```

Each hierarchy step is added to the schema as a blank node (`_:hsi`) by `qb4o:HierarchyStep` property, in which the cardinality and topological relationships are represented in between the child and parent levels as follows;

```
_:hs1 a qb4o:HierarchyStep; qb4o:inHierarchy gnw:geography;
   qb4o:childLevel gnw:customer, gnw:supplier;
   qb4o:parentLevel gnw:city; qb4o:cardinality qb4o:ManyToOne;
   qb4so:hasTopologicalRelation qb4so:Within.
```

The components of the facts are described at the schema level in the cube definition. The dimension level for `gnw:customer` is given with `sdmx-dimenson:refArea` property, which indicates the spatial characteristic of the dimension. Measures require the specification of the aggregate functions in the cube definition. As there are only numeric measures in the use case data,

---

[11] For our tests we used Virtuoso Universal Server and `virtrdf:Geometry` is a special RDF typed literal which is used for geometry objects in Virtuoso. Normally, WGS84 (EPSG:4326) is the SRID of any such geometry.

aggregate function for the sample measure gnw:quantity is given as qb4o:sum. The general overview of the cube schema $\mathcal{CS}$ which is given with the related components as follows:

```
### Cube definition ###
gnw:GeoNorthwind rdf:type qb:DataStructureDefinition;
### Lowest level for each dimension in the cube ###
qb:component [qb4o:level gnw:customer, sdmx-dimension:refArea;
qb4o:cardinality qb4o:ManyToOne].
### Measures in the Cube ###
qb:component [qb:measure gnw:quantity; qb4o:aggregateFunction qb4o:sum].
```

A spatial fact cube may contain spatial measure components besides spatial dimension according to QB4SOLAP. The implementation scope of this work covers only spatial facts, with spatial dimension and numerical measure components.

6   Querying the GeoNorthwind DW in SPARQL

We show next how some of the spatial OLAP queries from Sect. 4.3 can be expressed in SPARQL[12].

Query 1 (S-Roll-Up): Total sales to customers by city of the closest suppliers.

```
SELECT ?city (SUM(?sales) AS ?totalSales)
WHERE {?o a qb:Observation; gnw:customerID ?cust;
        gnw:supplierID ?sup; gnw:salesAmount ?sales.
        ?cust qb4o:inLevel gnw:customer;gnw:customerGeo ?custGeo;
        gnw:customerName ?custName; skos:broader ?city.
        ?city qb4o:inLevel gnw:city.?sup gnw:supplierGeo ?supGeo.
#Inner Select:Distance to the closest supplier of the customer
    {SELECT ?cust1 (MIN(?distance) AS ?minDistance)
    WHERE{?o a qb:Observation; gnw:customerID ?cust1;
        gnw:supplierID ?sup1. ?sup1 gnw:supplierGeo ?sup1Geo.
        ?cust1 gnw:customerGeo ?cust1Geo.
    BIND (bif:st_distance( ?cust1Geo, ?sup1Geo ) AS ?distance)}
    GROUP BY ?cust1 }
    FILTER (?cust = ?cust1 && bif:st_distance(?custGeo, ?supGeo)=
    ?minDistance)} GROUP BY ?city ORDER BY ?totalSales
```

The query above shows the spatial roll-up operation example from Sect. 3.2 with the actual use case data. We have explained the semantics of s-roll-up operator in Sect. 4.3. The inner select verifies the spatial condition in order to find the closest distance to suppliers from the customers. The outer select prepares the traditional roll up of the total sales from customer (child) level to the city (parent) level. Filter on customer and supplier distance creates the aforementioned dynamic spatial hierarchy based on the proximity of the suppliers.

Query 2 (S-Slice): Total sales to the customers located in the city within a 10 km buffer area from a given point.

---

[12] SPARQL endpoint is available at: http://extbi.ulb.ac.be:8890/sparql.

```
SELECT ?custName ?cityName (SUM(?sales) AS ?totalSales)
WHERE {?o rdf:type qb:Observation; gnw:customerID ?cust;
   gnw:salesAmount ?sales. ?cust gnw:customerName ?custName;
   skos:broader ?city. ?city gnw:cityGeo ?cityGeo;
         gnw:cityName ?cityName.
FILTER(bif:st_within(?cityGeo, bif:st_point(2.3522,48.856),10))}
GROUP BY ?custName ?cityName ORDER BY ?custName
```

The semantics of the above (s-slice) operator is given in Sect. 4.3. Traditional slice operator removes a dimension, by fixing a single value in a level of dimension with a given fixed value (*i.e.* CityName = "Paris"). On the other hand, s-slice dynamically defines the city level, by a specified buffer area around a given custom point in the city. Thus, s-slice removes the dimension customer and its instance in city Paris, but only the customer instances within 10 km buffer area of the desired location. The project content with corresponding data sets and full query examples are available at: http://extbi.cs.aau.dk/QB4SOLAP/index. php.

7   Conclusion and Future Work

In this paper, we studied the modeling issues of spatially enhanced MD data cubes in RDF, defined the concept of SOLAP operators and implemented them in SPARQL. We showed that in order to model spatial DWs on the SW, an extended representation of MD cube elements was required. We based our representation on the most recent QB4OLAP vocabulary and make it viable for spatially enhanced MD data models through the new QB4SOLAP metamodel. This allows users to publish spatial MD data in RDF format. Then, we define well-known OLAP operations on data cubes with spatial conditions, in order to introduce spatial OLAP query classes and formally define their semantics. Subsequently, we present a use case and implement real-world SOLAP queries in SPARQL, to validate our approach.

Future work will be conducted in two areas: (1) defining complete formal techniques and algorithms for generating SOLAP queries in SPARQL based on a high-level MD Cube Algebra as in [5], and extending the coverage of SOLAP operations over multiple RDF cubes in SPARQL, *i.e.*, to support *S-Drill-Across*; (2) implement our QB4SOLAP approach on a more complex case study with spatial measures and facts which can support spatial aggregation (*S-Aggregation*) operator over measures with geometries. In order to support this *S-Aggregation* operator in SPARQL we will also investigate on creating user-defined SPARQL functions.

Acknowledgment. This research was partially funded by The Erasmus Mundus Joint Doctorate in "Information Technologies for Business Intelligence – Doctoral College (IT4BI-DC)".

References

1. Abelló, A., Romero, O., Pedersen, T.B., Berlanga Llavori, R., Nebot, V., Aramburu, M., Simitsis, A.: Using semantic web technologies for exploratory OLAP: a survey. TKDE **99**, 571–588 (2014)
2. Andersen, A.B., Gür, N., Hose, K., Jakobsen, K.A., Pedersen, T.B.: Publishing danish agricultural government data as semantic web data. In: Supnithi, T., Yamaguchi, T., Pan, J.Z., Wuwongse, V., Buranarach, M. (eds.) JIST 2014. LNCS, vol. 8943, pp. 178–186. Springer, Heidelberg (2015)
3. Battle, R., Kolas, D.: GeoSPARQL: enabling a geospatial SW. Seman. Web **3**(4), 355–370 (2012)
4. Bimonte, S., Johany, F., Lardon, S.: A first framework for mutually enhancing chorem and spatial OLAP systems. In: DATA (2015)
5. Ciferri, C., Gómez, L., Schneider, M., Vaisman, A.A., Zimányi, E.: Cube algebra: a generic user-centric model and query language for OLAP cubes. IJDWM **9**(2), 39–65 (2013)
6. Cyganiak, R., Reynolds, D., Tennison, J.: The RDF Data Cube Vocabulary. W3C (2014)
7. Deb Nath, R.P., Hose, K., Pedersen, T.B.: Towards a programmable semantic extract-transform-load framework for semantic data warehouses. In: DOLAP (2015)
8. Diamantini, C., Potena, D.: Semantic enrichment of strategic datacubes. In: DOLAP (2008)
9. Etcheverry, L., Vaisman, A., Zimányi, E.: Modeling and querying data warehouses on the semantic web using QB4OLAP. In: Bellatreche, L., Mohania, M.K. (eds.) DaWaK 2014. LNCS, vol. 8646, pp. 45–56. Springer, Heidelberg (2014)
10. Gómez, L.I., Gómez, S.A., Vaisman, A.A.: A generic data model and query language for spatiotemporal OLAP cube analysis. In: EDBT (2012)
11. Han, J., Stefanovic, N., Koperski, K.: Selective materialization: an efficient method for spatial data cube construction. In: Wu, X., Kotagiri, R., Korb, K.B. (eds.) PAKDD 1998. LNCS, vol. 1394, pp. 144–158. Springer, Heidelberg (1998)
12. Kämpgen, B., O'Riain, S., Harth, A.: Interacting with statistical linked data via OLAP operations. In: Simperl, E., Norton, B., Mladenic, D., Valle, E.D., Fundulaki, I., Passant, A., Troncy, R. (eds.) ESWC 2012. LNCS, vol. 7540, pp. 87–101. Springer, Heidelberg (2012)
13. Koubarakis, M., Karpathiotakis, M., Kyzirakos, K., Nikolaou, C., Sioutis, M.: Data models and query languages for linked geospatial data. In: Eiter, T., Krennwallner, T. (eds.) Reasoning Web 2012. LNCS, vol. 7487, pp. 290–328. Springer, Heidelberg (2012)
14. Kyzirakos, K., Karpathiotakis, M., Koubarakis, M.: Strabon: a semantic geospatial DBMS. In: Cudré-Mauroux, P., et al. (eds.) ISWC 2012, Part I. LNCS, vol. 7649, pp. 295–311. Springer, Heidelberg (2012)
15. Le Grange, J.J., Lehmann, J., Athanasiou, S., Rojas, A.G., et al.: The GeoKnow generator: managing geospatial data in the linked data web. In: Linking Geospatial Data (2014)
16. Malinowski, E., Zimányi, E.: Advanced Data Warehouse Design: From Conventional to Spatial and Temporal Applications. Springer, Heidelberg (2008)
17. Nebot, V., Berlanga, R., Pérez, J.M., Aramburu, M.J., Pedersen, T.B.: Multidimensional integrated ontologies: a framework for designing semantic data warehouses. In: Spaccapietra, S., Zimányi, E., Song, I.-Y. (eds.) Journal on Data Semantics XIII. LNCS, vol. 5530, pp. 1–36. Springer, Heidelberg (2009)

18. Revesz, P.: Introduction to Databases: From Biological to Spatio-Temporal. Springer, Heidelberg (2010)
19. Stadler, C., Lehmann, J., Hffner, K., Auer, S.: Linkedgeodata: a core for a web of spatial open data. Semant. Web **3**(4), 333–354 (2012)
20. Vaisman, A.A., Zimányi, E.: A multidimensional model representing continuous fields in spatial data warehouses. In: ACM SIGSPATIAL (2009)

RDF Graph Visualization by Interpreting Linked Data as Knowledge

Rathachai Chawuthai[1,2(✉)] and Hideaki Takeda[1,2]

[1] SOKENDAI (The Graduate University for Advanced Studies),
Kanagawa, Japan
{rathachai, takeda}@nii.ac.jp
[2] National Institute of Informatics, Tokyo, Japan

Abstract. It is known that Semantic Web and Linked Open Data (LOD) are powerful technologies for knowledge management, and explicit knowledge is expected to be presented by RDF format (Resource Description Framework), but normal users are far from RDF due to technical skills required. As we learn, a concept-map or a node-link diagram can enhance the learning ability of learners from beginner to advanced user level, so RDF graph visualization can be a suitable tool for making users be familiar with Semantic technology. However, an RDF graph generated from the whole query result is not suitable for reading, because it is highly connected like a hairball and less organized. To make a graph presenting knowledge be more proper to read, this research introduces an approach to sparsify a graph using the combination of three main functions: graph simplification, triple ranking, and property selection. These functions are mostly initiated based on the interpretation of RDF data as knowledge units together with statistical analysis in order to deliver an easily-readable graph to users. A prototype is implemented to demonstrate the suitability and feasibility of the approach. It shows that the simple and flexible graph visualization is easy to read, and it creates the impression of users. In addition, the attractive tool helps to inspire users to realize the advantageous role of linked data in knowledge management.

Keywords: Graph simplification · Knowledge representation · Linked data · RDF visualization · Semantic web application · Triple ranking

1 Introduction

It is known that Semantic Web and Linked Open Data (LOD) technologies aim to enable the connection among pieces of data around the world, and turn them into a global knowledge space [1]. For this activity, Resource Description Framework (RDF) becomes a standard for representing explicit knowledge. Thus, all pieces of knowledge in every repository are expected to be stored in RDF format in order to have data be exchangeable and linkable across repositories via the Internet. At the moment, many organizations such as research institutes, governments, and industries start opening their own data. More local data are continuously interconnected through the LOD cloud. Thus, it can say that we are in the age of the growing world knowledge management system [1–3].

© Springer International Publishing Switzerland 2016
G. Qi et al. (Eds.): JIST 2015, LNCS 9544, pp. 23–39, 2016.
DOI: 10.1007/978-3-319-31676-5_2

Many pieces of research regularly manage RDF at the data tier in order to improve searching ability, because the advantages of knowledge representation and knowledge reasoning can construct rich machine-readable data in the form of a graph of knowledge [4]. Large amount of connected data are required, however, RDF data are mostly provided by tech users [5] or ones who know Semantic Web. Encouraging lay users [5], or ones who have less knowledge about Semantic Web, to contribute RDF data is very challenging, because they never realize how linked data work and RDF syntax itself is not user-friendly [6, 7]. It is resulted in a barrier between human and linked data.

For this reason, RDF data should be located not only at the data tier but also at the presentation tier in order to have users be familiar with Semantic Web. In this case, we question, "How users can access linked data in a suitable way?" Since a concept-map or a node-link diagram can enhance the learning ability from beginner to professional level, RDF graph visualization becomes a suitable way for enabling users to learn knowledge described in RDF and making them appreciate the role of linked data in knowledge management [8–10]. However, converting RDF data into an easily-readable graph visualization is difficult due to a lot of issues caused by the behaviors of RDF data together with the reasoning results. As we analyzed data, we found some significant issues. First, the graph is highly connected like a hairball due to inferred data, so it is too hard to be read by users. Second, since there is no ordering to the triples in an RDF graph, it is interrupting the flow of reading of users who pay attention to gain knowledge from a graph. Last, users are not convenient to focus what they want due to a large number of data presented.

This research aims to offer an approach to the presentation of RDF graph visualization as a learning tool by interpreting RDF data as knowledge structures. The following features are initiated to address the mentioned problems.

– Graph Simplification: To simplify a graph by removing some redundant triples that are resulted from ontological reasoning processes.
– Triple Ranking: To give a ranking score of each triple from common information (background content) to topic-specific information (main content), and to allow users to filter a graph based on this score.
– Property Selection: To allow users to filter a graph by selecting some properties in order to display or hide some triples.
– User Interaction: To control the above operations according to user demand.

This paper is organized as follows. The background and motivation are introduced in this section. Related work is reviewed in Sect. 2. The data are analyzed in Sect. 3. Our approach is described in Sect. 4. The prototype is demonstrated in Sect. 5. The outcome is discussed in Sect. 6. Last, the conclusion is drawn in Sect. 7.

2 Literature Review

There are pieces of research that work on the issue about RDF visualization. They aimed to operate a complex network in any visualization canvas to be friendly for general users.

We first reviewed some network visualization tools. Motif Simplification [11] considered some topologies of subgraphs, and replaced them with basic shapes such as diamonds, crescent, and tapered diamonds. It intended to give a big picture of a network rather than the detail of node-link. Gephi Open Viz Platform [12] is a powerful visualization tool that generated a well-shaped layout of network, allowed users to filter nodes and links, and had an option to set colors according to user preference. Both tools are suitable for general networks, but they are not designed for dealing with RDF data.

One important issue of RDF data is a large number of inferred links creating a hairball-like graph, so the tools should consider this behavior in order to simplify a graph. RDF Gravity [13] provided an interactive view. Users could zoom a graph to view much more detail, and get details of nodes in the focus area using text overlay. Next, Fenfire [14] gave an alternative view of RDF. It displayed the full details of the focused node and its immediate neighbors, but the other links were faded away according to the distance from the focus node. Both RDF Gravity and Fenfire offered well-organized displays, but they do not point out the issue of redundant data from inferred triples. Moreover, IsaViz [15] is an interactive RDF graph browser that used graph style sheets to draw a graph. It provided meaningful icons describing the type of each node such as *foaf:Person*, and grouped its metadata into a table in order to reduce highly interlinked data. It also allowed users to filter some nodes or properties to sparsify a dense graph, but this task required human effort to select some preferred URIs one by one.

The other issue is about the readability of RDF data, because RDF data are not well arranged for reading from introduction to main contents. Some works target to rank query triples. Several approaches used Term Frequency- Inverse Document Frequency (TF-IDF) to extract keywords from a content [16, 17]. PageRank [18] gave a score to each page by estimating the number of links and the quality of neighbors. TripleRank [19] ranked query result by applying a decomposition of a three-dimensional tensor that is originated by HITS [20] to retrieve relevant resources and predicates. Ichinose [21] employed the idea of TF-IDF to identify how important of resources and predicates of each subject under the same classification for ranking the query result. Nevertheless, they did not discuss about how to order triples for supporting the readability of users.

3   Preliminary Data Analysis

This research views that besides storing RDF data at the data tier, they should be presented at the visualization tier in order to have users to realize how importance of linked data in knowledge management. Using graph visualization for presenting knowledge is a suitable way for users to read and understand Semantic Web data [8, 10].

The well-displayed graph visualization should be simple and sparse [22]. In other words, it should be similar to the original RDF data; however, a query result contains both raw RDF data and inferred data because of the manner of a SPARQL engine. In this case, querying a graph by accessing the whole neighborhood of a given node within two hops is recommended to be a general input for this research due to the following scenario. Let *bt* be a short-hand writing of the transitive property named *skos:broaderTransitive*, if raw RDF data are ⟨Dog, bt, Mammal⟩, ⟨Mammal, bt, Animal⟩, and ⟨Animal, bt, LivingThing⟩; the inferred triples can be ⟨Dog, bt, Animal⟩,

⟨*Dog, bt, LivingThing*⟩, and ⟨*Mammal, bt, LivingThing*⟩. The well-displayed graph should be like Fig. 1(a), but in practice, it is hardly possible to obtain this result directly. Querying the information of the given node within one hop does not provide enough triples for constructing an informative structure of a graph, because the original triples are rarely found as shown in Fig. 1(b). In contrast, querying within two hops can maintain the mostly complete structure of the raw data as shown in Fig. 1(c), so it has an opportunity to be transformed into a simple graph by removing some inferred triples out of the query graph. The following expression can be used as a guideline to query the whole neighborhood of a given node *(uri)* within two hops.

```
CONSTRUCT { ?s ?p ?o.   ?o ?p1 ?o1. }
WHERE     { ?s ?p ?o.   ?o ?p1 ?o1.   FILTER(?s = <uri>) } .
```

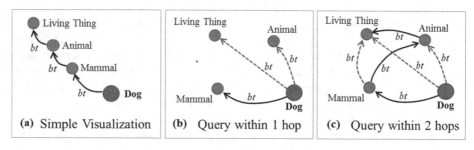

(a) Simple Visualization   **(b)** Query within 1 hop   **(c)** Query within 2 hops

Fig. 1. Query result of the given term "Dog". (Note: *"bt"* denotes *skos:broaderTransitive*, a black solid line indicates an original triple, a blue dashed line specifies an inferred triple, and a big yellow node represents the given node.)

Due to the nature of linked data and reasoning output, the according SPARQL statement usually forms some giant components in a graph. A hairball-like graph, as shown in Fig. 2, gives a bad impact to the readability of users and makes them be unsatisfied the way of learning and teaching using linked data. As we analyzed the query data from DBpedia [23] and LODAC [24] databases, we found two major issues.

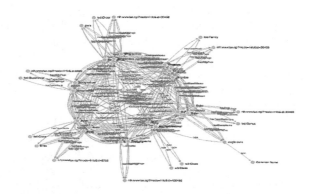

Fig. 2. Original RDF graph visualization from whole query result

Data Redundancy. A closer look at the data indicates that most inferred triples make a graph be highly complex. More than half of query triples are mainly formed by the reasoning results of *owl:sameAs*, *rdf:type* together with *rdfs:subClassOf*, and transitive properties. This behavior increases the average degree of the network and leads to have giant components, which have users be inconvenient to read.

Low Readability. In general, most well-organized articles such as academic papers prepare background knowledge of some essential concepts before bringing readers to the main content. Thanks to a well-outlined paper, beginners can understand it by reading from the beginning to the end, while experts of its domain can skip the introduction part and go to the main content directly. However, it is hardly possible to do with RDF data, because triples have no ordering. Thus, to give a ranking score to each triple is necessary.

In this case, we observed the distribution of URIs. It is found that the frequency of each URI in a query result *(fQ)* are distributed as shown in Fig. 3(a), where the horizontal axis shows individual URIs and the vertical axis shows the frequency of them. Several URIs have high degree. As we valued, most of them are important to display in a graph as key concepts. For example, if we query a term *dbpedia:Tokyo*, the high-degree URIs are *dbpedia:Tokyo*, *dbpedia:Japan*, *dbpedia:Honchu* (The island where Tokyo located), *rdf:type*, *owl:sameAs*, *dc:subject*, etc. The first three URIs are interesting because they are key concepts of "Tokyo", whereas the last three URIs are not much important for domain experts. Thus, we learned that using the frequency of each term in a query result alone is not enough. Next, we analyzed the frequency of every URI in a dataset *(fD)*, and compared each to the *fQ* chart one-by-one as shown in Fig. 3(b). This chart was drawn on a logarithmic scale because its distribution is extremely high variance. As *fD* of every URI found in the query result are estimated, a lot of high frequent ones are common properties such as *rdf:type*, *owl:sameAs*, *dc:subject*, etc. while *dbpedia:Tokyo*, *dbpedia:Japan*, and *dbpedia:Honchu* are not much high.

This characteristic of the data is meaningful. As query results are carefully analyzed, we found that URIs having high *fQ* can be treated as key concepts in the graph, while URIs having high *fD* indicate common information of the key concepts. This fundamental analysis will be utilized for ranking triples in the next section.

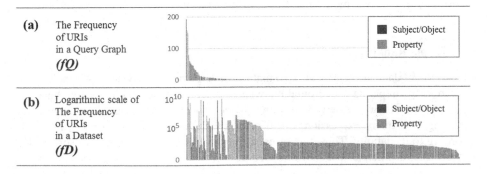

Fig. 3. Statistical analysis of URIs in the query result from DBpedia.

4 Proposed Approach

As we discussed, to motivate users to consume and contribute RDF data, they have to familiar with the knowledge representation of linked data. In this case, graph visualization is a suitable way to reduce a gap between human and Semantic Web. Understanding knowledge from a graph is quite challenging, because a graph is just a mathematical graph containing nodes and edges. In order to deliver graph-based knowledge to readers, an application should interpret all nodes and links as knowledge structures and make decision to maintain or eliminate some triples. To achieve this goal, we have to address the issues that are mostly discussed in the previous section. Thus, this research is initiated to serve the following purposes.

- To simplify a complex graph by removing redundant triples which are resulted from ontological reasoning.
- To serve different subgraphs on the basis of reading levels from common to topic-specific information.
- To filter a graph based on user preference.

4.1 Graph Simplification

As we mentioned, some well-prepared RDF repositories did reasoning on ontologies in order to support a SPARQL service, however, the inferred triples resulted in having giant components in a graph. As we investigate, equivalent or same-as instances (*owl: sameAs*), transitive properties (e.g. *skos:broaderTransitive*), and hierarchical classification (*rdf:type* together with *rdfs:subClassOf*) are commonly found in any complex RDF graph. Thus, this method aims to remove some redundant triples automatically by using rules that are defined in Table 1 and some description as follows:

- *R.1 – R.3:* To merge some same-as nodes into one and remain only unique links.
- *R.4:* To remove implicit links that resulted by the chain of transitive links.
- *R.5:* To remove inferred links that caused by hierarchical classification.

Several rules use the occurrence number of a URI counted across data repositories in order to choose the most popular node from a same-as pair, because it has high opportunity to discover more knowledge in the next query.

4.2 Triple Ranking

The section of data analysis mentioned that the arrangement of any content is necessary for readers by preparing background knowledge at first in order to understand the main content well. As we reviewed, existing works focused on seeking relevant data according to a query expression, but they less mentioned about how to order them according to readability. Thus, this research introduces a simple method to sort triples on the basis of different levels of knowledge.

Table 1. A set of rules used to simplify a highly connected graph resulted by inferred triples. (Note: The term *fD(uri)* is a frequency of a URI occurred in datasets.)

Rule	Triples	Condition	Display only
To merge nodes			
R.1	s_1 —p_1→ o_1; s_1 owl:sameAs s_2; s_2 —p_2→ o_1	$fD(s_1) > fD(s_2)$	s_1 —p_1, p_2→ o_1
	s_1 —p_1→ o_1; o_1 owl:sameAs o_2; s_1 —p_2→ o_2	$fD(o_1) > fD(o_2)$	s_1 —p_1, p_2→ o_1
R.2	s_1 —p_1→ o_1; s_1 owl:sameAs s_2; s_2 —p_2→ o_2	$fD(s_1) > fD(s_2)$	s_1 —p_1→ o_1; s_1 —p_2→ o_2
	s_1 —p_1→ o_1; o_1 owl:sameAs o_2; s_2 —p_2→ o_2	$fD(o_1) > fD(o_2)$	s_1 —p_1→ o_1; s_2 —p_2→ o_1
R.3	s_1 —p_1→ o_1; o_1 owl:sameAs o_2; s_2 —p_2→ o_2	$fD(o_1) > fD(s_2)$	s_1 → o_1; o_1 —p_2→ o_2
To remove links			
R.4	s_1 —p_1→ o_2; s_1 —p_1→ o_1; o_1 —p_1→ o_2	:p1 rdf:type owl:TransitiveProperty .	s_1 —p_1→ o_1; o_1 —p_1→ o_2
R.5	x —rdf:type→ C_2; x —rdf:type→ C_1; C_1 —rdfs:subClassOf→ C_2		x —rdf:type→ C_1; C_1 —rdfs:subClassOf→ C_2

A general article contains different roles of concepts. **General Concepts** are terms that are commonly known such as "name", "address", and "class", and they are always found in a corpus. Besides, **Key Concepts** are important terms that are always found in the given article but rarely found in a dataset. The key concepts are more relevance to the given article rather than the general ones. In terms of RDF, concepts are resources (including subjects and objects) and properties. The key concepts always present thorough the article, while general concepts are used as composition information for giving background knowledge of the key concepts as shown in Fig. 4(a).

In addition, different levels of information are defined. **Common Information** explains background knowledge that supports readers to understand the main content. It generally gives introduction of key concepts by using general terms. It means that triples being common information consist of general concepts rather than key concepts as shown in Fig. 4(b). In contrast, **Topic-Specific Information** contains specific terms that are highly relevance to the article. Thus, some triples acting as topic-specific information comprise of key concepts rather than general concepts as shown in Fig. 4(c).

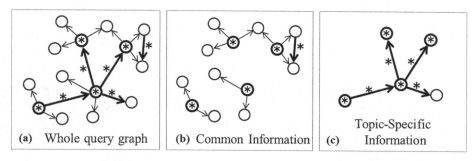

(a)   Whole query graph   (b) Common Information   (c)   Topic-Specific Information

Fig. 4. Graphs display the idea of common information and topic-specific information. Nodes and links with stars (*) indicate key concepts, whereas the others are general concepts.

The level of each concept is valued based on a query result. As we analyzed, the key concepts are commonly found in the query result but they are rarely found in the dataset, while the general concepts are frequently appeared in the dataset. This manner is consistent with the TF-IDF method, however an RDF dataset contains only separated triples but not documents of many words, so this method has to be adapted for RDF data.

In this research, we intend to define that a key concept has higher score than a general concept, so the scoring function of a URI *(w(uri))* is the occurrence number of a URI in a query result *(fQ(uri))* weighted by the its occurrence number found in datasets *(fD(uri))*. Since the data analysis informed that the variance of *fD(uri)* is extremely high, logarithm is taken for this term. The function *w* is defined by the equation:

$$w(uri) = \frac{fQ(uri)}{log(fD(uri) + 1)}$$

Next, a function named Visualization-Weight *(vw)* is defined to measure a triple $(\langle s,p,o \rangle)$ that should be in the direction of common or topic-specific information. It is the summary of weighting scores of subject (s), predicate (p), and object (o) of a triple as presented by the following equation.

$$vw(s,p,o) = \frac{\alpha \cdot w(s) + \beta \cdot w(p) + \gamma \cdot w(o)}{\alpha + \beta + \gamma}$$

The coefficients (α, β, and γ) of these terms are 1.0 by default; however, they can be adjusted if some domains place important to each term differently.

For example, the scores, $vw(\langle :Aves,:hasTaxonName,:Birds\rangle) = 0.33$ and $vw(\langle :Aves,:hasParentTaxon,:Coelurosauria\rangle) = 0.59$, show that the former is in the direction of common information, while the latter is more likely to be topic-specific information.

4.3 Property Selection

Moreover, although the problems discussed in the previous part can be solved, there are much more triples remained in the visualization. These data contain both necessary and unnecessary triples for readers. Since users have their own expectation to view a graph, they should customize the graph based on their interest by themselves. They always prefer to filter a graph by selecting only properties that they are interested.

This additional method named "Property Selection" is lastly described in this paper. The method helps users to focus on information that they desire to view. It is a simple technique that is always found in any visualization tool. In addition, we learn that most triples related to RDF, RDFS, and OWL are not needed by readers, for example, $\langle foaf:$ $Person, rdf:type, rdfs:Class\rangle$, $\langle foaf:Person, rdfs:subClassOf, foaf:Agent\rangle$, $\langle foaf:Person,$ $owl:disjointWith, foaf:Organization\rangle$, etc. In this case, filter out some of these properties and resources consume much user effort. Thus, the triples containing some vocabularies from RDF, RDFS, and OWL can be removed from a graph by considering the namespaces of subjects, predicates, and objects.

5 Prototype

The proposed approach originates an idea to organize RDF data for graph visualization. In order to verify the suitability and the feasibility of the proposed methods, a prototype has been developed.

5.1 User Requirement

Apart from the data analysis, we have gathered requirements from different users who have different levels of experience with Semantic Web and domain knowledge. In this part, the requirements from users are summarized into the following topics.

General Requirements.

- An application should provide different input interfaces for different types of users. A simple interface allows users to enter a URI, and then a graph is automatically queried. Besides, tech users are allowed to input a SPARQL expression for the advanced query.
- It is known that URIs are fundamental components in Semantic Web, and they are used as identifiers for machine-readable data on the web. However, most of them are difficult to be read by lay users. Thus, in a visualization, it should display human-readable labels in a graph for general users by default, and also provide an option to display URIs for tech users.

- Users are possible to move any node in the graph.
- In an RDF statement, its subject and property are URI, and object can be either URI or literals. Pairs of a datatype property and a literal node are used to be metadata of only one resource, and literal nodes may be long string, so they are not suitable to display in the limit area of the node-link diagram. Since these data are useful for readers, they should be displayed in another panel that users can access them easily.

Graph Simplification.

- Users can simplify a graph by merging same-as nodes and removing transitive links.

Triple Ranking.

- Since users have different background knowledge in a specific topic, beginners may interested in reading common information before getting topic-specific information, while experts may prefer to read only topic-specific information. Thus, the application should dynamically alter a graph according to the level of knowledge that users can customize and access on demand.

Property Selection.

- Users can select only properties that they prefer to view.
- Some triples containing vocabularies from RDF, RDFS, and OWL can be ignored.

5.2 Implementation

According to the user requirements, the prototype is implemented on the basis of the following features.

- Graph Simplification: To simplify a graph by removing redundant triples.
- Triple Ranking: To give ranking scores to triples based on common and topic-specific information.
- Property Selection: To filter a graph by selecting preferred properties.
- User Interaction: To control a graph according to user demand.

For this prototype, the functional diagram that described user actions and system workflows is shown in Fig. 5, the user interface is demonstrated in Fig. 6, and example graphs that are resulted from user actions are displayed in Fig. 7. The prototype is accessible at the following URL.

http://rc.lodac.nii.ac.jp/rdf4u/

This prototype is a web application that is mainly developed using the force layout of the D3 JavaScript library[1]. The main user scenarios are described in the following topics together with each step in Figs. 5, 6 and 7.

---

[1] Data-Driven Documents (D3) http://d3js.org/.

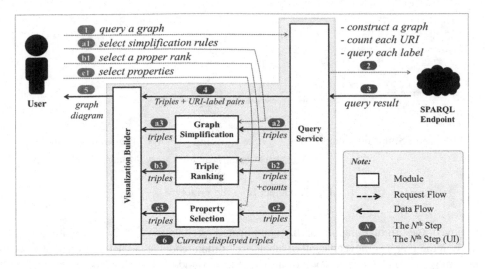

Fig. 5. Functional diagram

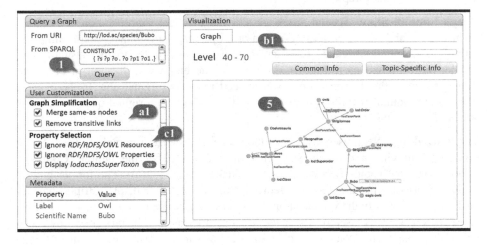

Fig. 6. User interface

General Requirements. First, the main flow visualization is the query of graph. Users request a graph (Step 1 in Fig. 5) by giving either a single URI or a SPARQL CON-STRUCT statement (Step 1 in Fig. 6). After that, the module "Query Service" forwards the query statements for getting a graph, counting the number of each URI, and inquiring the label of each URI to any SPARQL endpoint (Step 2 in Fig. 5). When the result is returned to the Query Service (Step 3 in Fig. 5), it forwards to the module "Visualization Builder" (Step 4 in Fig. 5). Then, the Visualization Builder generates graph visualization to users (Step 5 in Figs. 6 and 7). Since inferred triples are also

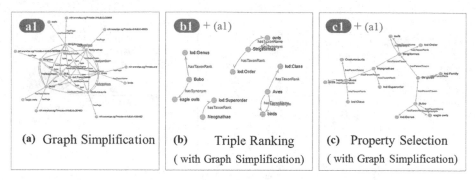

(a) Graph Simplification	**(b)** Triple Ranking	**(c)** Property Selection
	(with Graph Simplification)	(with Graph Simplification)

Fig. 7. Output from the prototype

retrieved, the original graph is highly complicated as shown in Fig. 2. In addition, each node in the graph is moveable, all of labels are human readable, a URI is shown when a user move a pointer over a node or a link. When a node is double-clicked, the literal information of a node is shown (In panel "Metadata" in Fig. 6). Moreover, every displayed triple is synchronized to the Query Service to be input data for other modules in next user actions (Step 6 in Fig. 5).

Graph Simplification. Second, users are allowed to select simplification rules (Step a1 in Figs. 5 and 6). When users click on any options, the module "Graph Simplification" executes some related rules and forwards result triples to the Visualization Builder (Step a2 and a3 in Fig. 5). As a result, the graph visualization in Fig. 7(a) shows that the simplified graph is easier to read than the original one. In the experiment, some redundant triples that are about 50–70 % of the original query graph are removed during this process.

Triple Ranking. Next, users can select the range of visualization ranking (Step b1 in Fig. 5) by moving a two-way slider bar or clicking either the button "Common Information" or "Topic-Specific Information" (Step b1 in Fig. 6). The former button displays triples having the lower vw score, while the latter one displays triples having the higher vw score. In the visualization tier, the integer number indicated the percentile of vw score is used as visualization level, because it is easier to be recognized by users rather than using the floating number of the vw value. Then, the module "Triple Ranking" computes and returns the triples that satisfy user input (Step b2 and b3 in Fig. 5). The result of this action together with the graph simplification is shown in Fig. 7(b). It displays common information that contains some key concepts and some general concepts in order to give background knowledge of the key concepts.

Property Selection. Last, users can customize a graph by selecting only preferred properties (Step c1 in Fig. 5). The interface allows users to hide resources and predicates that are vocabularies of RDF, RDFS, and OWL; and to show triples containing selected properties (Step c1 in Fig. 6). Then, the module "Property Selection" filters the triples according to the user input, and forwards the result to the Visualization Builder (Step c2 and c3 in Fig. 5). An example result of this scenario together with the graph simplification is shown in Fig. 7(c).

In summary of this section, the prototype demonstrates that our approach is pos-sible and suitable to implement. The features that we provide satisfy all requirements that we have previously discussed.

6   Discussion

This research aims to provide a suitable RDF graph visualization that users are easily to consume knowledge by learning from relationship among concepts. Thus, the three main methods: graph simplification, triple ranking, and property selection, are proposed to deliver an easily-readable graph to readers. The first and the second methods are major contribution, while the last one is an additional method used for fulfilling some minor requirements. In this paper, we intend to introduce the according methods rather than a new fully-functioned visualization tool. Thus, this section points to the dis-cussion about the usefulness, uniqueness, novelty, and prospect of this research.

6.1   Usefulness

Since a graph generated by RDF data is complicated by nature, users are not con-vention to read and understand knowledge from a graph. The analysis of mathematical features of a graph alone is not enough to simplifying the complexity of an RDF graph, because the RDF graph has semantic relationships that should be interpreted as knowledge structures. We carefully examined the actual behavior of RDF datasets, and found that the semantic structure of the datasets is meaningful in terms of knowledge representation, and it is useful for our research. The observation includes data redun-dancy such as same-as nodes and inferred relationships. When same-as nodes are merged and some inferred triples are filtered out by the simplification rules, some giant components in a network are eliminated, so the interactive graph on two-dimensional canvas becomes more sparse and convenient for users to control and read.

In addition, the degree of importance of triples such as distinction between common and topic-specific information was also investigated. For this reason, we have to realize the importance of triples depended on the expertise level of users. For domain experts, only topic-specific information is needed to show, while common information should be more emphasized for beginners. A case of multiple links between two nodes caused by the hierarchy of property demonstrates how this method is suitable for arranging data for readers. In general, a super property in an upper ontology is labeled by a common vocabulary describing the broader meaning, while a sub property is used by a specific domain. After reasoning, the number of a super property is certainly greater than the number of a sub property, so the super property trends to be displayed at the common level while the sup property often appears at the topic-specific level.

6.2   Uniqueness

The uniqueness of this research is discussed by functional comparison. The functionality of some visualization tools: Motif [11], Gephi [12], RDF Gravity [13], Fenfire [14], and IsaViz [15]; are studied according to the key methods of this research.

Graph Simplification. There are several works support this feature but the strategies are different. Motif replaces a dense component by an abstract shape, so a graph seems simple, but its detail is omitted. Gephi uses mathematical characteristics of a graph such as a node degree and a weight on edge, but it does not employ the knowledge structure of Semantic Web to reduce some redundant links. FenFire fades away some far nodes, but the subgraph including the focused node and its neighbors can produce giant components. Next, RDF Gravity and IsaViz can simplify a graph by having users to query inside the graph or select some URIs to be visible or hidden. However, they less discuss about options to merge same-as nodes and remove transitive links, which are the main issues of having dense parts in a graph. Unlike these existing tools, our approach adopts Semantic Web rules to interpret data and eliminate this issue automatically.

Triple Ranking. The according visualization tools do not mention about a way to arrange contents in a graph for serving different levels of knowledge to different users. A workaround is to filter some resources or properties based on user interest, but users have to put their effort to learn what they want to view and how to filter. Thus, in this case, our work provides a smart way to solve this issue by analyzing the statistical feature of data and then automatically ordering a graph from common to topic-specific information.

Property Selection. Filtering a graph by selecting preferred properties is a common feature that most visualization tools provide. Our work was implemented in the same way. In addition, we added an option to show or hide triples containing some vocabularies from RDF, RDFS, and OWL automatically, so users do not have to remove them one by one.

In summary, considering these three features, our solution has advantage over the existing visualization tools because our approach does not only allow users to customize a graph but also automatically deliver an easy-readable graph based on the knowledge interpretation and the statistical analysis of Semantic Web data.

6.3   Novelty

Due to the contradictory requirements from different types of users, we adapted TF-IDF method for ordering triple from common to topic-specific levels. The degree of commonness versus specificity is calculated by evaluating the nature of the dataset with the algorithm. After that, the RDF visualization application is designed to allow users to choose how common or domain-specific information that they need by clicking a button or controlling a two-way slider bar. The prototype was demonstrated and it got positive impression from users. Moreover, it can be considered that this work is a novel approach because it operates a graph at the knowledge level by concerning domain independent, so this approach is applicable to any domain datasets.

6.4 Prospect

Since the arrangement of triples for reading is a novel approach, it has opportunity to be value-added by the community of Semantic Web researchers. This approach can be extended by applying various algorithms in order to satisfy diverse characteristics of data in other domains. We are going to apply this system as a learning and teaching tool for a specific domain such as biodiversity informatics [25, 26], because a graph diagram can enhance the learning of biology [10], and it can be clearly to identify the level of users from beginners (e.g. high school students) to experts (i.e. researchers).

Moreover, as our observation, although this RDF graph visualization application does not give technical knowledge of RDF to lay users directly, it makes them appreciate and understand the role of linked data for future knowledge management. This one important task that attempts to break a barrier between human and Semantic Web.

7   Conclusion

This paper aims to deliver a suitable RDF graph visualization for every level of users, because a node-link diagram can enhance the learning ability of users and the amount of LOD is positively growing. Since the nature of RDF data makes a graph be complicated, it is difficult for users to read and understand. As we analyze, the root causes are data redundancy due to inferred triples, and low readability due to lacking of reading flow in a RDF graph. This research initiates three main methods to support readers. First, Graph Simplification executes the proposed Semantic Web rules to remove some inferred data. Second, Triple Ranking prepares different sections of a graph from common to topic-specific information for different levels of users by adapting TF-IDF algorithm for an RDF graph. Last, Property Selection is additionally developed to allow users to display or hide triples by selecting some properties, and to help users to filter some triples containing some vocabularies from RDF, RDFS, and OWL. These methods mostly use the statistical feature of a RDF graph together with the interpretation of RDF data as knowledge structures in order to produce an easily-readable node-link diagram for readers. The prototype is implemented by including interactive RDF visualization in order to verify the suitability and feasibility of our approach. It demonstrates that our methods can be developed on the basis of today's technologies and the prototype enables users to realize the power of Semantic Web and LOD for enhancing the ability of knowledge management.

In future, we plan to measure the expertise level of users, and allow system to adjust the visualization by applying various algorithms for different domain-specific datasets in order to deliver a more appropriate RDF graph for all levels of users.

References

1. Heath, T., Christian, B.: Linked data: evolving the web into a global data space. Synth. Lect. Semant. Web: Theory Technol. **1**(1), 1–136 (2011)
2. Suchanek, F., Weikum, G.: Knowledge harvesting in the big-data era. In: Proceedings of the 2013 ACM SIGMOD, pp. 933–938. ACM (2013)
3. Bizer, C., Heath, T., Berners-Lee, T.: Linked data-the story so far. In: Semantic Services, Interoperability and Web Applications: Emerging Concepts, pp. 205–227 (2009)
4. Hitzler, P., Krotzsch, M., Rudolph, S.: Foundations of Semantic Web Technologies. CRC Press, Boca Raton (2009)
5. Dadzie, A.-S., Rowe, M.: Approaches to visualising linked data: a survey. Semant. Web J. **2**(2), 89–124 (2011)
6. Bezerra, C., Freitas, F., Santana, F.: Evaluating ontologies with competency questions. In: Web Intelligence (WI) and Intelligent Agent Technologies (IAT), vol. 3 (2013)
7. Zemmouchi-Ghomari, L., Ghomari, A.: Translating natural language competency questions into SPARQLQueries: a case study. In: The First International Conference on Building and Exploring Web Based Environments, pp. 81–86 (2013)
8. Schwendimann, B.: Concept maps as versatile tools to integrate complex ideas: from kindergarten to higher and professional education. Knowl. Manage. E-Learn. **7**(1), 73–99 (2015)
9. Edelson, D., Gordin, D.: Visualization for learners: a framework for adapting scientists' tools. Comput. Geosci. **24**(7), 607–616 (1998)
10. Liu, S., Lee, G.: Using a concept map knowledge management system to enhance the learning of biology. Comput. Educ. **68**, 105–116 (2013)
11. Dunne, C., Shneiderman, B.: Motif simplification: improving network visualization readability with fan, connector, and clique glyphs. In: SIGCHI, pp. 3247–3256 (2013)
12. Mathieu, B., Sebastien, H., Mathieu, J.: Gephi: an open source software for exploring and manipulating networks. In: AAAI 2009 (2009)
13. Goyal, S., Westenthaler, R.: RDF Gravity (Rdf Graph Visualization Tool). Salzburg Research, Austria (2004)
14. Tuukka, H., Cyganiak, R., Bojars, U.: Browsing linked data with Fenfire. In: LDOW 2008 at WWW 2008 (2008)
15. Pretorius, J., Jarke, J., Van, W.: What does the user want to see? What do the data want to be?. Inf. Vis. **8**(3), 153–166 (2009)
16. Lee, S., Kim, H.J.: News keyword extraction for topic tracking. In: NCM 2008 (2008)
17. Li, J., Zhang, K.: Keyword extraction based on tf/idf for Chinese news document. Wuhan Univ. J. Nat. Sci. **12**(5), 917–921 (2007)
18. Brin, S., Page, L.: The anatomy of a large-scale hypertextual web search engine. In: WWW 1998 (1998)
19. Franz, T., Schultz, A., Sizov, S., Staab, S.: TripleRank: ranking semantic web data by tensor decomposition. In: Bernstein, A., Karger, D.R., Heath, T., Feigenbaum, L., Maynard, D., Motta, E., Thirunarayan, K. (eds.) ISWC 2009. LNCS, vol. 5823, pp. 213–228. Springer, Heidelberg (2009)
20. Kleinberg, J.: Authoritative sources in a hyperlinked environment. J. ACM (JACM) **46**(5), 604–632 (1999)
21. Ichinose, S., Kobayashi, I., Iwazume, M., Tanaka, K.: Ranking the results of DBpedia retrieval with SPARQL query. In: Kim, W., Ding, Y., Kim, H.-G. (eds.) JIST 2013. LNCS, vol. 8388, pp. 306–319. Springer, Heidelberg (2014)

22. Novak, J., Cañas, A.: The theory underlying concept maps and how to construct and use them. In: Florida Institute for Human and Machine Cognition (2006)
23. Lehmann, J., et al.: DBpedia – a large-scale, multilingual knowledge base extracted from Wikipedia. Seman. Web J. **6**(2), 167–195 (2015)
24. Minami, Y., Takeda, H., Kato, F., Ohmukai, I., Arai, N., Jinbo, U., Ito, M., Kobayashi, S., Kawamoto, S.: Towards a data hub for biodiversity with LOD. In: Takeda, H., Qu, Y., Mizoguchi, R., Kitamura, Y. (eds.) JIST 2012. LNCS, vol. 7774, pp. 356–361. Springer, Heidelberg (2013)
25. Chawuthai, R., Takeda, H., Hosoya, T.: Link prediction in linked data of interspecies interactions using hybrid recommendation approach. In: Supnithi, T., Yamaguchi, T., Pan, J.Z., Wuwongse, V., Buranarach, M. (eds.) JIST 2014. LNCS, vol. 8943, pp. 113–128. Springer, Heidelberg (2015)
26. Chawuthai, R., et al.: A logical model for taxonomic concepts for expanding knowledge using Linked Open Data. In: Workshop on Semantics for Biodiversity (2013)

Linked Open Vocabulary Recommendation Based on Ranking and Linked Open Data

Ioannis Stavrakantonakis[✉], Anna Fensel, and Dieter Fensel

University of Innsbruck, STI Innsbruck, Technikerstr. 21a, 6020 Innsbruck, Austria
{ioannis.stavrakantonakis,anna.fensel,dieter.fensel}@sti2.at

Abstract. The vocabulary space of the Semantic Web includes more than 500 vocabularies according to the Linked Open Vocabularies (LOV) initiative that maintains the directory list and provides search functionality on top of the curated data. Domain experts and researchers have populated it to facilitate the interpretation and exchange of information in the Web of Data. The abundance of vocabularies and terms available in the LOV space, on one hand aims to cover the major knowledge management needs, but on the other hand it could be cumbersome for a non-expert or even a vocabulary expert to find the correct way through the collection. To address this problem, we present an approach that helps to identify the most appropriate set of LOV vocabulary terms for a given Web content context by leveraging the existing dynamics within the LOV graph and the usage patterns in the LOD cloud. The paper describes the framework architecture that enables the discovery of vocabularies; it focuses on the corresponding metrics and algorithm, and discusses the outcomes of the applied experiments.

1 Introduction

Existing crawling based surveys about the Web of Data and the semantic annotations presence show that there is a long distance to cover in order to see structured data broadly provided by the websites. According to Common Web Crawler[1] only ca. 17 % (2,722,425 out of 15,668,667) of domains were found with triples in Q4 2014. Based on the corpus analysis by [3,11], the growth rate year over year could be considered significant, however, the vocabulary discovery and application ease still remains questionable. Focusing on specific domains, we realise that local businesses in business areas heavily relying on the Web presence, are far behind the structured data paradigm with triples existence ratio to be significantly lower as presented in [17], which shows only a 5 % triples inclusion in hotel websites.

Further investigating, it is relevant for the motivation frame of the proposed approach to analyse the level of misuse of vocabularies. Exposed structured data by using semantic annotations could be wrongly realised by using vocabularies in the wrong way or simply having wrong format. In both cases, the result

---

[1] http://www.webdatacommons.org/structureddata/.

© Springer International Publishing Switzerland 2016
G. Qi et al. (Eds.): JIST 2015, LNCS 9544, pp. 40–55, 2016.
DOI: 10.1007/978-3-319-31676-5_3

would not be accessible by a search engine or other parsing processes. Hogan et al. in [8] present some of the issues that they discovered while systematically examining existing Linked Data. They discovered many error types including syntax errors, "ontology-hijacking" referring to misuse of the terms semantics, while ca. 15 % of triples were using undeclared property URIs. Meusel et al. in [10] demonstrate their findings about common errors in deployed microdata that implement schema.org annotations and provide heuristics for fixing them at the consumer side. Proof of the complexity in creating annotations is the fact that according to their study over half of the examined sites (56.58 %) confuse object properties (OP) with datatype properties, at least once, by using OPs with a literal value instead of referring to an instance of a class.

Taking the above mentioned figures in consideration, it is apparent that there is a strong need in assisting the data and content publishers with methodologies and tools not only to weave annotations, but also to generate them in the correct way. Moving forward in addressing this problem we make the hypothesis that it is possible to automatise steps in the process of producing annotations that can positively affect both the results of it and the learning curve. In brief, the presented approach is positioned in the wider direction of facilitating the uptake of vocabularies and creation of structured data out of the already published content in the Web. In this context, we leverage the dynamics between the vocabularies in the LOV space [18], taking in consideration the collective knowledge as that is expressed through the Linked Open Data (LOD) cloud in order to provide a methodology and algorithm that will help to produce annotations on an existing website or data node.

The remainder of the paper describes related approaches that aim to address similar obstacles in the process of creating semantic annotations in Sect. 2; presents the proposed approach, namely the LOVR framework in Sect. 3; evaluates the approach in Sect. 4; and closes with our conclusions in Sect. 5.

2   Related Work

The Linked Open Vocabulary initiative [18] has a fundamental role in lowering the barriers of the Semantic Web adoption in the plateau of productivity by providing a curated directory of vocabularies. Each vocabulary is accompanied with a profile page[2] that provides useful metadata about the vocabulary itself, e.g. the namespace, the number of classes and properties of the latest version, the number of incoming and outgoing links, the versions history, the authors and the raw vocabulary schema in N3 notation. In addition, a number of discovery interfaces have been implemented including a search for the vocabularies and the terms in order to find the most relevant resources for a given keyword; a SPARQL endpoint to query the data repository; and a JSON based REST API that facilitates the integration of the search functionality with external applications, like the approach that we present in this paper.

---

[2] E.g. the schema.org profile is available under the URL: http://lov.okfn.org/dataset/lov/vocabs/schema.

Defining the related work space of the presented approach, we mainly consider work that is relevant to vocabularies recommendation, ranking of vocabularies and usages of the LOV data in general. To the best of our knowledge based on our research there are not many endeavours aiming to tackle the problem that was defined in the introduction section.

The closest approach to our topic, as presented by Atemezing and Troncy in [1], examines the problem of vocabularies recommendation based on a ranking metric that has been developed by introducing the concept of Information Content (IC) to the context of LOV. In comparison to the methodology proposed in [1], we follow a more holistic approach by starting earlier in the funnel of searching for vocabulary terms to annotate a given web page content, while the IC approach aims only to rank the vocabularies. However, we still consider it relevant to the *LOVRank* method that we present later in Subsect. 3.1.

Furthermore, Butt et al. in [4,5] aim to address the ontology ranking problem by introducing the DWRank algorithm. DWRank consists of two main scores the Hub score and the Authority score, which measure the centrality of the concepts within the ontology and the ontology authoritativeness (i.e. the importance of the ontology in the ontologies space), respectively. DWRank and the proposed approach share the same perspective of ranking the concepts defined within the ontologies in order to find the best match to a keyword search. One of the main differences in comparison to the proposed approach is related to the consideration of the LOD cloud as an input to the algorithm, which is included in our approach as presented later.

Researching about related work in a wider perspective that covers the vocabulary reuse, the work presented by Schaible et al. in [13], aims to support the ontology engineer by providing a methodology that guides the creation of Linked Data through the best practices for modelling new entities [7]. The approach is mainly based on Swoogle[3] and the SchemEX index and consists of an iterative process, where each iteration cycle finishes with the definition of one or several mappings of data to vocabulary terms. Compared to our approach it differs at fundamental decisions, like using the Swoogle service instead of the LOV directory, which has a more up to date coverage as Swoogle seems to have stopped being updated before the launch of schema.org [6]. Moreover, it is intended to be used by ontology experts and not by non-specialised developers in ontologies; and also it does not introduce a ranking method.

Moreover, studying the survey presented in [14] about various existing strategies to vocabulary reuse, the reader can extract a few useful insights that hail from the collective intelligence of ontology experts that is captured in the study via questions and tasks about vocabularies reuse. One of the interesting facts that is highlighted as relevant to our methodology is the fact that using popular terms from popular vocabularies is preferred over using mainly one domain specific vocabulary that covers the needs of the given data.

Part of our proposed methodology employs the vocab.cc service [15], which provides metrics about the LOD usage patterns of the various vocabulary terms

---

[3] http://swoogle.umbc.edu/.

that appear in the LOD cloud. Vocab.cc has analysed the crawled dataset of the Billion Triples Challenge Dataset (BTCD) [9], and applied two frequency measures to the identified class and property URIs, i.e. (a) overall frequency of URI use in the BTCD; and (b) document frequency of URI use (how many different documents refer to them). Apart from the vocab.cc stats, another source of related metrics is the LOD stats [2], which aims to provide a comprehensive picture of the current state of the Data Web. The algorithm presented later in Subsect. 3.3 leverages this usage information to filter and rank the candidate terms that will qualify to the final recommendation set of terms.

3   The LOVR Framework

The Linked Open Vocabulary Recommendation (LOVR) framework as described within this section, proposes an approach to address the above mentioned problem by suggesting a set of vocabularies and terms that can be used to annotate a given webpage, relying on scores related to the importance of the various vocabulary terms in the Web of Data. The usage of the LOVR framework is relatively simple and aims to facilitate the selection of vocabulary terms by web developers. The tangible goal of the framework is to give a list of candidate terms to the end user and save her time in the process of finding the appropriate terms from the abundance of terms across the more than 500 vocabularies of the Linked Open Vocabularies space.

The overall workflow starts by giving the target webpage at the input and then receiving at the output a set of vocabulary terms from the LOV space. The framework's architecture includes all the necessary modules to map the webpage content to a set of vocabulary terms. As shown in Fig. 1, a natural language processing (NLP) module initiates the process by enabling the extraction of concepts and important tokens from the webpage content. Those tokens are then used as input to the *LOV search* module to find related terms within the vocabularies of the LOV directory. The LOV search stands for both the discovery of vocabulary terms; and the exploration of the LOV metadata about the vocabularies. The results are enriched with useful metadata about the terms, including two important for our approach: (a) the relevance score of the vocabulary term compared to the query token, and (b) the usage of the term within the Linked Open Data cloud according to the datasets that LOV is aware. Furthermore, the LOV vocabulary profile pages, as it has already been said in Sect. 2, provide useful metadata about the vocabulary itself, from which we leverage the metrics about the number of incoming and outgoing links, which is explained in details later in Subsect. 3.1, in order to compute a ranking score for the vocabulary within the LOV space. This ranking is computed within the *LOV Ranking* module of Fig. 1.

Figure 1 uses dashed lines to mark the nodes that represent external systems or services, i.e. the *NLP, NER* node, the *LOV* service and the *vocab.cc* service. The *LOD Search* step in Fig. 1 of the architecture refers to the retrieval of usage patterns about the specific vocabulary terms of the T set in the Linked Open Data cloud.

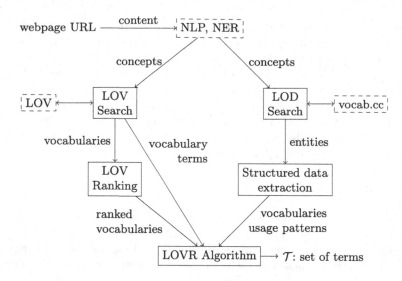

Fig. 1. The LOVR architecture.

The above described results of the *LOV Seach* and the *LOV Ranking* components in conjunction with the *LOD usage patterns* are passed to the core module, the *LOVR algorithm* presented in Subsect. 3.3, to compute a further score for each proposed term. This score is used to decide on the relevance of the term to the given webpage context. The score of each term, or the *Linked Open Term Ranking (LOTR)* as we call it, is returned as part of the output with the list of the candidate terms. Before demonstrating the algorithm in detail, the two dimensions that comprise the fundamental pillars of the algorithm, i.e. the LOV ranking methodology and the LOD terms usage patterns, are presented respectively in the Subsects. 3.1 and 3.2. The rationale behind considering those two dimensions as important for the result term's score is the need of parameters that reflect the importance of the shortlisted candidate terms within the vocabulary ecosystem. The LOV ranking of a vocabulary is definitely a way to evaluate how important and widely accepted it is by the vocabularies creators. Moreover, the usage of a term within the LOD cloud demonstrates the acceptance that a specific terms has by the data producers in LOD. Thus, in a nutshell the idea is to put those two perspectives at the ground base of the framework and incorporate insights extracted from the existing contributions of ontology experts (both vocabulary creators and data developers).

3.1   LOV Ranking

The purpose of introducing a ranking score across the LOV space is to facilitate the exploration of the most used vocabularies and reuse those that are widely accepted by the community. Inspired by the work of Page and Brin in [12] about the PageRank method, which measures the relative importance of web pages, we introduce the *LOVRank* that aims to unveil the importance of vocabularies within the LOV graph.

Definition 1. *If $B_V(t)$ is the number of the backlinks to the vocabulary V of the term t, $V \in LOV$, then the B_V is divided by the total number of vocabularies to represent the ranking of the vocabulary in the LOV:*

$$VR_{LOV,V}(t) = \frac{B_V(t)}{|LOV|}$$

The number of backlinks to the vocabulary B_V, is marked as "incoming links" in the vocabulary LOV profile page. We divide B_V with the number of available vocabularies to normalise it, therefore $VR_{LOV,V} \in [0,1]$. According to the proposed formula, the number of the backlinks of a vocabulary are proportional to the LOV ranking score, to materialise the reasoning that was explained at the beginning of the section, which assumes that the more central in the LOV graph a vocabulary is, the more acceptance and impact it has.

Table 1. LOV vocabulary ranking examples.

Vocabulary V	B_V	$VR_{LOV,V}$
dbpedia-owl:	7	0.01
dcterms:	403	0.78
event:	36	0.07
foaf:	307	0.60
gr:	37	0.07
og:	0	0.00
schema:	42	0.08
sioc:	20	0.04
skos:	83	0.16
vcard:	10	0.02

Table 1 presents the score for a few popular vocabularies. At the time of the manuscript authoring, the number of the registered vocabularies in LOV is $|LOV| = 512$, which is used to calculate the figures by applying the formula of Definition 1. Studying the $VR_{LOV,V}$ scores, it is evident that the most reused vocabularies are those that provide terms for basic concepts, like *foaf:* that is widely used to describe personal details, while others that are more specialised do not have that many incoming links as it is difficult to reuse (extend).

```
SELECT ?p ?b {
GRAPH <http://lov.okfn.org/dataset/lov>{
  ?vocab a voaf:Vocabulary.
  ?vocab vann:preferredNamespacePrefix ?p.
  ?vocab voaf:reusedByVocabularies ?b.
}} ORDER BY desc (?b)
```

Listing 1.1. LOV SPARQL for vocabularies reuse.

Unavoidably, there will always be niches that are not supported by the overall highest ranked vocabularies as they may not cover concepts that are needed for a given domain. Taking that in consideration the formula presented later in Definition 4, that makes use of the ranking score, it includes in addition a relevance score that mainly refers to the distance between the given search keyword and the retrieved term. Using the SPARQL endpoint of LOV[4] to run the query shown in Listing 1.1, we were able to calculate the frequency distribution of the $VR_{LOV,V}$ metric as depicted in Fig. 2. As it is shown in the distribution diagram, most of the vocabularies are assigned a low score in the window between 0 and 0.1, while only a few make it beyond the 0.2 score, which could be considered a high score as it reflects a reusability of the vocabulary from the 20 % of the registered vocabularies. Those that are very close to the maximum score value, are considered to be outliers as they are mostly vocabularies that include very basic terms, like *rdf:*, *rdfs:*, etc.

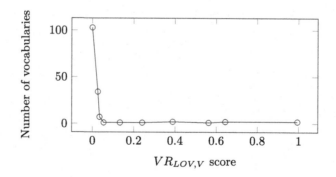

Fig. 2. $VR_{LOV,V}$ score frequency distribution.

3.2   LOD Terms Usage

Leveraging the data of vocab.cc to our ranking equation could have been implemented either by using the overall occurrences of a term and the number of documents that refer to a term, or by using the ranking metrics of the term in the corpus. We decided to use the ranking that is provided by vocab.cc as it reflects relatively the importance of the term among the examined documents

---

[4] http://lov.okfn.org/dataset/lov/sparql.

instead of introducing variables in the equation for the absolute figures. In this way, it is easier to combine the ranking variables with other factors and produce a better equation.

The usage of the vocab.cc service is straightforward and simple as it provides a RESTful way via content negotiation[5] to access the metrics for each term that has been crawled. The vocab.cc endpoint is based on the Billion Triples Challenge Dataset [9]. On the other hand, incorporating the ranking scores in our approach we ensure that the variables reflect an evaluation result that has the same reference point, thus it can be introduced in a formula with other ranking factors without the prerequisite that all the factors should refer to same datasets. For example, the schema:Place (referring to the http://schema.org/Place URI) is in position 51 of the overall ranking and in position 41 of the document ranking. In this case, the $TR_{BTCD,t}$ would be 45.72 as shown in Table 2.

Definition 2. *If OR_t is the overall ranking and DR_t the document ranking of a term t, $t \in V$ and $V \in LOV$, then:*

$$TR_{BTCD,t} = (OR_t \cdot DR_t)^{1/2}$$

The formula of Definition 2 will obviously return a lower value for those terms with a better ranking position within vocab.cc, thus we have $TR_{BTCD,t} \in [1, +\infty)$. We use the geometric mean as it gives a meaningful average between the overall ranking and the document ranking, while the arithmetic mean would not help to normalise the ranges and the one with the greater values would dominate the weighting between the two factors.

Beyond the metrics of the vocab.cc service, we have considered incorporating the statistics that have been computed as part of the LODStats initiative [2], which was introduced in the related work section. The LODStats is a great source of comprehensive statistical metrics (32 different metrics) due to the stream based computation that uses, which enables its processes to handle millions of triples and scale up better than other approaches. For the time being, we assume that the integrated data from the LODStats into the LOV infrastructure is adequate to give a sound picture of the vocabularies and terms usage within the LOD cloud as that is realised by the LODStats algorithms. This assumption is supported by the metadata that accompany the search results of vocabularies and terms, e.g. it is stated that the schema:Place occurs 3,164,782 times in 4 LOD datasets. Furthermore, browsing various terms in the LOV directory, we realised that the number of occurrences are distributed with great outliers. Thus, there are terms with millions of occurrences, e.g. schema:Place, while others are met a few thousand times, e.g. gr:Offering 22,584, or only a few times like schema:isRelatedTo, which has only 2 occurrences. However, for the context of the presented approach the usage of a term 17M times or 8M times does not make any significant difference about the popularity of the term, while it would make

---

[5] By notifying the server that the client accepts content in the *application/rdf+xml* format.

a difference if a term is used only once, 1000 times or 1M times. In this respect, the formula in Definition 3 is based on the logarithmic function with base-2 as in principle logarithmic scales reduce wide-ranging quantities to smaller scopes. Thus, this formula will reflect better the profile of the previous mentioned data within a smaller range, i.e. 0 for $log_2(1)$ to ca. 24 for $log_2(17 \cdot 10^6)$. Base-2 is preferred over base-10 or ln due to the nature of the values that we deal with, causing the $\log_2$ results to spread more on the axis than the rest; allowing a better comparison of the calculated metric.

Definition 3. *If OC_t is the number of occurrences of a term t, $t \in V, V \in LOV$, and $OC_t \in [0, +\infty)$ then:*

$$TR_{LOD,t} = \log_2(OC_t + 1)$$

According to the aforementioned definition, the $TR_{LOD,t}$ metric is higher for those terms that are more popular, with worst value to be the 0, i.e. $TR_{LOD,t} \in [0, +\infty)$. The formula adds 1 before computing the log base-2 in order to circumvent the issue of computing the ranking for a term with zero occurrences, which is not a real number. In addition, adding 1 to the occurrences does not change the impact of the final result.

Table 2. LOD term ranking examples.

Term t	$TR_{BTCD,t}$	$TR_{LOD,t}$
schema:Place	45.72	21.59
dbpedia-owl:Place	46.73	5
foaf:Person	6.93	21.14
schema:Person	54.41	19.90
owl:sameAs	10.39	24.01

Table 2 uses the prefixes that are used in LOV, without mentioning the complete URIs for space savings. As shown in Table 2, foaf:Person has significantly better score compared to schema:Person based on the $TR_{BTCD,t}$, while they seem to be on par according to the $TR_{LOD,t}$. The reason is that the former takes in consideration the number of datasets that a term occurs, while in the latter we include the occurrences decoupled from the number of documents that include them.

3.3   The LOVR Algorithm

The LOVR algorithm aims to generate accurate suggestions about the best candidate terms for semantic annotations on a webpage based on two aspects: (a) the ranking of the vocabulary, that the term belongs to, in the LOV space; and

(b) the ranking of the term in the Linked Open Data cloud taking in consideration the usage of it. Those two parameters define the overall ranking of the term, namely the Linked Open Term Rank (LOTR) as presented in Definition 4.

The *LOTR* ranking metric aims to assign a score to each single term t of the extracted vocabulary terms set by accounting the importance of the vocabulary across the LOV space as demonstrated in Sect. 3.1; the usage of the term by the existing LOD datasets; and the relevance of the term to the related concept c of the input content or webpage. The range of *LOTR* is the set of positive real numbers including zero, $LOTR \in \mathbb{R}_{\geq 0}$ and a higher score denotes a higher position in the ranking sequence.

Definition 4. *If $re_{t,c}$ is the relevance of the term t to a given concept c, $re_{t,c} \in (0,1]$, $VR_{LOV,V}$ is the ranking of the vocabulary V within LOV, $VR_{LOV,V} \in (0,1]$, $TR_{BTCD,t}$ is the ranking of t in the BTCD and $TR_{LOD,t}$ is the ranking of t in the Linked Open Data cloud based on its usage, then:*

$$LOTR_t(c) = (\frac{TR_{LOD,t}}{TR_{BTCD,t}} + \alpha \cdot VR_{LOV,V}) \cdot re_{t,c}$$

The $TR_{BTCD,t}$ factor of the formula is at the denominator of the fraction due to the range of it as described in Definition 3 and the fact that a smaller value represents a better ranking position of the term in BTCD. In addition, the α constant (set to $\alpha = 100$ throughout the experiments) has been introduced in the formula in order to address the problem of having the multiplication of two factors with values below 1, i.e. $VR_{LOV,V}$ and $re_{t,c}$, which would have a product lower than each of the two factors.

Table 3. LOTR concept ranking example.

Term t	Concept c	$re_{t,c}$	$LOTR_{t,c}$
schema:Place	place	0.784	2.94
event:Place		0.751	0.21
dbpedia-owl:Place		0.458	0.05

Table 3 depicts the LOTR scores for various candidate terms about the concept *place*. According to the assigned scores, the *schema:Place* and *event:Place* are the highest ranked terms while the former wins over the latter mainly due to the number of occurrences in the LOD datasets as the LOVRank $VR_{LOV,V}$ scores are very close to each other.

Moving forward to the declaration of the *LOVR* algorithm, it can be described as an iterative process that starts from the extraction of the concepts c from a webpage $\mathcal{W} = \{c_1, c_2, ..., c_n\}$ and then iterates through all of them. The goal is to find the best matching vocabulary term t for each concept c. For each one of the extracted concepts a search in the LOV directory is performed in order to find the most relevant terms, which form the set $V_{LOV,c}$. All those terms are accompanied with a relevance score re, provided by the LOV search; $V_{LOV,c} = \{[t_1, re_1], [t_2, re_2], ..., [t_n, re_n]\}$. Thus, at this stage, for each concept c we have mapped a set $V_{LOV,c}$ with those vocabulary terms that best match based on the string similarity metrics that LOV search materialises. For each term t of this set, the algorithm executes the computations for the Linked Open Term Rank $LOTR_c(t)$ formula as described in Definition 4. The result is kept together with the results for the rest of the terms of the $V_{LOV,c}$ set and then the term with the highest score is marked as the most appropriate match for the given concept c. Putting together the best candidate term t with the concept c, a pair is created, which is one of the items of the set $\mathcal{T}$ that will be returned as output of the algorithm.

Let $\mathcal{T}$ be a set of pairs [concept,term]. The goal of the LOVR algorithm is to return a set $\mathcal{T}$ for a given webpage.

Data: Webpage content
Result: Set of vocabulary terms to annotate the given content

1 extract the concepts of the webpage: $\mathcal{W} = \{c_1 \ ... \ c_n\}$;
2 **foreach** *concept* $c \in \mathcal{W}$ **do**
     /* $V_{LOV,c}$ keeps a set of terms from LOV   */
3      $V_{LOV,c} \leftarrow$ LOVSearch(c);
4   **foreach** *term* $t \in V_{LOV,c}$ **do**
5          $r_{c,t} \leftarrow LOTR_c(t)$;
        /* $r_{c,t}$: ranking of t for the c */
6   add $r_{c,t}$ to the $R_{c,t}$ set ;
7   **end**
8      $r_{max} \leftarrow \max(R_{c,t})$;
9   add the pair c, t that corresponds to the r_{max} to the $\mathcal{T}$ set;
10 **end**
11 **return** $\mathcal{T}$;

Algorithm 1. The LOVR algorithm.

Figure 1 demonstrates the workflow of the LOVR architecture, which employs the *LOVR algorithm* after the extraction of concepts and the LOV querying step in order to shortlist the gathered terms and rank them based on the aforementioned metrics.

The proof of concept for the effectiveness of the LOVR algorithm described in Algorithm 1 is implicitly provided by the scores presented in Table 3 regarding the LOTR rank. At the core of the algorithm stands the LOTR formula which is applied to every term that has been correlated with the concepts of a webpage. Therefore, the LOTR scores are those that steer the decisions within the algorithm and support the formation of the $\mathcal{T}$ set. Apart from the above presented results, we experimented on real world web content in order to see the strengths and weaknesses of the approach. The next section focuses on those experiments.

4   Experiment and Discussion

The experiment section aims to bring together the various metrics and evaluate the proposed approach by testing the results of the LOVR architecture as depicted in Fig. 1 from an aggregated perspective. Evidence about the effectiveness of the individual formulas and the LOVR algorithm has already been presented in the respective aforementioned sections by applying the metrics on example cases of terms and concepts.

Thus, the approach in the experiment that we conducted follows the simplest usage scenario, giving as input a webpage and getting the set of terms T at the output. Then, the output terms are checked against the initial input to evaluate if they are eligible to be used in order to transform the content to structured data content by weaving semantic annotations. For this experiment, we chose a webpage from the allrecipes.com[6] that describes a recipe preparation.

Table 4. Extracted concepts.

Concept c	Candidate term t	$LOTR_{c,t}$
Ingredients	schema:ingredients	19.016
	food:ingredientListAsText	0.498
	food:percent	0.332
Nutrition	schema:nutrition	14.176
	food:NutritionData	0.623
	schema:NutritionInformation	4.296
Calories	schema:calories	19.824
	dbpedia-owl:approximateCalories	1.239
	spfood:MaxCaloriesPreference	0.000
Recipe	food:Recipe	1.716
	sio:SIO_001042	0.000
	schema:Recipe	18.184
Servings	schema:servingSize	9.504
	schema:recipeYield	2.206
Review	schema:review	45.921
	schema:reviews	32.28
	schema:reviewBody	28.33
	schema:itemReviewed	27.41

It is worth mentioning that the chosen webpage for the evaluation includes already semantic annotations, albeit they were ignored throughout the process as we used only the text of the page. However, the existing annotations can be used for comparison, in terms of comprehensiveness, with the vocabulary terms that our methodology proposed.

---

[6] http://allrecipes.com/Recipe/Veggie-Pizza.

After running the methodology for the given input URL, the output $\mathcal{T}$ set included the terms that are presented in the right column of Table 5. All of the terms are part of the popular vocabulary schema.org, which is the most used one in the context of recipes description on the Web, according to the best of our knowledge. The LOTR scores, that the terms are accompanied with, are presented as part of Table 4, where they are compared with the rest of the candidate terms.

Having a closer look in the rest of the scores of Table 4 we realise that there are cases, for which the winner term does not have a large distance from the rest of the group, e.g. the candidates for the *review* concept. Furthermore, all the shortlisted terms for this specific concept are considered in principle relevant to describe a review entity. Thus, one of the future improvements of the approach would be to return not only the highest ranked term but the top three ranked terms. In addition, this feature would facilitate one more long term goal, which is the education of the end user (i.e. developer or ontology engineer) about the existing vocabularies and terms.

During the evaluation phase, we encountered a challenge with the vocab.cc results, as it is based on the BTCD, which does not include all the terms that appear in the examined shortlist. This fact, forced us to apply a default value to the $TR_{BTCD,t}$ for those cases in order not to break the calculations. However, experimenting with various constants, we realised that it did not affect the outcome significantly as the occurrences in LODStats were also 0 or very low. Additionally, it turned to be a good idea to have a high default value (ca. 1000) for the $TR_{BTCD,t}$ metric as a way to penalise the terms that did not appear within the BTCD, which means that they are not mature enough in the LOD cloud.

Table 5. Allrecipes.com annotations vs $\mathcal{T}$.

Webpage annotation	$\mathcal{T}$ term
schema:ingredients	schema:ingredients
schema:nutrition	schema:nutrition
schema:calories	schema:calories
schema:recipe	schema:recipe
schema:recipeYield	schema:servingSize
schema:review	schema:review

According to the results of Table 5, the overall success score of the approach is more than 80 %, taking in consideration that five out of the six proposed vocabulary terms match with the annotations that the examined concepts had already been implemented with. One of the reasons that we considered a recipe for the experiment is the structured data of the allrecipes.com, which we consider it as a good example for the domain of recipes as far as the recipes vocabulary is concerned. Furthermore, a recipe description has a lot of fields that need to be annotated and that makes it a good example to work with while building an algorithm for vocabulary recommendations. On the other hand, the recipes vocabulary domain is mostly dominated by the schema.org initiative, and as a consequence most of the output vocabulary terms of Table 5 belong to the schema.org namespace.

Coming back to the implementation details of the approach, there are a few notes that we would like to mention in this section, including the NLP part and the overall performance. Regarding the Natural Language Processing step in Fig. 1, we used the AlchemyAPI[7] endpoint to extract concepts from the given input text. The NLP step could be completely skipped if we were requesting more input from the user, i.e. the main concepts of the given webpage. However, such a decision would affect negatively the simplicity in the usage of it.

Regarding the speed performance of the approach, we consider as the most time consuming parts those that are related to external API calls. Precomputing metrics that are not related to specific input parameters but based on data that is available in advance, we achieve a significant optimisation at the running stage. For example, the *LOVRank* presented in Definition 1 is basically a ranking score for the vocabularies of the LOV space, therefore, it is possible to compute it for all the vocabularies and prepare the values that the *LOVR algorithm* will need for the respective vocabularies of the discovered terms. On the other hand, the metrics described in the LOD terms usage, under Subsect. 3.2, cannot be precomputed due to the abundance of terms available. However, the results for each term is stored to be reused in the future if it reoccurs in the recommended terms set, which has a high probability when applying the proposed methodology in a set of webpages within the same context (i.e. a website).

Finally, taking in consideration the initiative of the schema.org roadmap to allow the introduction of extensions to it, we foresee a challenge for the web developers related to the vocabularies selection. The problems of finding the best way to build on top of an existing vocabulary could be addressed up to a significant extent by the proposed approach based on the evaluation outcomes and success that have been presented above.

5   Conclusions

The goal of the presented work is threefold: (a) propose a ranking approach for the vocabularies of the LOV space based on their importance; (b) propose a ranking approach for the vocabulary terms based on the LOVRank and the usage patterns in the LOD cloud; and (c) provide a methodology to identify the most appropriate set of LOV vocabulary terms for a given webpage by leveraging the existing dynamics within the LOV graph and the usage patterns in the LOD cloud. The presented approach has been designed in order to be easily deployable on the Web as an application or service and support the broader initiative of facilitating the creation of semantic annotations and bridging the gap between web developers and theoretical semantic web methodologies.

---

[7] http://www.alchemyapi.com/.

The proposed algorithm and formulas are agnostic about the existence period of the vocabulary, which has as a consequence to implicitly penalise those vocabularies that are newer in the LOV space, thus probably less used in the LOD cloud by the various datasets and less (if not at all) reused by other vocabularies (also known as the Zipf's law). Thus, the time dimension is a parameter that is considered to be relevant for the *LOVR algorithm* calculations and part of the outlook. The inclusion of a temporal dimension will allow the recommendation of new terms to the final output set. Furthermore, based on the results of the *LOVRank*, we find interesting to leverage the formula to consider the outgoing links as well apart from the incoming in order to measure how integrated vs isolated is the vocabulary in the LOV space. In this context, it could be also useful to incorporate the ranking of the incoming links of a vocabulary in a recursive way.

Moreover, part of the future work is to conduct larger evaluations that employ both automatic and manual processes. The aim of a larger evaluation on one hand is to further test the effectiveness of the approach but more important is the feedback loop that it will help to build with the presented metrics and formulas. This feedback data will lead to adjustments and possibly further granular parametrisation.

Inspired by the work of Meusel et al. [10] regarding the quality of the deployed microdata, we consider as valid and interesting future work direction of the presented approach, the recommendation of the annotations with the content weaved together in JSON-LD format in conjunction with the shortlist of vocabulary terms. This approach will help the data producers to publish content with a higher quality in the annotations, which will benefit the visibility and the machine readability of the websites. Moreover, we plan to integrate into the LOVR algorithm another work in progress related to annotations specific about possible actions on websites as showcased briefly in [16], where we presented a simplified approach to recommend annotations from the schema.org actions based on the existing annotations of a given webpage. Finally, the described approach is planed to be provided as an online service endpoint, which can be used by developers and ontology engineers to assist the discovery of relevant terms for their content.

Acknowledgments. This work has been partially supported by the EU projects BYTE, ENTROPY, EUTravel, FWF project OntoHealth, as well as FFG projects OpenFridge and TourPack.

References

1. Atemezing, G.A., Troncy, R.: Information content based ranking metric for linked open vocabularies. In: Proceedings of the 10th International Conference on Semantic Systems, pp. 53–56. ACM (2014)
2. Auer, S., Demter, J., Martin, M., Lehmann, J.: LODstats – an extensible framework for high-performance dataset analytics. In: ten Teije, A., Völker, J., Handschuh, S., Stuckenschmidt, H., d'Acquin, M., Nikolov, A., Aussenac-Gilles, N., Hernandez, N. (eds.) EKAW 2012. LNCS, vol. 7603, pp. 353–362. Springer, Heidelberg (2012)

3. Bizer, C., Eckert, K., Meusel, R., Mühleisen, H., Schuhmacher, M., Völker, J.: Deployment of RDFa, microdata, and microformats on the web – a quantitative analysis. In: Alani, H., Kagal, L., Fokoue, A., Groth, P., Biemann, C., Parreira, J.X., Aroyo, L., Noy, N., Welty, C., Janowicz, K. (eds.) ISWC 2013, Part II. LNCS, vol. 8219, pp. 17–32. Springer, Heidelberg (2013)

4. Butt, A.S.: Ontology search: finding the right ontologies on the web. In: Proceedings of the 24th International Conference on World Wide Web Companion, pp. 487–491. International World Wide Web Conferences Steering Committee (2015)

5. Sahar Butt, A., Haller, A., Xie, L.: Relationship-based top-k concept retrieval for ontology search. In: Janowicz, K., Schlobach, S., Lambrix, P., Hyvönen, E. (eds.) EKAW 2014. LNCS, vol. 8876, pp. 485–502. Springer, Heidelberg (2014)

6. Guha, R.: Introducing schema.org: search engines come together for a richer web (2011). http://insidesearch.blogspot.com/2011/06/introducing-schemaorg-search-engines.html

7. Heath, T., Bizer, C.: Linked data: evolving the web into a global data space. Synth. Lect. Semant. Web: Theory Technol. 1(1), 1–136 (2011)

8. Hogan, A., Harth, A., Passant, A., Decker, S., Polleres, A.: Weaving the Pedantic Web (2010)

9. Käfer, T., Harth, A.: Billion Triples Challenge data set (2014). Downloaded from http://km.aifb.kit.edu/projects/btc-2014/

10. Meusel, R., Paulheim, H.: Heuristics for fixing common errors in deployed schema.org microdata. In: Gandon, F., Sabou, M., Sack, H., d'Amato, C., Cudré-Mauroux, P., Zimmermann, A. (eds.) ESWC 2015. LNCS, vol. 9088, pp. 152–168. Springer, Heidelberg (2015)

11. Meusel, R., Petrovski, P., Bizer, C.: The WebDataCommons microdata, RDFa and microformat dataset series. In: Mika, P., Tudorache, T., Bernstein, A., Welty, C., Knoblock, C., Vrandečić, D., Groth, P., Noy, N., Janowicz, K., Goble, C. (eds.) ISWC 2014, Part I. LNCS, vol. 8796, pp. 277–292. Springer, Heidelberg (2014)

12. Page, L., Brin, S., Motwani, R., Winograd, T.: The PageRank citation ranking: bringing order to the web. Technical report 1999–66, Stanford InfoLab, November 1999

13. Schaible, J., Gottron, T., Scheglmann, S., Scherp, A.: Lover: support for modeling data using linked open vocabularies. In: Proceedings of the Joint EDBT/ICDT Workshops, pp. 89–92. ACM (2013)

14. Schaible, J., Gottron, T., Scherp, A.: Survey on common strategies of vocabulary reuse in linked open data modeling. In: Presutti, V., d'Amato, C., Gandon, F., d'Aquin, M., Staab, S., Tordai, A. (eds.) ESWC 2014. LNCS, vol. 8465, pp. 457–472. Springer, Heidelberg (2014)

15. Stadtmüller, S., Harth, A., Grobelnik, M.: Accessing information about linked data vocabularies with vocab. cc. In: Li, J., Qi, G., Zhao, D., Nejdl, W., Zheng, H.-T. (eds.) Semantic Web and Web Science, pp. 391–396. Springer, New York (2013)

16. Stavrakantonakis, I., Fensel, A., Fensel, D.: Matching web entities with potential actions. In: SEMANTICS (2014)

17. Stavrakantonakis, I., Toma, I., Fensel, A., Fensel, D.: Hotel websites, web 2.0, web 3.0 and online direct marketing: the case of Austria. In: Xiang, Z., Tussyadiah, L. (eds.) Information and Communication Technologies in Tourism, pp. 665–677. Springer, Switzerland (2014)

18. Vandenbussche, P.-Y., Vatant, B.: Linked open vocabularies. ERCIM News 96, 21–22 (2014)

Heuristic-Based Configuration Learning
for Linked Data Instance Matching

Khai Nguyen[1,2](✉) and Ryutaro Ichise[1,2]

[1] The Graduate University for Advanced Studies, Hayama, Japan
{nhkhai,ichise}@nii.ac.jp
[2] National Institute of Informatics, Tokyo, Japan

Abstract. Instance matching in linked data has become increasingly important because of the rapid development of linked data. The goal of instance matching is to detect co-referent instances that refer to the same real-world objects. In order to realize such instances, instance matching systems use a configuration, which specifies the matching properties, similarity measures, and other settings of the matching process. For different repositories, the configuration is varied to adapt with the particular characteristics of the input. Therefore, the automation of configuration creation is very important. In this paper, we propose *cLink*, a supervised instance matching system for linked data. *cLink* is enhanced by a heuristic algorithm that learns the optimal configuration on the basic of input repositories. We show that *cLink* can achieve effective performance even when being given only a small amount of training data. Compared to previous configuration learning algorithms, our algorithm significantly improves the results. Compared to the recent supervised systems, *cLink* is also consistently better on all tested datasets.

Keywords: Instance matching · Schema-independent · Supervised · Linked data

1 Introduction

The general problem of instance matching was first employed in the 1950s. Up to now, researchers have conducted numerous related studies. The targets of the solutions for this problem are diverse, from the general areas of data mining, to more specific application of data cleansing and integration. Instance matching is very important in integrating information from multiple sources because it maintains the consistency and integrity of the final data. This study focuses on instance matching for linked data repositories. Instance matching on linked data has been a topic of extensive research [3]. However, developing a perfect instance matching algorithm remains an open research problem.

The main difficulties of instance matching originate from the ambiguity and heterogeneity of input repositories. The ambiguity exists because many instances of a repository can share similar descriptions (e.g., *Tokyo bay* and *Tokyo station*), especially when the number of instances is high. Meanwhile, the heterogeneity

© Springer International Publishing Switzerland 2016
G. Qi et al. (Eds.): JIST 2015, LNCS 9544, pp. 56–72, 2016.
DOI: 10.1007/978-3-319-31676-5_4

relates to the inconsistent representations of the identical information, including the instance-level (e.g., *Tokyo* and *the capital of Japan*) and schema-level[1] (e.g., *name* and *label*). Compared to other type of data, web-based data like linked data is more heterogeneous and ambiguous due to its high scale and the open-access mechanism.

Basically, in order to match two instances of different repositories, instance matching algorithms have to estimate the similarities between the values of the same attributes. After that, these similarities are aggregated into one matching score. This score is used to determine whether two instances are co-referent or not. The mechanisms of those steps can be described by a matching configuration [12,25].

The construction of configuration defines the approach of an instance matching system. Basic methods [2,19] manually define the configuration and thus are limited to only the repositories whose schema are well-documented and realized by user. Advanced methods create the matching configuration automatically using the observation on the input repositories. They are also called schema-independent methods. For obtaining the schema-independent goal, many techniques have been proposed, such as pure statistic [1,17], unsupervised learning [18], and supervised learning [6–8,11,13,21]. Among them, the supervised learning though requires a number of labeled co-referent instances, delivers the best accuracy.

Recently, supervised learning of configuration has been investigated with genetic algorithm [8,11] and information-gain based selection [7]. The reported results of using these methods are promising. Unfortunately, the genetic algorithm is not supported by a clear strategy as it is based on random search principle. Meanwhile, the information-gain based selection ignores the evaluation of combining different similarity estimation methods. We are motivated in constructing an instance matching system that can take these issues into account.

We propose *cLink*, an instance matching system that is enhanced by a novel configuration learning algorithm named *cLearn*. From two input repositories, *cLink* performs a pre-learning stage to automatically generate initial similarity functions and co-referent candidates. Given that some candidates are labeled and input into *cLearn*, the algorithm uses a heuristic search method to optimize the combination of similarity functions as well as all other settings of the instance matching process. The learning outcome of *cLearn* is an optimal configuration for discovering the co-references from unlabeled candidates. *cLearn* is previously presented in its initial form in [15]. In this paper, we present the *cLink* system whose core learning algorithm is *cLearn* and report in detail many important descriptions, analyses, and evaluations.

The remaining parts of this paper are organized as follows. Section 2 reviews remarkably related work. Section 3 describes the details of *cLink*. Section 4 reports the experiments. Section 5 summarizes the paper and future work.

---

[1] In this paper, the schema of a linked data repository is defined as the list of all predicates, which is part of the RDF statements ($< subject, predicate, object >$).

2 Related Work

SILK [25] and LIMES [12] are among the frontiers of linked data instance matching frameworks. These frameworks define a declarative structure to represent a manually-defined matching configuration. Zhishi.Links [19], AggrementMaker [2], and RiMOM [10] are more advanced system of manual approach, which focus on both accuracy and scalability. Zhishi.Links leverages domain knowledge while AggrementMaker and RiMOM combine many strategies to improve matching accuracy. RiMOM is also one of the current state-of-art automatic instance matching systems.

SERIMI [1] and SLINT+ [17] attempted to eliminate user involvement by automatically detecting property alignments using overlap measure. KnoFuss [18] is an unsupervised system that learn the matching configuration using genetic algorithm. The advantage of Knofuss is that this system does not require labeled data for training. However, since the fitness value is calculated using a pseudo value of the actual accuracy, the learned configuration can contain incorrect alignments and thus reduce the performance. PARIS [24] is an automatic system that consider the similarities of RDF objects, predicates, instances, and classes as related components. The matching process is done by iteratively computing and propagating the similarities over these components. Also based on probabilistic theory but combining with crowdsourcing technology, Zencrowd offers a large scale instance matching framework. Although these automatic systems obtained good results in experiment, when the high accuracy is the first priority and the existing co-references are available, the supervised learning approach is still the first option.

ADL [7] and ObjectCoref [6] learn discriminative properties that are expected to effectively separate the positive and negative co-references. These systems include the domain knowledge in order to improve the accuracy. Instance matching also can be treated as a binary classification problem. A benchmark of many different classifiers for linked data instance matching can be found in [23]. Recently, Rong et al. used Adaboost algorithm to enhance the performance of a random forest classifier [21]. The disadvantage of using classifier is that the matching scores of instances are not explicitly calculated and hence the further post-processing steps, whose these scores are the demand, cannot be implemented. For example, adaptive filtering in SLINT+ greatly improves the result [17].

Most related to our work are RAVEN [11], EAGLE [13], EUCLID [14], and ActiveGenLink [8], which learn the matching configuration using supervised approach. While RAVEN and EUCLID use deterministic methods, EAGLE and ActiveGenLink apply genetic algorithm in order to improve the efficiency, especially when many attributes are input. However, genetic algorithm still has to check many configurations to reach the convergence. In addition, active learning is used for RAVEN, EAGLE, and ActiveGenLink. Compared to these systems, *cLink* is different at the novel learning algorithm. *cLink* also supports many similarity aggregation strategies and matching score post-filtering. In the next section, we describe the detail of *cLink*.

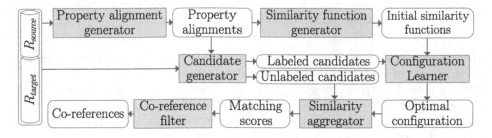

Fig. 1. The workflow of *cLink*.

3   The *cLink* System

cLink consists of six components, property alignment generator, similarity function generator, candidate generator, configuration learner, similarity aggregator, and co-reference filter. The workflow of *cLink* is illustrated in Fig. 1. Given two input repositories, R_{source} and R_{target}, *cLink* first creates the property alignments. Then, these alignments are used to build the initial similarity functions and to select the possibly co-referent candidates. Using the existing co-references, some candidates are labeled and input into the configuration learner to find an optimal configuration. The optimal configuration defines the appropriate similarity functions selected from the initial similarity functions. Each similarity function computes the similarity of two instances on one property. The similarity aggregator combines many similarities into a final matching score. Finally, co-references filter considers the matching score of all candidates and produces the final co-references. Next, we describe the details of each component.

3.1   Property Alignment Generator

This component generates the property alignments between R_{source} and R_{target}. An appropriate property alignment is expected to describe the same information of two instances (e.g., 'born in' and 'home town'). In linked data, as properties are represented by RDF predicates, the outputs of this component are the alignments between RDF predicates.

Property alignments are generated in two steps. First, *cLink* selects in the source repository the predicates that satisfy two conditions. Second, *cLink* aligns each selected predicate with the corresponding ones in the target repository using an overlap measure. The conditions used in the first step are the discriminability and the coverage, which are constructed by a modification on the basic of [22]. The discriminability $discr(p_k)$ expresses the diversity of the RDF objects declared by p_k, while the coverage $cover(p_k)$ represents the instance-wise frequency of p_k. Eqs. 1 and 2 describe $discr(p_k)$ and $cover(p_k)$, respectively:

$$discr(p_k) = \frac{|\{o|x \in R_{source}, <s, p_k, o> \in x\}|}{|\{<s, p_k, o> |x \in R_{source}, <s, p_k, o> \in x\}|} \qquad (1)$$

$$cover(p_k) = \frac{|\{x|x \in R_{source}, < s, p_k, o >\in x\}|}{|R_{source}|} \qquad (2)$$

where R stands for repository, $< s, p, o >$ is a RDF triple, and x is an instance, which is a set of RDF triples sharing the same subject s. We separate the predicates by their data type: $string, number, date$, and URI. The type of a predicate is assigned with the most frequent type of its RDF objects. For each type, we select the predicates having the discriminability $discr$ higher than α and then retain only the K_{cover} predicates with the highest coverages $cover$.

One predicate selected from the source repository can be aligned with many predicates having the same type in the target. The confidence $overlap$ of an alignment $[p_{source}, p_{target}]$ is measured as follows:

$$overlap([p_{source}, p_{target}]) = \frac{|O_{p_{source}} \cap O_{p_{target}}|}{|O_{p_{source}}|} \qquad (3)$$

$$O_{p_k} = \{E(o)|x \in R_k, < s, p_k, o >\in x\}$$

where E is a pre-processing function. E is applied for each RDF object. E works in flexible ways depending on the property type. For $string$, E collects the tokens. For $number$, E returns the rounded value at two decimal digits. For $date$, E keeps the original value. For URI, E extracts the remaining string after stripping away the domain. For each predicate of the source repository, we select the top K_{align} alignments having the highest $overlap$. The confidence is not enough to guarantee the correctness of generated alignments. Therefore, a medium or large number of property alignments is safer to retain the useful alignments. Removing incorrect alignments is the mission of configuration learner. In $cLink$, we set $\alpha = 0.5, K_{cover} = 4$ and $K_{align} = 4$ for all types as the default parameter. In total, by default this component produces at most 64 property alignments.

Equation 3 works under the assumption that the target repository contains many co-references with the source repository. Therefore, the denominator is related only to the source repository instead of both repositories like a Jaccard measure, which is used in SERIMI [1] and SLINT+ [17]. Equation 3 is reasonable because currently there are many large repositories that cover a wide range of instances (e.g., DBpedia, Freebase, and Wikidata). Furthermore, as the predicate alignments are generated using only RDF objects, $cLink$ does not require the specification of the schemata. In other words, $cLink$ is schema-independent.

3.2 Similarity Function Generator

This component assigns to each property alignment the suitable similarity metrics and the result of one assignment is one similarity function. In other word, each similarity function computes the similarity of two instances on one designated property. The final matching score is computed by multiple similarity functions. The detail is described in Sect. 3.5. Equation 4 is the definition of a similarity function sim:

$$sim_{\{[p_{source}, p_{target}], metric\}}(x, y) = \max_{o_x, o_y}(metric(E(o_x), E(o_y))) \qquad (4)$$

$$< s, p_{source}, o_x > \in x, < s, p_{target}, o_y > \in y$$

A similarity function is specified by two pieces of information: a property alignment $[p_{source}, p_{target}]$ and a similarity measure $metric$. For two instances: $x \in R_{source}$ and $y \in R_{target}$, a similarity function returns the similarity of the most similar RDF objects declared by p_{source} and p_{target}. In other words, the max operator effects if p_{source} or p_{target} appears many times in x or y. The function E defined in Sect. 3.1 is used to pre-process the RDF objects.

Table 1. Example of property alignments and assigned similarity measures

Source property	Target property	Similarity measures
Label	name	Levenshtein, TFIDF-Cosine
	leader	Levenshtein, TFIDF-Cosine
Description	comment	TFIDF-Cosine
	abstract	TFIDF-Cosine
Establish	named date	ExactMatch
Population	area size	ReversedDifference
	elevation	ReversedDifference

The similarity measure $metric$ is assigned to the similarity function in accordance with the type of p_{source} and p_{target}, as illustrated in Table 1. $cLink$ supports five similarity measures. For 'date' and 'URI', $cLink$ uses the exact matching, which returns 1.0 if two values are identical and 0.0 otherwise. For 'number', $cLink$ uses the reversed difference (Eq. 5), which calculates how much close two numbers a and b are to each other:

$$diff(a, b) = (1 + |a - b|)^{-1} \qquad (5)$$

For $string$, $cLink$ supports two measures having different characteristics. For short string[2], we select Levenshtein because of its robustness. For long string, we select $cosine$ similarity on TFIDF vectors calculated from the token sets. In order to obtain the TFIDF weight for each token, we adapt the original weighting calculation for being compatible with linked data instances. We calculate the normalized term frequency for TF and probabilistic inverse frequency for IDF, which are described by Eqs. 6 and 7, respectively.

$$TF(w, x, R) = \frac{|\{< s, p, o > | w \in E(o), < s, p, o > \in x\}|}{\max_{y \in R} |\{< s, p, o > | w \in E(o), < s, p, o > \in y\}|} \qquad (6)$$

$$IDF(w, R) = \log \frac{|R| - |\{y | < s, p, o > \in y, w \in E(o), y \in R\}|}{|\{y | < s, p, o > \in y, w \in E(o), y \in R\}|} \qquad (7)$$

---

[2] The strings whose length is less than 100 characters.

where, w is the token of interest, x is an instance belonging to the repository R. Function E extracts the tokens of the input RDF object.

All property alignments generated previously are input into this component. In parallel, the alignments of string properties are input into the candidate generator, which is described in the following section.

3.3   Candidate Generator

This component detects the pairs of potentially co-referent instances. Such pairs are called candidates. *cLink* uses a simple token-based prefix blocking approach [20], by which a pair of two instances is considered as a candidate if they share a number of first tokens of the RDF objects declared by designated properties. In *cLink*, we use only one token so that *cLink* can retain as many correct candidates as possible. In addition, only the generated string property alignments are used for comparing the tokens.

The mission of the candidate generator is to limit the huge number of pairwise alignments between instances, which is $|R_{source}| \times |R_{target}|$. Therefore, this component is very important because it is impractical to check all pairs of instances, especially when the repositories are large.

There is possibility to use weighting schemes for reducing the number of candidates [9,17,25]. However, such reduction is companied with a drop in the number of correct candidates. For that reason, we recommend using the simple token-based blocking without weighting.

Given that positive and negative labels are assigned (e.g., manually or automatically by using some existing co-references) to some candidates for creating labeled ones, they are divided into training set T and validation set V, which are required by the learning algorithm in the next step.

3.4   Configuration Learner

The input of this component are the labeled candidates, including a training set T and a validation set V; initial similarity functions I_{sim} (Sect. 3.2); and similarity aggregators I_{agg} (Sect. 3.5). The output of this component is the optimal configuration C_{opt} that is most suitable to the input repositories. A configuration specifies the combination of similarity functions F_{sim}, the similarity aggregator Agg, the parameters δ_{sim} associated with each similarity function sim, and the parameter δ of the co-reference filter (Sect. 3.6). For solving the configuration learning problem, we use *cLearn* algorithm. *cLearn* uses a heuristic search method to learn the combination of the similarity functions, the similarity aggregator, and other parameters. The pseudo code of *cLearn* is given in Algorithm 1. Note that in the pseudo code, we use dot ('.') notation as the member accessing function. The *Init* function creates a configuration by assigning Agg, F_{sim}, and δ with given values. The *Evaluate* function first executes the similarity aggregator and the co-reference filter specified by a configuration, on a set of candidates. Then, based on the label of those candidates, it computes the performance, $F1$ score. $F1$ is

---

Algorithm 1. *cLearn*

Input: Training set T, validation set V, integer paramerter K_{top}
list of similarity functions I_{sim}, list of similarity aggregators I_{agg}
Output: Optimal configuration C_{opt}

1 $C_{agg} \leftarrow \emptyset$
2 **foreach** $A \in I_{agg}$ **do**
3    $\quad$ $visited \leftarrow \emptyset$
4    $\quad$ **foreach** $sim \in I_{sim}$ **do**
5    $\quad\quad$ $c \leftarrow Init(Agg \leftarrow A, F_{sim} \leftarrow sim, \delta \leftarrow 0)$
6    $\quad\quad$ $[c.\delta, F1] \leftarrow EvaluateAndAssignThreshold(c, T)$
7    $\quad\quad$ $c.\sigma_{sim} \leftarrow c.\delta$
8    $\quad\quad$ $visited \leftarrow visited \cup \{[c, F1]\}$
9    $\quad$ $candidate \leftarrow TopHighestF1(visited, K_{top})$
10   $\quad$ **while** $candidate \neq \emptyset$ **do**
11   $\quad\quad$ $next \leftarrow \emptyset$
12   $\quad\quad$ **foreach** $g \in candidate$ **do**
13   $\quad\quad\quad$ **foreach** $h \in candidate$ **do**
14   $\quad\quad\quad\quad$ $c \leftarrow Init(Agg \leftarrow A, F_{sim} \leftarrow g.c.F_{sim} \cup h.c.F_{sim}, \delta \leftarrow 0)$
15   $\quad\quad\quad\quad$ $[c.\delta, F1] \leftarrow EvaluateAndAssignThreshold(c, T)$
16   $\quad\quad\quad\quad$ $visited \leftarrow visited \cup \{[c, F1]\}$
17   $\quad\quad\quad\quad$ **if** $g.c.F_{sim} \neq h.c.F_{sim}$ **and** $F1 \geq g.F1$ **and** $F1 \geq h.F1$ **then**
18   $\quad\quad\quad\quad\quad$ $next \leftarrow next \cup \{[c, F1]\}$
19   $\quad\quad$ $candidate \leftarrow next$
20   $\quad$ $F1 \leftarrow Evaluate(argmax_{v \in visited}(v.F1).c, V)$
21   $\quad$ $C_{agg} \leftarrow C_{agg} \cup \{[c, F1]\}$
22 $[C_{opt}, F1] \leftarrow argmax_{v \in C_{agg}}(v.F1)$
23 **return** C_{opt}

---

the harmonic mean of the recall *rec* and the precision *pre*, which are calculated as follow:

$$rec = \frac{\text{Number of correctly detected co-references}}{\text{Number of actual co-references}} \quad (8)$$

$$pre = \frac{\text{Number of correctly detected co-references}}{\text{Number of all detected co-references}} \quad (9)$$

The *EvaluateAndAssignThreshold* function works similarly to *Evaluate*, but at the same time, it finds a value for δ. This function selects the top N candidates with highest matching score, where N is the number of the actual co-references in L. After that, it assigns the lowest score of the correctly detected co-reference to δ. In *cLearn*, K_{top} controls the maximum size of F_{sim} in the learned configuration and is fixed to 16 by default. This parameter also limits the number of configurations that the algorithm has to check to $2^{K_{top}}$ and thus guarantees a constantly maximum response time.

cLearn begins with each single similarity function and then considers the combinations of those individuals. *cLearn* is based on a greedy enhancement heuristic (line 17), which assumes that the performance of a new combination of similarity functions must not be less than that of the combined components. The heuristic is reasonable because a series of similarity functions that reduces the performance has little possibility of generating a further combination with improvement.

The global optimum of configuration can be found using exhaustive search. However, such method is extremely expensive in term of computational cost and thus has not been used. Some systems try to use genetic search to solve the issue [8,13,18] but this algorithm is still time-consuming and contains many parameters. We use a heuristic search to reduce the complexity and minimize the parameters.

cLearn uses validation set V (line 20) in order to increase the generality of the final configuration C_{opt}. Each iteration controlled by line 2 finds an optimal configuration associated to a similarity aggregator A. Therefore, C_{agg} contains $|I_{agg}|$ configurations that is optimized for training set T. In order to identify the most general configuration, using V, a different set from T, is a better option.

cLearn algorithm is generic because the function $EvaluateAndAssignThre-$ *shold* can be replaced by other functions having the similar purpose. Therefore, this algorithm not only works in *cLink*, but also is compatible with any configuration-based matching system [7,12,14,25].

3.5   Similarity Aggregator

The similarity aggregator computes the final matching score for each candidate using the similarity functions F_{sim} and their parameter δ_{sim}, which are specified by the configuration. The computation of the matching score $mScore(x,y)$ for two instances x and y is defined as follows:

$$mScore(x,y) = weight(y) \times combine_{F_{sim}}(x,y) \tag{10}$$

where *weight* is a function weighting the target instance y and *combine* is a similarity combination function. *cLink* provides *non-weighting* and *weighting* versions. For *non-weighting*, $weight(y)$ simply returns 1.0. For *weighting*, the weight is calculated by Eq. 11:

$$weight(y) = \log_{\max_{t \in R_{target}} size(t)} size(y) \tag{11}$$

where $size(y)$ counts the number of RDF triples existing in y. By using Eq. 10, we assume that instances containing many triples are more prioritized. The logarithmic scale is used to reduce the weight of instances whose *size* is particularly large. This weighting method is effective when the target repository is very ambiguous, such as large repositories.

For similarity combination, $cLink$ uses the following equation:

$$combine_{F_{sim}}(x,y) = \frac{1}{valid(U_{F_{sim}}(x,y))} \sum_{v \in U_{F_{sim}}(x,y)} v^k \qquad (12)$$

$$U_{F_{sim}}(x,y) = \{sim(x,y)|sim(x,y) \geq \sigma_{sim}, sim \in F_{sim}\}$$

where $k \in \{1,2\}$, $valid$ is a counting function, and σ_{sim} is the parameter for each similarity function sim, which is determined automatically by $cLearn$ (line 7 of Algorithm 1). k controls the transformation for each similarity v. When $k = 1$, $combine$ acts as a first order aggregation. When $k = 2$, we have a quadratic aggregation. There are also two variations of $valid$, which return the number of elements in $U_{F_{sim}}(x,y)$ and 1.0 always. The difference between these variations is that the latter penalizes the (x,y) pair having similarities $sim(x,y) < \sigma_{sim}$ while the former does not. In addition, $cLink$ provides a $restriction$ mechanism to enable or disable σ_{sim}. When disabling, all σ_{sim} are set to zero instead of their original value. In total, there are 16 combinations of $weight$, $valid$, k, and $restriction$. Consequently, there are 16 different aggregators supported by $cLink$. All aggregators are used to initialize I_{agg} and let $cLearn$ select the best one.

3.6   Co-reference Filter

This component produces the final predictive co-references on the basis of matching scores of all candidates. $cLink$ reuses the adaptive filter of in SLINT+ [17]. This filter follows the idea of the stable marriage problem [4]. A pair of instances (x,y) is co-referent if its matching score $mScore(x,y)$, where $x \in R_{source}$ and $y \in R_{target}$, satisfies the conditional statement of Eq. 13:

$$mScore(x,y) \geq \max(\max_{z \in R_{source}} mScore(z,y), \max_{t \in R_{target}} mScore(x,t)) \qquad (13)$$

In addition, this component uses a cut-off filter to eliminate the incorrect candidates but satisfying Eq. 13. A threshold δ is used for this task. δ is assigned by the learning algorithm. Only instance pairs whose scores satisfy the condition statement of the filter and threshold δ are promoted to be a final co-reference.

Above we have described the details of $cLink$. The next section reports the experiments and the results.

4   Experiment

We report in total three experiments. The first experiment is to evaluate the candidate generator. The second experiment evaluates learning algorithm $cLearn$. In this experiment, we compare $cLearn$ with other configuration learning algorithms. The third experiment compares $cLink$ with other systems, including the systems that use supervised learning approach. We use the default parameters for the settings of $cLink$ for every experiment. The details of the experiments are described from Sect. 4.3 to 4.5.

$cLink$ in implemented as part of ScSLINT framework [16] and is available at http://ri-www.nii.ac.jp/ScSLINT/.

Because $cLink$ uses training data for the learning algorithm, before reporting the experiments, we describe the annotation process, which is compatible not only for $cLink$ but also for other learning-based instance matching systems.

4.1   Candidate Labeling and Separation

From the input of source repository R_{source}, a target repository R_{target}, and the actual co-references R, we execute the first three components of $cLink$ to obtain the candidates. After that, we annotate the candidates using the actual co-references. Concretely, if a candidate (x, y) appears in R, where $x \in R_{source}$ and $y \in R_{target}$, we assign positive label to (x, y) and candidate label to (x, z), where $y \neq z \in R_{target}$.

When splitting the candidates into different sets, we propose to put all the candidates sharing the source instance into one set. In other words, each separated set does not share any source instance with each other. This separation strategy is used because it is compatible with the practical annotation process. In order to keep the quality of annotated data, the following process can be applied. Given an instance from source repository, a ranked list of instance pairs is created using a simple matching method. Annotators are asked to assign the positive labels for the top ranked pairs. Because the list is sorted, the remaining pairs can be assumed to be negative. By following this manner, such positive label assignments have tiny possibility to be incorrect. That is, the quality of training data is guaranteed. In previous experiments [7,8,13,21], other supervised systems randomly select training examples from the candidates set and thus fail to address the issue of annotation quality in real matching tasks.

For our experiments, we first randomly separate the candidates into two sets using the above strategy. The size of these sets is determined differently for each experiment. The first set is reserved for learning, and the second set (test set) is used for evaluating the performance. After that, the first set is similarly separated into training set T (80 %) and validation set V (20 %). Furthermore, since the source repository frequently contains many instances that are not co-referent with any instance in the target, an arbitrary randomization possibly creates the sets that are drastically different from each other in term of the ratio between the positive and the negative candidates. Therefore, in our experiment, we remain the same ratio r for all separated sets, where r is the actual ratio between the number of the source instances related to at least one co-reference and the ones that do not involve with any co-reference.

4.2   Datasets

We use two the datasets provided by the instance matching track of OAEI 2010 and OAEI 2012, which have been considering to be among the most challenging datasets. There are five subsets (D1 to D5) for OAEI 2010, and seven subsets (D6 to D12) for OAEI 2012. The details of these subsets are given in Table 2.

Table 2. Summary of OAEI 2010 (D1 to D5) and OAEI 2012 (D6 to D12) datasets

ID	Source repository			Target repository			Co-references
	Name	#Instances	#Properties	Name	#Instances	#Properties	
D1	Sider	2,670	10	Drugbank	19,689	118	1,142
D2	Sider	2,670	10	Diseasome	8,149	18	344
D3	Sider	2,670	10	Dailymed	10,002	27	3,225
D4	Sider	2,670	10	DBpedia	4,183,461	45,858	1,449
D5	Dailymed	10,002	27	DBpedia	4,183,461	45,858	2,454
D6	NYT loc	3,837	22	DBpedia	4,183,461	45,858	1,917
D7	NYT org	5,967	20	DBpedia	4,183,461	45,858	1,922
D8	NYT peo	9,944	20	DBpedia	4,183,461	45,858	4,964
D9	NYT loc	3,840	22	Freebase	40,358,162	2,455,627	1,920
D10	NYT org	6,045	20	Freebase	40,358,162	2,455,627	3,001
D11	NYT peo	9,958	20	Freebase	40,358,162	2,455,627	4,979
D12	NYT loc	3,785	22	Geonames	8,514,201	14	1,729

Table 3. Result of candidate generation.

	D1	D2	D3	D4	D5	D6
#cans	5,771	4,258	5,013	482,605	987,856	38,201,823
rec	0.9721	0.9535	0.9939	0.9538	0.9780	0.9718
	D7	D8	D9	D10	D11	D12
#cans	61,702,166	46,942,099	222,686,524	357,365,003	620,073,101	32,161,659
rec	0.9880	0.9970	0.9875	0.9770	0.9912	0.9676

For OAEI 2010, we use the data provided by the OAEI website. The OAEI 2010 dataset contained five repositories related to healthcare domain: Sider, Diseasome, Drugbank, Dailymed, and DBpedia. The OAEI 2012 dataset contains four repositories NYTimes (NYT), DBpedia, Freebase, and Geonames with three domains location (loc), organization (org) and people (peo). For this dataset, we use the dump of NYTimes 2014/02, DBpedia 3.7 English, Freebase 2013/09/03, and Geonames 2014/02. There are a few slight inconsistencies between the ground-truth provided by OAEI 2012 and our downloaded dump data, because of the difference in the release dates. Therefore, we exclude 130 (0.298 %) source instances which are related to such inconsistencies.

We use these datasets because the purpose of our experiments is to know the performance of $cLink$ on the real data with large size. Although there are some newer datasets, they are either small, artificial, or focus on the benchmarks with some special targets (e.g., reasoning-based, string distortion, language variation). Therefore, we do not use such datasets. In addition, using OAEI 2010 dataset provides us the opportunity to compare our proposed system with other learning-based systems, which were already tested on this dataset.

Table 4. F1 score of *cLink* when using *cLearn* and other algorithms.

	cLearn	genetic	naive	none
D1	0.9365	**0.9375**	0.8767	0.8137
D2	0.8679	0.8450	**0.8733**	0.8430
D3	**0.8686**	0.7665	0.6741	0.6841
D4	**0.6676**	0.6673	0.6407	0.5535
D5	0.4725	**0.4727**	0.3100	0.2080
D6	**0.8835**	0.8249	0.8289	0.7336
D7	**0.9058**	**0.9058**	0.8594	0.4508
D8	**0.9645**	0.9635	0.9581	0.9506
D9	**0.8781**	**0.8781**	0.8609	0.1934
D10	**0.9116**	0.9089	0.8105	0.2223
D11	0.9465	**0.9477**	0.9325	0.4640
D12	**0.9163**	0.9106	0.8908	0.8852
H.Mean	**0.8191**	0.8022	0.7236	0.4249

4.3 Experiment 1: Candidate Generation

We evaluate the recall of the generated candidates by using Eq. 8. In order to use Eq. 8, we consider all the candidates as the detected co-references. We report this experiment because candidate generation is a very important step of instance matching system. The recall of this step is the upper bound of the final recall of the full instance matching process. In addition, we report the number of generated candidates because it reflects the complexity of further tasks.

Table 3 lists the number of the generated candidates #cans and the recall *rec* by each subset. According to this table, the recall is very high although it cannot reach 1.0, even when *cLink* uses the token sharing, a very relaxed constraint. Such a lose constraint leads to high number of candidates, compared to the expected co-references (Table 2), especially on OAEI 2012 dataset (D6 to D12). However, compared to the number of all possible instance pairs between the source and the target repository, for all subsets, more than 99.9 % of pairwise alignments are excluded. Shortly, candidate generator shows its important role as it considerably reduces the complexity of further components, such as the similarity aggregator and the configuration learner.

4.4 Experiment 2: Evaluation of *cLearn* Algorithm

We report the result of *cLink* when using *cLearn* and other alternatives, so that we can simultaneously evaluate the effectiveness of *cLearn* and compare it with other algorithms. We implement two baseline algorithms, non-optimization (*none*) and naive (*naive*). *none* does accept all generated similarity functions without any validation. This approach is similar to that of SLINT+ [17] and

Table 5. F1 score of the compared systems on OAEI 2010.

Training data	System	D1	D2	D3	D4	D5	H.mean
5 %	*cLink*	**0.911**	**0.824**	**0.777**	0.6414	**0.424**	**0.663**
	Adaboost	0.903	0.794	0.733	0.641	0.375	0.628
Varied by subset	*cLink*	**0.894**	**0.829**	**0.722**			**0.799**
	ObjectCoref	0.464	0.743	0.708			0.611
Reference systems							
RiMOM		0.504	0.458	0.629	0.576	0.267	0.445
PARIS		0.649	0.108	0.149	0.502	0.219	0.207

SERIMI [1], which totally rely on the accuracy of property alignment generator. *naive* selects the K_{top} similarity functions that offer the highest $F1$ on training data. For other alternatives, we select genetic algorithm (*genetic*) as the state-of-the-art in configuration learning [8, 13]. *genetic* represents the combination of similarity functions by binary array, in which an element indicates the active or inactive status of the associated similarity function. We choose exponential ranking for fitness selection, 0.7 for single point cross-over probability, 0.1 for single point mutation probability, and 50 for the population size.

5-folds cross validation is used for this experiment. We choose cross-validation because this option in turn puts all candidates into training as well as testing.

In Table 4, we report the average F1 scores on each subset of the tested algorithms. According to this table, *cLearn* consistently outperforms other algorithms in term of harmonic mean of all subsets. Although *genetic* is competitive with *cLearn* on some subsets, when considering each fold separately, so that there are 60 test cases for 12 subsets, the paired t-test over all tests at 0.05 significant level shows that *cLearn* is significantly better than *genetic*, as well as all other algorithms. In addition, the efficiency of *cLearn* is much better than *genetic*. *genetic* spends averagely 7,231 s³ for learning on one subset of OAEI 2012 while *cLearn* only takes 2,977 s³ (See Footnote 3). The average numbers of configurations that *cLearn* and *genetic* have to check is 137 and 263, respectively. That is, almost 50 % configurations are skipped by using *cLearn* compared to *genetic*. This fact supports the efficiency of using our heuristic against the random convergence principle of *genetic*.

4.5   Experiment 3: Compare *cLink* with Other Systems

We compare *cLink* with two supervised instance matching systems, Object-Coref [6] and the work in [21], which uses Adaboost to train a classifier. In this paper, we temporarily refer to this work with the name Adaboost. In addition, we also report the result of RiMOM [10] and PARIS [24] for reference. RiMOM and PARIS are the two state-of-the-art systems among non-learning based ones.

---

³ Tested on a computer equipped with two Intel E5-2690 CPUs and 256 GB memory.

OAEI 2010 dataset is used for this comparison. According to the reported result of Adaboost and ObjectCoref, Adaboost used 5 % candidates for training on all subsets, while ObjectCoref uses 20 actual co-references, which is equivalent to 2.3 %, 11.6 % and 1.2 % candidates on D1, D2, and D3, respectively[4]. Therefore, we feed the same amount of training data into *cLink*, on each respective subset and for each comparison. The results of *cLink* in this experiment are the average of 10 times repeat.

Table 5 reports the F1 scores of *cLink* and other systems. According to this table, *cLink* consistently outperforms the others. In detail, *cLink* improves the results against ObjectCoref and *cLink* is remarkably better than Adaboost on D3 and D5. These subsets are difficult for other systems because they involve with DailyMed, a repository containing the highest number of co-references inside. The comparison also reveals that *cLink* can achieve promising results by being given a small amount of training data (at most 5 %). Compared to the reference systems, RiMOM and PARIS, all of *cLink*, Adaboost, and ObjectCoref are far better. This fact supports the necessity of supervised resolution systems when effectiveness is the first priority.

5   Conclusion and Future Work

We presented *cLink*, an effective and efficient schema-independent entity resolution system. *cLink* supports many aggregation methods and uses a reasonable filtering technique to refine the final resolution results. The matching configuration is optimized by a novel and efficient learning algorithm using only a small amount of training data. The *cLearn* algorithm significantly improves the result of *cLink* against other algorithms, including the ones that are used by other state-of-the-art systems. *cLink* also achieves better performance compared to recent systems.

In future work, we are interested in experimenting with learning goals that are different from F1 for each particular property (e.g., prioritizing the recall or precision). Since there is labeled data, a learning algorithm for optimizing the candidate generator is also very useful to reduce the candidates. In addition, transfer learning and active learning are the two technologies that are applicable in *cLink* in order to even reduce more the training data.

References

1. Araujo, S., Tran, D.T., DeVries, A., Hidders, J., Schwabe, D.: SERIMI: Class-based disambiguation for effective instance matching over heterogeneous web data. In: 15th ACM SIGMOD Workshop on the Web and Databases, pp. 25–30 (2012)
2. Cruz, I.F., Antonelli, F.P., Stroe, C.: AgreementMaker: efficient matching for large real-world schemas and ontologies. VLDB Endowment **2**, 1586–1589 (2009)
3. Ferrara, A., Nikolov, A., Scharffe, F.: Data linking for the semantic web. Int. J. Semant. Web Inf. Syst. **7**(3), 46–76 (2011)

---

[4] Only the results on D1, D2, and D3 are available for ObjectCoref [5].

4. Gale, D., Shapley, L.S.: College admissions and the stability of marriage. Am. Math. Mon. **96**(1), 9–15 (1962)

5. Hu, W., Chen, J., Cheng, G., Qu, Y.: ObjectCoref & Falcon-AO: results for OAEI 2010. In: 5th ISWC Workshop on Ontology Matching, pp. 158–165 (2010)

6. Hu, W., Chen, J., Qu, Y.: A self-training approach for resolving object coreference on the semantic web. In: 20th International Conference on World Wide Web, pp. 87–96 (2011)

7. Hu, W., Yang, R., Qu, Y.: Automatically generating data linkages using class-based discriminative properties. Data Knowl. Eng. **91**, 34–51 (2014)

8. Isele, R., Bizer, C.: Active learning of expressive linkage rules using genetic programming. Web Semant.: Sci. Serv. Agents World Wide Web **23**, 2–15 (2013)

9. Isele, R., Jentzsch, A., Bizer, C.: Efficient multidimensional blocking for link discovery without losing recall. In: 14th ACM SIGMOD Workshop on the Web and Databases (2011)

10. Li, J., Tang, J., Li, Y., Luo, Q.: RiMOM: a dynamic multistrategy ontology alignment framework. IEEE Trans. Knowl. Data Eng. **21**(8), 1218–1232 (2009)

11. Ngomo, A.C.N., Lehmann, J., Auer, S., Höffner, K.: RAVEN - active learning of link specifications. In: 6th International Semantic Web Conference Workshop on Ontology Matching, pp. 25–36 (2011)

12. Ngomo, A.C.N., Auer, S.: LIMES: a time-efficient approach for large-scale link discovery on the web of data. In: 22nd International Joint Conference on Artificial Intelligence, pp. 2312–2317 (2011)

13. Ngomo, A.C.N., Lyko, K.: EAGLE: efficient active learning of link specifications using genetic programming. In: Simperl, E., Cimiano, P., Polleres, A., Corcho, O., Presutti, V. (eds.) ESWC 2012. LNCS, vol. 7295, pp. 149–163. Springer, Heidelberg (2012)

14. Ngomo, A.C.N., Lyko, K.: Unsupervised learning of link specifications: deterministic vs. non-deterministic. In: 8th International Sematic Web Conference Workshop on Ontology Matching, pp. 25–36 (2013)

15. Nguyen, K., Ichise, R.: A heuristic approach for configuration learning of supervised instance matching. In: 14th International Semantic Web Conference Posters and Demonstrations Track (2015)

16. Nguyen, K., Ichise, R.: ScSLINT: time and memory efficient interlinking framework for linked data. In: 14th International Semantic Web Conference Posters and Demonstrations Track (2015)

17. Nguyen, K., Ichise, R., Le, B.: Interlinking linked data sources using a domain-independent system. In: Takeda, H., Qu, Y., Mizoguchi, R., Kitamura, Y. (eds.) JIST 2012. LNCS, vol. 7774, pp. 113–128. Springer, Heidelberg (2013)

18. Nikolov, A., d'Aquin, M., Motta, E.: Unsupervised learning of link discovery configuration. In: Simperl, E., Cimiano, P., Polleres, A., Corcho, O., Presutti, V. (eds.) ESWC 2012. LNCS, vol. 7295, pp. 119–133. Springer, Heidelberg (2012)

19. Niu, X., Rong, S., Zhang, Y., Wang, H.: Zhishi.Links results for OAEI 2011. In: 6th ISWC Workshop on Ontology Matching, pp. 220–227 (2011)

20. Papadakis, G., Ioannou, E., Palpanas, T., Niederée, C., Nejdl, W.: A blocking framework for entity resolution in highly heterogeneous information spaces. IEEE Trans. Knowl. Data Eng. **25**(12), 2665–2682 (2013)

21. Rong, S., Niu, X., Xiang, E.W., Wang, H., Yang, Q., Yu, Y.: A machine learning approach for instance matching based on similarity metrics. In: Cudré-Mauroux, P., et al. (eds.) ISWC 2012, Part I. LNCS, vol. 7649, pp. 460–475. Springer, Heidelberg (2012)

22. Song, D., Heflin, J.: Automatically generating data linkages using a domain-independent candidate selection approach. In: Aroyo, L., Welty, C., Alani, H., Taylor, J., Bernstein, A., Kagal, L., Noy, N., Blomqvist, E. (eds.) ISWC 2011, Part I. LNCS, vol. 7031, pp. 649–664. Springer, Heidelberg (2011)
23. Soru, T., Ngomo, A.C.N.: A comparison of supervised learning classifiers for link discovery. In: Proceedings of the 10th International Conference on Semantic Systems, pp. 41–44. ACM (2014)
24. Suchanek, F.M., Abiteboul, S., Senellart, P.: PARIS: probabilistic alignment of relations, instances, and schema. VLDB Endowment 5(3), 157–168 (2011)
25. Volz, J., Bizer, C., Gaedke, M., Kobilarov, G.: Discovering and maintaining links on the web of data. In: Bernstein, A., Karger, D.R., Heath, T., Feigenbaum, L., Maynard, D., Motta, E., Thirunarayan, K. (eds.) ISWC 2009. LNCS, vol. 5823, pp. 650–665. Springer, Heidelberg (2009)

Alignment Aware Linked Data Compression

Amit Krishna Joshi[✉], Pascal Hitzler, and Guozhu Dong

Wright State University, Dayton, OH, USA
{joshi.35,pascal.hitzler,guozhu.dong}@wright.edu

Abstract. The success of linked data has resulted in a large amount of data being generated in a standard RDF format. Various techniques have been explored to generate a compressed version of RDF datasets for archival and transmission purpose. However, these compression techniques are designed to compress a given dataset without using any external knowledge, either through a compact representation or removal of semantic redundancies present in the dataset. In this paper, we introduce a novel approach to compress RDF datasets by exploiting alignments present across various datasets at both instance and schema level. Our system generates lossy compression based on the confidence value of relation between the terms. We also present a comprehensive evaluation of the approach by using reference alignment from OAEI.

1 Introduction

Linked data has experienced accelerated growth in recent years due to its interlinking ability across disparate sources, made possible via machine processable non-proprietary RDF data. Today, large number of organizations, including governments and news providers, publish data in RDF format, inviting developers to build useful applications through re-use and integration of data. This has led to tremendous growth in the amount of RDF data being published on the web. Although the growth of RDF data can be viewed as a positive sign for semantic web initiatives, it also causes performance bottlenecks for RDF data management systems that store and provide access to it [8]. As such, the need for compressing structured data is becoming increasingly important.

Various RDF representations and compression techniques have been developed to reduce the size of RDF data for storage and transmission. Representation like N3, Turtle and RDF/JSON offer compactness while maintaining readability by reducing verbosity of original RDF/XML format. Earlier RDF compression studies [3,4,18] focused on dictionary encoding and RDF serialization techniques. [4] proposed a new compact representation format called Header-Dictionary-Triples (HDT) that takes advantage of skewed data in large RDF graphs. [9] introduced the notion of a lean graph which is obtained by eliminating triples with blank nodes that specify redundant information. [17] and [14] studied the problem of redundancy elimination on RDF graphs in the presence of rules and constraints. [12] introduced rule based compression technique that exploits the semantic redundancies present in large RDF graph by mining frequent patterns and removing triples that can be identified by association rules.

© Springer International Publishing Switzerland 2016
G. Qi et al. (Eds.): JIST 2015, LNCS 9544, pp. 73–81, 2016.
DOI: 10.1007/978-3-319-31676-5_5

These techniques perform compression by identifying syntactic and semantic redundancies in a dataset but do not exploit alignments present across various datasets. In this paper, we propose a novel technique to incorporate alignments for compressing datasets. This is the first study investigating lossy RDF compression based on alignments and application context.

This work was supported by the National Science Foundation under award 1017225 III: TROn (Tractable Reasoning with Ontologies).

2   Motivation

Ontology Alignment [16] refers to the task of finding correspondences between ontologies. It's a widely explored topic and numerous applications have been developed that perform the task of ontology alignment and mapping for schemas and instances. It finds a number of use cases including schema integration, data integration, ontology evolution, agent communication and query answering on the web [13,15].

In this study, we utilize ontology alignments for compressing RDF datasets. Below, we list a few properties that could be exploited while compressing multiple datasets.

Schema Heterogeneity and Alignment: Linked datasets cater to different domains, and thus require different modeling schemas. Even when datasets belong to the same domain, they could be modeled differently depending on the creator. For instance, Jamendo[1], and BBC Music[2] both belong to music domain but they use different ontologies[3,4]. Different ontologies, whether belonging to the same domain or not, often share some similarities and the terms can be aligned. Based on the resulting alignment, individual datasets can be rewritten using fewer schema terms before further processing. Many studies have been focused on schema alignment using various approaches such as sense clustering [6], instance similarity [10,19] and structural/lexical similarities [11]. Factforge [1] uses an upper level PROTON[5] as a reference layer and has more than 500+ mapping across various datasets.

Datasets rewritten using a set of mapping terms lead to increased occurrences of same terms, resulting in a better compression.

Entity Co-reference and Linking: The purpose of entity co-reference is to determine if different resources refer to the same real world entity. Often datasets have overlapping domains and tend to provide information about the same entity [5]. One of the approach include using similarity properties such as owl:sameAs or skos:exactMatch. For instance, LinkedMdb provides information about the

---

[1] http://dbtune.org/jamendo/.
[2] http://www.bbc.co.uk/music.
[3] http://musicontology.com/.
[4] http://www.bbc.co.uk/ontologies/bbc.
[5] http://www.ontotext.com/proton-ontology/.

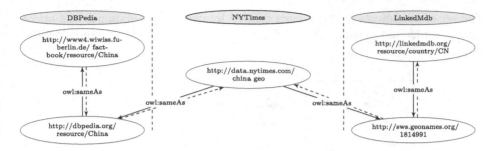

Fig. 1. LinkedMdb connects to DBPedia via NYTimes

Romeo and Juliet movie and provides direct reference to DBPedia using the owl:sameAs property.

However, there are cases where the two instances might not be directly connected but a path exists for such a co-reference as shown in Fig. 1. Here, the Geonames resource for China is linked to the CIA Factbook concept and the DBPedia concept for China, using an "owl:sameAs" link from the NYTimes dataset.

Varied Alignments: Alignment results vary greatly among different ontology matching systems (see [7]). Some of these work best for one set of ontologies while perform low in a different set of ontologies. The alignments can differ even when manually performed among group of experts for the same set of ontologies. For instance, conference track of OAEI provides the reference alignments[6] with a confidence score of 1 (signifying exact match) for all mappings within a collection of ontologies describing the domain of organizing conferences. On the contrary, [2] introduced a new version of the Conference reference alignment for OAEI that includes the varying confidence values reflecting expert disagreement on the matches.

3   Alignment Aware Linked Data Compression

In this section, we elaborate on the internals of our compression system. The main task involves identification of alignments across various datasets. The alignments can be manual or generated using existing Ontology matching systems[7].

Given two ontologies O_i and O_j, we can compute multiple mappings between the ontology terms, t_i and t_j.

Alignment, μ is defined as $\mu = < t_i, t_j, r, s >$ where r denotes the relationship and $s \in [0, 1]$ is the confidence score that the relationship holds in the mapping.

Figure 5 represents the high level overview of our system. Given a set of input datasets, we first identify alignments present across these datasets. For this, we extract terms from each dataset and check for alignments with other participating datasets either manually or using automated ontology matching systems. It should be noted that the alignments can be in both schema and instance level.

---

[6] http://oaei.ontologymatching.org/2014/conference/data/reference-alignment.zip.
[7] http://www.mkbergman.com/1769/50-ontology-mapping-and-alignment-tools/.

The set of alignments are then consolidated by performing mapping to a set of *master* terms and pruning all mappings that have a confidence score below the threshold Fig. 2.

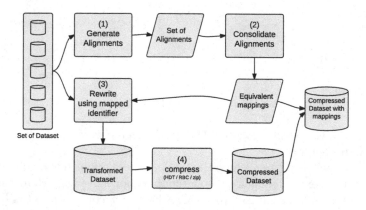

Fig. 2. Conceptual system overview

The resulting unique set of mappings, together with the original datasets go through a transformation phase where all datasets are merged and the equivalent terms are replaced with *master* terms. Once the transformation is complete, the combined dataset is then passed to existing compression systems such as HDT, RBC and Zip to generate compressed dataset.

Algorithm for the consolidation of alignments is listed in Algorithm 1. Given a threshold and a set of alignments, mappings with confidence score less than a threshold are pruned and a set of master items is generated. Each master item maps to a group of equivalent ontology terms. These master items are later used to rewrite the dataset to replace ontology terms with corresponding master item. The alignments can contain both instance and schema terms. Figure 3 shows two master items and corresponding consolidated alignments.

```
<http://ekaw#Regular_Paper>
-<http://cmt#PaperFullVersion>
-<http://confOf#Contribution>
-<http://iasted#Submission>
-<http://cmt#Paper>
-<http://ekaw#Regular_Paper>
-<http://edas#Paper>
-<http://confOf#Paper>
-<http://sigkdd#Paper>
-<http://conference#Paper>
-<http://ekaw#Paper>
```

```
<http://ekaw#Research_Topic>
-<http://cmt#SubjectArea>
-<http://edas#Topic>
-<http://confOf#Topic>
-<http://ekaw#Research_Topic>
-<http://conference#Topic>
```

Fig. 3. Sample grouping of equivalent terms for ekaw#Regular_Paper and ekaw#Research_Topic using OAEI reference alignment.

Algorithm 1. Consolidation of Alignments

Require: A Alignment set and θ threshold for alignments
        Pruning mappings with confidence lower than threshold value
1: Valid Mapping M as $\langle k, V \rangle \leftarrow \phi$
2: Term Mapping $G \leftarrow \phi$
3: Set $S \leftarrow \phi$
4: MasterItem Mapping $I \leftarrow \phi$
5: **for each** mapping, $< e1, e2, r, s >$ that occurs in A **do**
6:    **if** $r =' equivalence'$ and $s >= \theta$ **then**
7:       $M \leftarrow M \cup \langle e1, V \cup e1 \rangle$ ▷ add a new valid mapping
8:       $M \leftarrow M \cup \langle e2, V \cup e2 \rangle$
9:    **end if**
10: **end for**
        Grouping equivalent terms
11: **for all** $\langle k, V \rangle$ in M **do**
12:    **if** $k \notin keys(G)$ and $k \notin S$ **then**
13:       $G \leftarrow G \cup \langle k, V_k \rangle$ ▷ mark this k as master item
14:       $S \leftarrow S \cup k$ ▷ mark this k as processed item
15:       **for each** $t \in V_k$ **do**   ▷ group all items in V_k under k
16:          $G \leftarrow G \cup \langle k, V_t \rangle$ ▷ k maps to $V_k \cup V_t$
17:          $S \leftarrow S \cup t$ ▷ mark this t as processed item
18:       **end for**
19:    **end if**
20: **end for**
        One to One mapping with master item
21: **for each** $(k, V) \in G$ **do**
22:    **for each** $v \in V$ **do**
23:       $I \leftarrow I \cup \langle v, k \rangle$ ▷ map to master item
24:    **end for**
25: **end for**

---

4 Evaluation

For this paper, we built a prototype, LinkIt, in JAVA to test the validity of our approach. We experimented using reference alignments from OAEI.

4.1 Dataset Generation

Since our primary purpose is to validate that RDF data can be compressed in presence of alignments, we need a set of ontologies, reference alignment for those ontologies and RDF data large enough to be tested. For the evaluation, we generated large size of synthetic RDF data using SyGENiA[8] tool and a set of Conference ontologies and the reference ontologies available from OAEI[9]. Given a set of queries and an ontology, SyGENiA tool can automatically generate a

---

[8] https://code.google.com/p/sygenia/.
[9] http://oaei.ontologymatching.org/2014/.

large number of individuals. The set of queries that we use for generating RDF data is available from[10]. In order to test the compression against dataset of varying size, we created multiple queries and generated eight different dataset. The size of evaluation dataset size is shown in Fig. 4.

	Dataset size(MB) created using query							
ontology	Q1	Q2	Q3	Q4	Q5	Q6	Q7	Q8
Conference	113	261	257	123	195	213	113	727
confOf	107	152	149	77	137	129	98	546
iasted	84	161	157	74	129	108	84	670
sigkdd	98	158	146	92	137	126	88	390
cmt	67	149	140	79	97	99	56	658
edas	107	192	181	90	137	139	108	769
ekaw	94	181	177	63	146	147	92	704
Total	670	1254	1207	598	978	961	639	4464

Fig. 4. Dataset size for various set of queries.

4.2 Varied Alignments and Compression

We evaluated two versions of Conference reference alignment available from OAEI and [2]. These reference alignments include 16 ontologies related to the conference organization and they are based upon the actual conference series and corresponding web pages[11]. The mappings in the Conference:V1 are all set to be exact match. Figure 5 compares the distribution of valid mappings for various

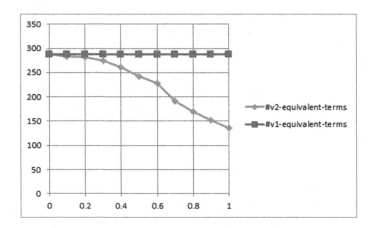

Fig. 5. Number of mappings at different thresholds for two versions of conference reference alignments

---

[10] http://bit.ly/1hgNsRv.
[11] http://oaei.ontologymatching.org/2014/conference/index.html.

	Alignment System					
threshold	AML	Xmap	RSDLWB	OMReasoner	MaasMtch	LogMap
1	9	134	162	154	219	194
0.9	146	134	162	154	482	204
0.8	170	143	162	154	532	213
0.7	194	145	162	154	532	218
0.6	213	145	162	154	532	225
0.5	220	146	162	154	532	230
0.4	220	148	162	154	532	238
0.3	220	148	162	154	532	239
0.2	220	148	162	154	532	240
0.1	220	148	162	154	532	240

Fig. 6. Comparison of various automated alignment systems demonstrating varying number of equivalent terms for same threshold

thresholds for both reference alignments. The number of mappings are generated after the consolidation of alignments. As expected, the number of mapping decreases with the increase of threshold in Conference:V2 reference alignment.

Furthermore, for the same set of datasets, various ontology matching systems can produce different set of alignments. Figure 6 shows a comparison of various alignment systems with varying number of equivalent terms for same threshold, as seen in the results of OAEI[12]. The alignments are generated for the same set of ontologies used in Conference:V1 and Conference:V2 reference alignments. As seen in Fig. 6, some alignment systems such as RSDLWB and OMReasoner generate all alignments with a confidence score of 1, while others like LogMap and XMap generate alignments with varying confidence score.

Since the alignment is not one to one, we cannot recover the original data once compressed and hence the compression is lossy.

The evaluation result for varying alignments is shown in Fig. 7 for one of the datasets which has original size of 670 MB. The compressed size can be compared against the output resulting from HDT alone which is 56 MB.

AlignmentSystem	Compressed size (MB)
V1	51
V2	53
AML	53
Logmap	51
OmReasoner	52
Maasmtch	51
rsdlwb	53
xmap	53

Fig. 7. Compressed size (in MB) against original size of 670 MB

---

[12] http://oaei.ontologymatching.org/2014/conference/eval.html.

5   Conclusion

In this paper, we have introduced a novel compression technique that exploits alignments present across various datasets at schema and instance level. We have explored lossy RDF compression, the area which has barely been researched in the semantic web field. The system extracts all mappings with confidence score greater or equal to the threshold and group them using single identifiers. Hence, our approach is flexible enough to cut-off alignments based on threshold that are context dependent.

In the future, our research can be directed towards finding applications of lossy RDF compression. We will also explore the effect of alignments in compressing RDF streams.

References

1. Bishop, B., Kiryakov, A., Ognyanov, D., Peikov, I., Tashev, Z., Velkov, R.: Fact-forge: a fast track to the web of data. Semant. Web **2**(2), 157–166 (2011)
2. Cheatham, M., Hitzler, P.: Conference v2.0: an uncertain version of the OAEI conference benchmark. In: Mika, P., Tudorache, T., Bernstein, A., Welty, C., Knoblock, C., Vrandečić, D., Groth, P., Noy, N., Janowicz, K., Goble, C. (eds.) ISWC 2014, Part II. LNCS, vol. 8797, pp. 33–48. Springer, Heidelberg (2014)
3. Fernández, J.D., Gutierrez, C., Martínez-Prieto, M.A.: RDF compression: basic approaches. In: Proceedings of the 19th International Conference on World Wide Web, pp. 1091–1092. ACM (2010)
4. Fernández, J.D., Martínez-Prieto, M.A., Gutiérrez, C., Polleres, A., Arias, M.: Binary RDF representation for publication and exchange. Web Semant.: Sci. Serv. Agents on the World Wide Web **19**, 22–41 (2013)
5. Glaser, H., Jaffri, A., Millard, I.: Managing co-reference on the semantic web (2009)
6. Gracia, J., d'Aquin, M., Mena, E.: Large scale integration of senses for the semantic web. In: Proceedings of the 18th International Conference on World Wide Web, pp. 611–620. ACM (2009)
7. Grau, B.C., Dragisic, Z., Eckert, K., Euzenat, J., Ferrara, A., Granada, R., Ivanova, V., Jiménez-Ruiz, E., Kempf, A.O., Lambrix, P., et al.: Results of the ontology alignment evaluation initiative 2013. In: Proceedings of the 8th ISWC Workshop on Ontology Matching (OM), pp. 61–100. No commercial editor (2013)
8. Huang, J., Abadi, D.J., Ren, K.: Scalable SPARQL querying of large rdf graphs. Proc. VLDB Endowment **4**(11), 1123–1134 (2011)
9. Iannone, L., Palmisano, I., Redavid, D.: Optimizing RDF storage removing redundancies: an algorithm. In: Ali, M., Esposito, F. (eds.) IEA/AIE 2005. LNCS (LNAI), vol. 3533, pp. 732–742. Springer, Heidelberg (2005)
10. Isaac, A., Van Der Meij, L., Schlobach, S., Wang, S.: An Empirical Study of Instance-Based Ontology Matching. Springer, Heidelberg (2007)
11. Jean-Mary, Y.R., Shironoshita, E.P., Kabuka, M.R.: Ontology matching with semantic verification. Web Semant.: Sci. Serv. Agents on the World Wide Web **7**(3), 235–251 (2009)
12. Joshi, A.K., Hitzler, P., Dong, G.: Logical linked data compression. In: The Semantic Web: Semantics and Big Data, pp. 170–184. Springer (2013)

13. Joshi, A.K., Jain, P., Hitzler, P., Yeh, P.Z., Verma, K., Sheth, A.P., Damova, M.: Alignment-based querying of linked open data. In: Meersman, R., et al. (eds.) OTM 2012, Part II. LNCS, vol. 7566, pp. 807–824. Springer, Heidelberg (2012)
14. Meier, M.: Towards rule-based minimization of RDF graphs under constraints. In: Calvanese, D., Lausen, G. (eds.) RR 2008. LNCS, vol. 5341, pp. 89–103. Springer, Heidelberg (2008)
15. Noy, N.F.: Semantic integration: a survey of ontology-based approaches. ACM Sigmod Rec. **33**(4), 65–70 (2004)
16. Noy, N., Stuckenschmidt, H.: Ontology alignment: an annotated bibliography. In: Semantic Interoperability and Integration 4391 (2005)
17. Pichler, R., Polleres, A., Skritek, S., Woltran, S.: Redundancy elimination on RDF graphs in the presence of rules, constraints, and queries. In: Hitzler, P., Lukasiewicz, T. (eds.) RR 2010. LNCS, vol. 6333, pp. 133–148. Springer, Heidelberg (2010)
18. Urbani, J., Maassen, J., Drost, N., Seinstra, F., Bal, H.: Scalable RDF data compression with MapReduce. Concurrency and Computation: Practice and Experience **25**(1), 24–39 (2013)
19. Wang, S., Englebienne, G., Schlobach, S.: Learning Concept Mappings from Instance Similarity. Springer, Heidelberg (2008)

ERA-RJN: A SPARQL-Rank Based Top-k Join Query Optimization

Zhengrong Xiao[1,2(✉)], Fengjiao Chen[1,2], Fangfang Xu[1,2], and Jinguang Gu[1,2]

[1] College of Computer Science and Technology, Wuhan University of Science and Technology,
Wuhan 430065, China
243881942@qq.com
[2] Hubei Province Key Laboratory of Intelligent Information Processing
and Real-Time Industrial System, Wuhan 430065, China

Abstract. With the wide use of RDF data, searching and ranking semantic data with SPARQL has become a research hot-spot. While there is no much research work on top-k join queries optimization in RDF native stores. This paper proposes a new rank-join operator algorithm ERA-RJN on the basis of SPARQL-RANK algebra, making use of the advantage of random access availability in RDF native storage. This paper implements the ERA-RJN operator on the ARQ-RANK platform, and performs experiments, verifies the high efficiency of ERA-RJN algorithm dealing with SPARQL top-k join query in RDF native storage.

Keywords: RDF native storage · Top-k join · Rank-join operator · SPARQL-RANK

1 Introduction

Ranking is an important process of the query processing [1]. According to the ranking criterion, the top-k join query is a process which obtains k results that mostly match conditions from multi database tables or data sets. In recent years, the research of top-k join query become popular with the widely use of RDF(Resource Description Framework) [2] data in different fields. During the querying processing, the multi complex joining and ranking led to high cost in obtaining exact result of top-k join query. The research of optimizing SPARQL top-k join query is not widespread. So optimizing SPARQL(Simple Protocol and RDF Query Language) [3] top-k join query with access characteristics has large significance in querying large semantic data.

2 Related Work

There are two existing methods of optimizing SPARQL top-k join query: the optimization of relational algebra [4–6] and the optimization of top-k join algorithm [7, 8].

The method based on optimization of relational algebra is embodied in the extension of SPARQL or relational algebra to meet the different kind of top-k query.

© Springer International Publishing Switzerland 2016
G. Qi et al. (Eds.): JIST 2015, LNCS 9544, pp. 82–88, 2016.
DOI: 10.1007/978-3-319-31676-5_6

RANKSQL system, which is achieved by Li, provides effective top-k query on relative database on the extension of relational algebra [9]. Magliacane proposed SPARQL-RANK algebra based on RANKSQL system [10].

The optimization of top-k join is a method that committed to speed up query connection. The typical example algorithm is Hash Rank Join Operator (HRJN) algorithm which is proposed by Elmagarmid [11]. RSEQ algorithm is the optimized algorithm of HRJN. RSEQ algorithm speeds up the query efficiency by using less sequential access and using random access only.

With the widely using in native storage of RDF, making further research on optimizing SPARQL top-k join with data storage is more and more important.

3   Top-k Join Query Optimization Based on SPARQL-RANK

The join operation of most systems is the 2-road join, and the SPARQL algebraic query plan tree is the left line form, the order of executing query is bottom-up sequentially, rank-join operator only deals with binary connection, so the top-k join algorithm described in this paper, are all based on two order input collection.

Sequential access is according to the base score descending sort of tuple to read, random access is to read tuples that conform to the connecting condition, regardless of whether the data has maximum or minimum base scores. Taking advantage of random access patterns, top-k join algorithm can have a lot ascension in efficiency, there are two main reasons: (1) compared with the large amount of data tuple that sequential access in reading, random access only need to read tuples that meet the requirements of connection, decrease the I/O overhead of reading tuples; (2) relative to the sequent access model, random access mode threshold calculation more accurate, in the top-k results determine phase, smaller threshold reduced the contrast range.

Primary storage is the mainstream way of RDF triples storage. RDF primary storage is a way that can support random access well, and through establish the index to achieve tuple sequential access, take advantage of this point, this paper extends the RSEQ algorithm, proposed the ERA-RJN (Extend the Random Access Rank the Join Operator) algorithm. The algorithm allows the two input collection support random access and sequential access at the same time. During reading tuples phase, parallel sequential read each input collection at a time, and then use random access patterns, two-way parallel to explore effective connecting mappings. At the same time, in view of the two-way exploration result in repeated connection, duplicate elimination strategy is proposed in this paper. Finally in order to further improving algorithm performance, the rapid termination strategy has been used, through accurate calculation threshold, the effective connecting mappings and threshold in the process of query contrast range decreases, in this way it's more conducive to output effective connecting mappings. Here are the strategies used in this paper:

(1) Random access two-way availability. When the two input sets L and R using sequential access pattern and random access pattern while reading tuples at the same time, threshold value calculated by formula [12]

$$T = f\left(L_{bottom}, R_{bottom}\right) \tag{2-1}$$

Among them L_{bottom} and R_{bottom} represent the latest collection of base scores come from the input tuples of L and R, also called the recent visit scores. ERA-RJN expand the RSEQ, allows two input collection support the sequential access and random access at the same time.

(2) Parallel sequential access. Parallel sequential read tuples, read tuples from the input sets L and R at once every time, and use two tuples, respectively, to explore another input collection to find effective connecting mappings. The two-way pattern that combine parallel access and random access, can achieve parallel operation. This way can discover all effective connecting mappings that under the minimum scanning depth, to improve the execution efficiency.

(3) Duplicate elimination strategy. In the process of connecting and exploring, all connecting mappings' total scores that over T(2-1) must have been produced before. Through comparing effective connecting mappings' composite score that produced each time with the threshold value, the redundant connecting mappings can be deleted in the process of exploring.

(4) Fast termination strategy. After exploring effective connecting mappings, if the current reading tuples do no contribute to subsequent effective connection, then using the scores of current reading tuples L_{bottom} or R_{bottom} to update the threshold, but it will overestimated the average threshold, increase the follow-up exploration contrast range. Aimed at this situation, this paper uses the method of demoting L_{bottom} or R_{bottom}, to set the next reading tuples' scores with L_{bottom} or R_{bottom}, which can calculate the threshold T(2-1) more accurately.

(5) Conflict Resolution. Because the different evaluation standards of duplicate elimination strategy and fast termination strategy, contradictory phenomena will be produced when they combined. To solve this problem, we use different evaluation standards to run these strategies.

4   ERA-RJN Algorithm Implementation

Based on RDF native data availability of random access storage mode, improved RSEQ algorithm and proposed the ERA - RJN algorithm, allow the two input support the sequential access and random access at the same time, and adds the repeated elimination strategy and quick end strategy. Table 1 describes the concrete realization of the algorithm:

Table 1. ERA - RJN algorithm

Input: two sort map sets the L and R, K results the number of constraints.

Initialization: the original threshold $T\_o = \infty$, precise threshold $T\_n = \infty$,
priority queue $Q = \varnothing$, k=K, $t_L = null$, $t_R = null$, $t_L' = null$, $t_R' = null$

Steps:

1) Judge k, if $k = 0$, then the top - k produced results, the algorithm is over;
   Otherwise enter the step (2);

2) Judge t_L' and t_R' whether they are empty, if not then $t_L = t_L'$, $t_R = t_R'$,
   $T\_o = T\_n$, else judge L and R whether they are finished reading, if both
   of them do not complete of reading, sequential access the L and R
   respectively, read the next tuple, t_L =L.getnext(), t_R =R.getnext();
   $L_{bottom} = t_L.score$, $R_{bottom} = t_R.score$; $T\_o = f(L_{bottom}, R_{bottom})$. If both
   inputs are complete reading, jump out of circulation, and end the algorithm;

3) Using t_L to explore R, using t_R to explore L, look for effective connection
   combination;

4) Judge L and R whether they are finished reading, if not, then continue reading
   the next tuple from the corresponding input t_L' and t_R', then update bottom
   scores and T_n; Or else then $T\_n = T\_o$;

5) For each active connection mapping *mapping*, using f to calculate a
   combination of score *mapping.score*, if *mapping.score* $\leq T\_o$,
   deposit effective linked map in Q with the combination of descending scores;

6) Analyzing priority queue Q whether is empty, if the queue is empty, go to (1),
   if not then enter (7);

7) Take out the first mapping of queue, judge *mapping.score* $>= T\_n$
   whether is established, if it is true then output *mapping*, k=k-1, remove
   from Q of *mapping*, return to step (6). If not satisfied return to step (1).

Output: top-k query results

5   Experimental Design and Analysis

5.1   Experimental Environment

In this paper, the machine is configured as a quad-core, 2.00GHz Intel® Xeon (R) E5504CPU, 7.00GB DDR2 system memory, the machine is running with CentOS release 6.4 (Final), the system kernel is 2.6.32-358.el6.x86_64, ARQ-RANK platform is running on Sun Java1.7.0, the maximum allowable stack memory is 4.0 GB.

This paper based on ARQ-RANK platform achieved the ERA-RJN operator. The BSBM (Berlin SPARQLBenchmark) data set is used for the experimental data set, as shown in Table 2 below.

Table 2. Experimental datasets

Data set	Dataset1	Dataset2	Dataset3	Dataset4	Dataset5
RDF triples number	107433	255240	500661	1036070	5122738

5.2   Experimental Results and Analysis

This paper compared the execution time of three algorithm HRJN, RSEQ and ERA - RJN under different data sets.

Figure 1 shows that in the same query on five datasets, with increasing number of data sets, the times needed to extract top-10 and top-100 are all increased. Since ERA-RJN algorithm supports bidirectional explore and parallel sequential read tuples, so the same query execution time is less, reflecting the efficiency of its performance. When facing large-scale data, for example run top-100 inquiries on Dataset5 dataset, the advantage of ERA-RJN algorithm is more obvious, the execution time of the algorithm

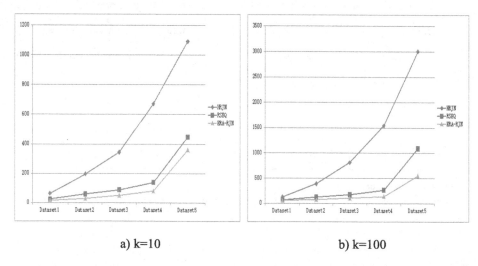

a) k=10   b) k=100

Fig. 1.   Under different data sets HRJN, RSEQ, ERA-RJN algorithm execution time (Unit: ms)

is the 1/5 of HRJN, and the 1/2 of RSEQ. So with the data set larger, compared HRJN and RSEQ, ERA-RJN algorithm is more efficient in connection.

6 Conclusion

This paper based on the top-k query optimization scheme of SPARQL-RANK algebra, analyzed the rank-join operator processing algorithm type on the basis of the native RDF storage mode. Compared the commonly used top-k join algorithm HRJN and its improved algorithm RSEQ, taking advantage of the RDF data availability of random access, then proposed ERA-RJN algorithm, improved the efficiency of the top-k join query. However the current ERA-RJN algorithm only consider advantages of random access to optimize the query performance, so it is not comprehensive. Based on predicate index optimization or any other strategies can also be considered to enhance the SPARQL top-k join query efficiency in the future.

Acknowledgements. This topic research is done under the guidance of Professor JinGuang Gu. Avail ourselves of this opportunity to express heartfelt thanks to Professor Gu. And student Fangfang Xu, YuWei Zou helped with some technical stuff. This work is supported by the Natural Science Foundation Project of Hubei Province (2013CFB334), the Scientific Research Project of Hubei Provincial Education Department (Q20101110, D2009110) and the Outstanding Young Scientific and Technological Innovation Team Project of Hubei Provincial Colleges and Universities (T201202), much thanks with it.

References

1. Ilyas, I.F., Beskales, G., Soliman, M.A.: A survey of top-k query processing techniques in relational database systems. ACM Trans. Comput. Surv. **40**(4), 198–205 (2008)
2. Decker, S., Melnik, S., Van Harmelen, F., et al.: The semantic web: the roles of XML and RDF. Internet Comput. IEEE **4**(5), 63–73 (2000)
3. Perez, J., Arenas, M., Gutierrez, C.: Semantics and complexity of SPARQL. ACM Trans. Database Syst. **34**(3), 341–352 (2009)
4. Schmidt M, Meier M, Lausen G. Foundations of SPARQL query optimization. In: Proceedings of the 13th International Conference on Database Theory, pp. 4–33. ACM (2010)
5. Nikolov, A., Schwarte, A., Hütter, C.: FedSearch: efficiently combining structured queries and full-text search in a SPARQL Federation. In: Alani, H., Kagal, L., Fokoue, A., Groth, P., Biemann, C., Parreira, J.X., Aroyo, L., Noy, N., et al. (eds.) ISWC 2013, Part I. LNCS, vol. 8218, pp. 427–443. Springer, Heidelberg (2013)
6. Bozzon, A., Della Valle, E., Magliacane, S.: Extending SPARQL algebra to support efficient evaluation of top-k SPARQL queries. In: Ceri, S., Brambilla, M. (eds.) Search Computing: Broadening Web Search. LNCS, vol. 7538, pp. 143–156. Springer, Heidelberg (2012)
7. Liu, J., Feng, L., Xing, Y.: A pruning-based approach for supporting top-k join queries. In: Proceedings of the 15th International Conference on World Wide Web, pp. 891–892. ACM (2006)
8. Martinenghi, D., Tagliasacchi, M.: Proximity measures for rank join. ACM Trans. Database Syst. (TODS) **37**(1), 1–45 (2012)

9. Li, C., Chen-Chuan, K., Ihab, C., et al.: RankSQL: query algebra and optimization for relational top-k queries. In: SIGMOD, pp. 131–142 (2005)
10. Magliacane, S., Bozzon, A., Della Valle, E.: Efficient execution of top-k SPARQL queries. In: Cudré-Mauroux, P., Heflin, J., Sirin, E., Tudorache, T., Euzenat, J., Hauswirth, M., Parreira, J.X., Hendler, J., Schreiber, G., Bernstein, A., Blomqvist, E. (eds.) ISWC 2012, Part I. LNCS, vol. 7649, pp. 344–360. Springer, Heidelberg (2012)
11. Ilyas, I.F., Aref, W.G., Elmagarmid, A.K.: Supporting top-k join queries in relational databases. VLDB J.—Int. J. Very Large Data Bases 13(3), 207–221 (2004)
12. Martinenghi, D., Tagliasacchi, M.: Cost-aware rank join with random and sorted access. IEEE Trans. Knowl. Data Eng. 24(12), 2143–2155 (2012)

Information Extraction

CNME: A System for Chinese News Meta-Data Extraction

Junbo Xia[1,2(✉)], Fei Xie[1,2], Mengdi Zhang[1,2], Yu Su[1,2], and Huanbo Luan[1,2]

[1] Knowledge Engineering Group, Department of Computer Science and Technology,
Tsinghua University, Beijing 100084, People's Republic of China
[2] Communication Technology Bureau, Xinhua News Agency, Beijing 100803, China
shiningbeer@gmail.com, mdzhangmd@gmail.com, luanhuanbo@gmail.com,
{xiefei,suyu}@xinhua.org

Abstract. News mining has gained increasing attention because of the overwhelming news produced everyday. Lots of news portals such as Sina (http://www.sina.com) and Chinanews (http://www.chinanews.com) develop tools to manage the billions of news and provide services to meet all kinds of needs. News analysis applications conduct news mining work and reveal valuable information. What they all need is news meta-data, the fundamental element to support news analysis work. To extract and maintain meta-data of news becomes an important and challenging task. In this paper, we present a system specialized for Chinese news meta-data extraction. It can identify 28 kinds of meta-data and provides not only a pipeline to extract them but also a systematic way for management. It facilitates the organizing and conducting of news mining processes and improves efficiency by avoiding duplication of work. More specifically, it introduces an innovative way to categorize news based on words' ability to represent category. It also adapts existing methods to extract keywords, entities and event elements. Integration of our system on news mining applications has proved its valuable contribution for news analysis work.

Keywords: News analysis · Meta-data extraction · Keyword extraction · Entity linking

1 Introduction

The web is overwhelmed by numerous daily news. Apart from news website, news has found various spreading ways to Internet users, such as Twitter, blogs and mobile phone applications. There is valuable information hidden in the ocean of news and it is a challenging task to mine the information. Lots of applications arose in response to tackle the problem, such as NewsMiner [3] and Eventsearch [11]. They are developed to query, analyze and mine millions of news to meet all kinds of needs. Regardless of the news generation, publication, spreading, or analysis, the basic information of the news is required in all processes to support their work. There is a term for such information: *meta-data*, which is described in Wikipedia[1] as "data about data".

---

[1] https://en.wikipedia.org/wiki/Metadata.

© Springer International Publishing Switzerland 2016
G. Qi et al. (Eds.): JIST 2015, LNCS 9544, pp. 91–107, 2016.
DOI: 10.1007/978-3-319-31676-5_7

Generally, there are lots of information to describe news, and in this paper, we identify the mostly-used 28 kinds of meta-data. As shown in Fig. 1, some are obvious and easy to obtain, such as publish time, source url and comment volume, while some need great efforts to extract, such as category, keywords, entities and event elements (i.e., at *when*, *who* did *what* at *where*, often referred as 5W1H[2]).

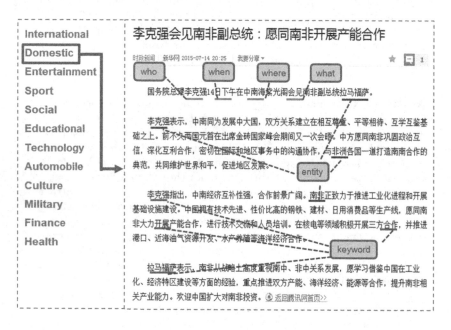

Fig. 1. Example of meta-data

The extracted meta-data provides plenty of information vital to news analysis related work, which presently, is troubled by the following problems:

- it may need to extract the same meta-data from the news in different stages, which leads to duplication of work and thus reduces the efficiency.
- it lacks a systematic way to organize the extracted meta-data.

It occurs to us that a system is needed to extract all the possible meta-data of news, so that other news applications can relieve themselves from the extraction work and concentrate on their own job. There will be no need for other news applications to deal with the raw news text and their input changes to well-structured meta-data as shown in Fig. 2. Such a system provides at least the following advantages to news mining: (1) it facilitates the data management and the mining processes; (2) it improves efficiency by avoiding duplicated work; (3) it provides an alternative way for quick access to news.

---

[2] When, Who, What, Where, Why, How.

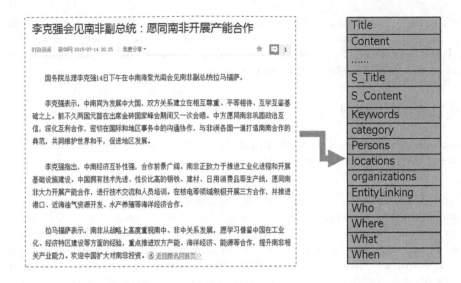

Fig. 2. Unstructured text to structured object

In this paper, Chinese News Meta-data Extraction (CNME), developed for extracting all kinds of meta-data of Chinese news, is presented. It extracts, preserves and maintains as many meta-data as possible to meet all the predictable needs of various further news querying, analyzing and mining processes. Presently, CNME extracts 28 kinds of meta-data, which describes the news from different aspects. In the paper, we only discuss the extraction details about those with more semantic information:

- *News category*: we introduce an innovative way to classify the news, which puts emphasis on word representative ability of category. Our method makes a great leap on efficiency compared with other machine learning methods.
- *Named entity*: we extract named entities, and then link them to the corresponding entries in public knowledge bases through a simple and effective method based upon context and co-mention.
- *Keyword*: we derive potential POS patterns for keywords, and design 7 features for computing word score as keyword.
- *Event element*: we extract 4W (i.e. who, where, when, what) since we believe "*who* does *what* at *when* at *where*" can describe the event of the news. The method is based on the topic sentence extraction.

Our main contributions include:

- We define a complete set of meta-data to describe the news from all aspects.
- We develop a pipeline to extract the defined meta-data.
- We validate the proposed method on real world data, and further apply the extracted meta-data in news analysis application and the results demonstrate the effectiveness and efficiency.

The rest of the paper is organized as follows: Sect. 2 reviews the related work and analyzes the main differences from our methods. We present the system overview and detailed methods in Sect. 3. Experimental results are reported in Sect. 4, and finally Sect. 5 concludes our work.

2 Related Work

There are several lines of researches that are related to our work, and we present some of the related as follows:

Manual Extraction. A common way for news meta-data extraction is by manual efforts. For example, the New York Times maintains a corpus with meta-data provided, which contained 1.8 million articles written and published by the New York Times. Most articles are manually summarized and tagged by a staff of library scientists. Articles are tagged for persons, places, organizations, titles and topics using a controlled vocabulary that is applied consistently across articles. This work can help researchers to test their algorithms, but it can not be widely applied due to its tedious and time-consuming nature.

Auto Extracting Systems. There exist many researches or projects that focus on automatical extraction and management of news data, such as IJS newsfeed [12] and NASS [2]. IJS is a RSS crawler and also provides user with extracted meta-data. It can extract named entities and keywords, but it mainly deals with English and Slovenian news and is not capable of Chinese news. Because Chinese words are continuous and not segmented by blank space, the pre-process of its sentences is very different from English-kind languages. NASS is a ontology-based classification system for Spanish news. It extracts keywords and entities from news to match the news to a node of the hierarchical tree of their ontology. It can not deal with Chinese news neither. Actually, we haven't encountered any meta-data extracting system specialized for Chinese news.

News Categorization. News categorization is a branch of text categorization. In the last 20 years, substantial studies have been done on text categorization. Recent years, Chinese text categorization also gains attentions. Lots of methods are compared in [9], and the improved visions including KNN [16], Bayes [8], Decision tree [4], and SVM [6]. These methods neither provide sufficient accuracy for news categorization nor meet the requirement of high efficiency for real-time system. Krishnalal et al. [5] claims more than 90 % accuracy with his method of HMM-SVM (Hidden Markov Model SVM), but his method was only tested on 3 categories, and we often deal with much more categories (e.g., in this paper, our system requires 12 categories).

Other Extractions. Meta-data extraction has received increasing attentions in recent years, but many works focus on *Information Retrieval* (IR), such as [1,13]. While our attention is settled on more semantic meta-data, which falls in the domain of *Information Extraction* (IE) [10]. The event (5W1H) extraction method in this paper is inspired by Wang's work [14]. He extracts 5W1H on the basis of topic sentence extraction. We improve his topic sentence extraction algorithm by adding more features and simplify his 5W1H extraction algorithm for higher efficiency. Our keywords extraction method is guided by Li's work [7], and our entity extraction method follows Zheng's work [15].

3 CNME System

In this section, we first present the overview of the system as well as the definition of news meta-data, and then talk about the implementation details of the extraction modules.

3.1 System Overview

The system mainly consists of a list of modules, each extracting one or more meta-data from the raw text of news or existing meta-data. To make the process easier, we construct an object named *NewsMeta* to combine all the meta-data and use it as the only medium between modules. To make it simple, we only demonstrate how the system extracts the 4 kinds of meta-data mentioned in Sect. 1. The input includes the news title and content, and our system will output an instance of *NewsMeta*, which is consisted of the properties in Table 1.

For one piece of news, we create one instance of *NewsMeta* with only title and content assigned with values. Then the instance will go through the assembling line of modules, each of which will assign the respective property with extracted values. At last, we store the completed instance of *NewsMeta* to a MongoDB database. Figure 3 shows the flow chart of the system.

Table 1. The description of *NewsMeta* (our system extracts meta data mainly from the news title and content, so other properties are omitted here although they should also be maintained for various purposes.)

Property	Data Type	Description
title	String	Title of the news
content	String	Content of the news
stitle	String	The segmented title
scontent	String	The segmented content
category	String	One of the pre-set category the news should be assigned to
keywords	List	Keywords and their respective scores.
entities	List	Entities in the news and their entries in the knowledge base
4W	List	Who does what at when at where

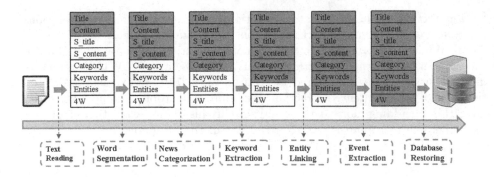

Fig. 3. Flow chart of the processing

As we can see from Fig. 3, apart from the input and output modules, currently we have 5 assembling modules, namely Word Segmentation, News Categorization, Keyword Extraction, Entity Linking, and Event Extraction. The modules are organized in plug-and-play manner so that we can add new modules or remove present modules freely.

3.2 The Input and Output Modules

The input module reads formatted news texts from disk or database, creates an instance of *NewsMeta* and assigns its two basic properties (title and content) with values. Then it hands the instance to the next module. After all the other modules complete their work, the output module will put the finished product, i.e. the *NewsMeta* instance with all properties filled, into the database.

3.3 Word Segmentation Module

Chinese sentences should be split into words, which is the first-of-all procedure of Chinese text processing. The segmentation tool we use in word segmentation module is called ICTCLAS, which can be freely downloaded from http://ictclas.nlpir.org. When receiving the *NewsMeta* instance, this module will segment sentences of the title and the content, and then assign the result to its two relevant properties named *sTitle* and *sContent*, which are the basis for any further process.

3.4 News Categorization Module

This module decides the category of the news. We studied the news portals of Xinhua[3], Chinanews[4], Renmin[5] and Tencent[6], and after combining their

---

[3] http://www.news.cn.

[4] http://www.chinanews.com.

[5] http://www.people.com.cn.

[6] http://www.tencent.com.

categorization methods, we divides all Chinese news into 12 categories, namely International, Domestic, Sports, Entertainment, Technology, Finance, Military, Culture, Education, Health, Social, and Automobile.

There are lots of machine learning methods for categorization, such as SVM, decision tree and Bayes. Experiments prove that common machine learning methods are not good at multi-category classification, not to mention that they are time-consuming, which is intolerable for our system. Therefore, we propose an innovative statistic-based algorithm, which is also an improved machine learning method.

The basic idea of the method is: for all the words in one category, we compute its score of its representative ability for the category. The words together with their scores form a vocabulary list. One list for one category, our categorization model is composed of all the 12 lists. Then, for a news waiting to be categorized, we use the model to compute its scores on each category, with the highest-score category to be its tag. Figure 4 describes the whole process.

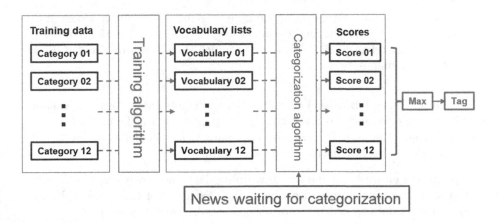

Fig. 4. Categorization method description

Model Training: We believe that, the higher the probability $F(w)$ of a word appears in a category, the stronger its ability to represent the category. Similarly, the more documents in a category contain a word, i.e., the higher document frequency $DF(w)$ of a word in a category, the stronger the word can represent the category. In this paper, we treat $F(w)$ and $DF(w)$ equally, and use their product as the final measuring metric. For every word in one category, we can compute its representative ability for the category as:

$$Score(w) = F(w) \times DF(w)$$
$$= \frac{category(w)}{corpus(w)} \times DF(w) \tag{1}$$

where:

- *category*(*w*) means the occurrence count of *w* in documents of the category
- *corpus*(*w*) means the occurrence count of *w* in all documents
- *DF*(*w*) denotes the number of documents that contains *w* in the category

For each category, we acquire a number of (e.g., 300) news as training data to train a categorization model, which is composed of 12 vocabulary lists. To achieve higher accuracy, we further setup several rules to screen the vocabulary lists by filtering out words:

- with too small $DF(w)$ (e.g., < 4);
- with too small occurrence count $C(w)$ (e.g., < 3);
- with too large document frequency in the entire corpora (including all 12 categories)(e.g., > 3000).

Note that the thresholds mentioned above should be decided by the scale of the training data.

There might be big gap between the highest score and smallest scores of the words. To avoid words acquiring too much weight, we normalize the score to an appropriate region (e.g., 1–10), the value of which should also be decided by the scale of the training data.

Vocabulary Lists Update: Due to the limit of the training data, the vocabulary lists should be updated regularly. In this paper, we use the newly classified data to update the lists. Because the new data is not as clean as the original one, we provide two alternative strategies to update:

- Update the entire vocabulary with training data and newly classified news.
- Retain the original words, use newly classified news of the same scale with the training data to add new words with equally weighting scores.

Categorization: Based on the established categorization model, we can compute a news' scores on all categories. We believe that the occurrence count of a word in a news is instructive for news categorization. So we set the score $S(w)$ of this word for news on a category to be the product of the score $Score(w)$ of this word in that category and its occurrence count $C(w)$ in news:

$$S(w) = Score(w) \times C(w) \tag{2}$$

Then the final score $Score(doc)$ of a news article with n words on a category is:

$$Score(doc) = \sum_{i=1}^{n} S(w_i) \tag{3}$$

The last step is to compare the news scores on the 12 categories and classify the news into the category with the highest score.

3.5   Keyword Extraction Module

Our keyword extraction algorithm is not based on machine learning, but a marriage of pattern and statistical information. The reasons are:

- Machine learning method requires a large training data, which needs tedious and time-consuming manual work.
- News event happens every day. It is impossible for the training data to follow the continuously emerging new keywords.
- The length of news content is suitable for the realization of frequency-computing methods.

Our proposed method is consisted of two parts, making rules and computing statistical features.

Making Rules. The input of this module is the segmented title and content of the news, i.e. the result of the splitting of ICTCLAS. ICTCLAS uses 22 kinds of part-of-speech (POS) for all Chinese words. After studying more than 1000 pieces of news articles, we derive the following POS-combining patterns for news keywords: uni-gram patterns (Table 2), bi-gram patterns (Table 3) and tri-gram patterns (Table 4).

Table 2. The uni-gram patterns

	Uni-gram potential POS patterns
Nouns	Common nouns (n), person names (nr), location (ns), organization (nt), proper nouns (nz)
	Temporal words (t)
	Abbreviation (j)
	Idiom (l)
Modifier	Adjectives (a), adnoun (an), noun-verb (vn), state adjectives (z)

Table 3. The bi-gram patterns

	Bi-gram potential POS patterns
Nouns	Noun(n)/verb(v)/noun-verb (vn)/state adjectives(z)/Abb(j) + suffix(k)/morpheme(ng, vg)
	Noun/verb/modifier + temporal morpheme(tg)
Noun phrase	Modifier + common noun
	Common noun + common noun
Verb phrase	Verb + common noun

Table 4. The tri-gram patterns

	Tri-gram potential POS patterns
Nouns phrase	Modifier + common noun + common noun
	Modifier + modifier + common noun
	Common noun + common noun + common noun

Compute Statistical Features. Taking consideration of frequency, POS, position and status, we define 7 features for each word:

- $w.tf$: the frequency of word w in the news
- $w.ctf$: the frequency of word w in the corpora
- $termSum$: the sum of frequencies of the words belonging to the word w's pos pattern
- $w.inTitle$: 0 or 1, indicating if the word w appears in title
- $w.inFirst$: 0 or 1, indicating if the word w appears in the first paragraph
- $w.quo$: 0 or 1, indicating if the word w is emphasized by quotation marks
- $w.sign$: a value set by the length of the word. $w.sign = g(x)$; where
  - $g(x) = 0 (x < 2)$
  - $g(x) = log_2 x (2 \leq x \leq 8)$
  - $g(x) = 3 (x > 8)$

Then we can compute the comprehensive score for each word by combining the features in accordance to Eq. 4.

$$score(w) = (w.tf)^{t_1} \times (1 + \sum_{fi \in F} w.fi \times t_{fi}) \times ln\frac{termSum^{t_2}}{w.ctf^{t_3}} \qquad (4)$$

where: F is the set containing $inTitle$, $inFirst$, quo and $sign$, with t_{fi} as their weights. t_1, t_2 and t_3 are the weights of $w.tf$, $termSum$ and $w.ctf$ respectively.

3.6   Entity Linking Module

Entity extraction can provide abundant extra information about news. Here it refers to the task of linking entity mentions in the news text with their referent entries in a knowledge base. So the task of this module is to extract named entities from the news text and link them to entries of our local knowledge base.

In this paper, we propose a simple and effective entity linking method based upon two features of the entity mention: context and co-mention.

- Context. Context similarity is proportional to the probability that an entity mention matches its candidate entity entry of the knowledge base.
- Co-Mention. The more co-mentions of an entity mention appear in its candidate entity entry, the higher probability that they matches.

We focus on three kinds of entities, namely person, location and organization, which provide much more important information than others. The extraction process is very easy thanks to ICTCLAS splitter, which adds POS tags to the words when splitting the Chinese sentences. Then the linking method is completed in two steps, candidate selection and entity disambiguation.

Candidate Selection. Candidates from the knowledge base are selected based on the name similarity, which is computed by edit-distance. Given the source string S_s and target string S_t, the edit distance is defined as:

$$Ed_{sim}(S_s, S_t) = 1 - \frac{|\{ops\}|}{max(length(S_s), length(S_t))} \tag{5}$$

where $|\{ops\}|$ means the minimum operation count to transform S_s to S_t , and length(*) means the length of string *.

Entity Disambiguation. We have entity mentions and their candidate entries. What left to do is to choose the right one from the candidates. We have two features at hand. One is context words, weighted by their frequency, and the other is co-mentioned entities, weighted by predefined values in Table 5.

Table 5. Weighting value of co-mention (d is a unit value to adjust the weighting ratio between the two features)

Contribution	Person	Organization	Location
Person	2d	2d	1d
Organization	3d	2d	1d
Location	1d	1d	1d

Table 5 is obtained based on the data observation. As you can see, organization contributes more to a person than location does. A simple example is that, *NBA* helps a lot to distinguish the mention of *Jordan* as a basketball player, while *Chicago* helps little.

A vector will be established through combining the two features, and the candidate with the largest cosine similarity will be the linking result.

3.7 Event Extraction Module

A common definition of event of a news is 5W1H. But "why" and "how" is too abstract to extract, and "at when who does what at where" is clearly enough to express the event of a news, so we only extract these 4W. For each news, we consider the title, as a sentence, has more important role than any sentence in the content. It contains lots of information related to the 5W1H. We also consider there are several topic sentences in the content that need more attention. To improve the efficiency and accuracy, we only pay attention to the title and topic sentences.

Topic Sentences Extraction. We set four features for a sentence, as listed below:

- $sNer$: score for having named entities
- $sLoc$: score for its location
- $sLen$: score for its length
- $sOverlap$: score for having overlapping words with the title

Then the total score for a sentence is computed as their weighted sum:

$$Score = wNer \times sNer + wLoc \times sLoc + wLen \times sLen + wOverlap \times sOverlap \quad (6)$$

where w leaded expression means the weight of the expression followed.

4W Extraction. For all the 4W, we only consider the words appear in the topic sentences to narrow the selecting spectrum.

We consider person and organization are proper candidates for WHO. The extracting steps are described as below:

1. Extract the persons and organizations from topic sentences to form the candidate set.
2. Calculate the frequency of each candidate in the news as their score.
3. Sort the candidates according to their scores.
4. Provide with the top N (a parameter set by user) candidates.

For WHERE and WHAT, the extracting steps are mostly the same as above, and the only variation takes place in the first step. For WHERE, we extract locations from topic sentences and for WHAT, we extract verbs.

For WHEN, the situation will be a little complicated due to its unique character that only expressions of time are applicable. First, we form a candidate set by using regular expression to match all the time expressions in the news, and transform them to a uniform format. Then we define 3 features for a time expression: whether it appears in topic sentence or not, its distance to the center time, and its index of sentences in the news.

We use logistic regression to assess the probability that a time could be a "WHEN" using Eq. 7.

$$P(Y = 1 | X = x_w) = \frac{1}{1 + e^{-(\alpha x_w + \beta)}} \quad (7)$$

where the x_w is the feature vector of time, α is the weight vector and β is the offset.

The training data consists of certain amount of news, in which the right WHEN is marked. The marked times are positive instances, and others are negative instances. Then we assess the optimal value α and β of by maximizing the likelihood:

$$O(\alpha, \beta) = \prod_{w \in W} P(Y = 1 \mid X = x_w)^{y_w} P(Y = 0 \mid X = x_w)^{1-y_w} \qquad (8)$$

where y_w is the tag of time, i.e., positive or negative.

4   Experiment

Experiment in this paper is mainly about news categorization. The methods of Keyword Extraction, Entity Extraction and 4W Extraction are modified from [7,14,15]. They have their own experimental results in their respective papers. Due to the space limitation, we do not present their performance on our dataset.

There is no official experimental dataset available for news categorization, so we crawled about 10,000 news from Xinhua, Tecent, Renmin and Chinanews. After filtering out unqualified ones, we choose 7,200 articles (600 articles for each category) as our training data. We also crawled another 2,400 articles (200 articles for each category) as the test data.

After completing the categorization model training, we first randomly choose 1,200 (100 articles per category) from the training data to test the training performance, then evaluate the model on the test data which is non-overlap with the training data. Table 6 shows the results on each category.

Table 6. Overall results for news categorization.

Category	Train		Test	
	Precision	Recall	Precision	Recall
International	93.8 %	92 %	89.3 %	87.2 %
Domestic	87.3 %	92 %	85.6 %	86.5 %
Social	86.2 %	83 %	84.3 %	79.3 %
Entertainment	90.6 %	87 %	87.4 %	83.8 %
Sports	95.8 %	93 %	91.7 %	88.3 %
Technology	98.9 %	90 %	90.8 %	86.4 %
Finance	85.6 %	97 %	82.3 %	87.2 %
Military	87.1 %	94 %	85.9 %	89.4 %
Culture	88.2 %	82 %	80.2 %	78.3 %
Education	89.4 %	90 %	85.7 %	86.2 %
Health	90.4 %	95 %	90.2 %	90.4 %
Automobile	96.1 %	99 %	91.8 %	91.5 %

Table 7. Precision comparison before and after vocabulary updating

	Before updating	After updating
Precision	87.1 %	87.8 %
Recall	86.2 %	87.1 %

As we can see from Table 6, the precisions of categories with more obvious features, such as Sports, Technology and Automobile, is comparatively higher, while the generalized categories suffer from lower precision. This is in accordance with reality, as Sports and Automobile are more distinctive from others.

Comparing the results on training and test data, we can find that test data gets lower precision than training data. That's because the vocabulary of our training model is limited. We also make a comparison for pre-and-post vocabulary updating. Table 7 shows the result. The precision does improve after updating the vocabulary lists. It is need to note that, the data used to update the vocabulary is not such high-quality as training data, and it is the result after our categorization without manual efforts, which means that, even though some data are not correct, the result of our categorization still can improve the vocabulary lists.

Because it is a problem of multi-category Chinese news categorization, we don't have any exiting experimental result to compare, so we conduct SVM categorization for comparison and present the result in Table 8. The result shows our method outperform SVM categorization (baseline) both on precision and efficiency.

Table 8. Efficiency comparison between our method and SVM (baseline)

	Our method	SVM
Precision	87.1 %	78.4 %
Recall	86.2 %	76.1 %
Efficiency	700 ms/1000 docs	3 s/1000 docs

The program of our method runs much faster than SVM. It is a significant advantage for a real-time system.

We have developed a demo web page to demonstrate the meta-data extraction result of a single news. As shown in Fig. 5, users can get the result of meta-data extraction after filling the text box with title and content of the news and clicking the "submit" button.

We also developed a demo web page (Fig. 6) to query the maintained meta-data. Due to the complete meta-data we preserved, the querying is easy and fast.

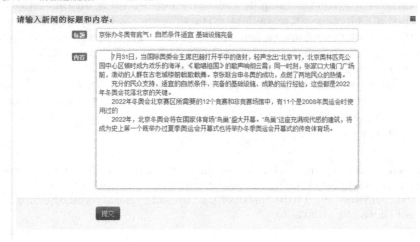

Fig. 5. Demo of extracting meta-data of one news article

Fig. 6. Demo of query by meta-data

In our MongoDB database, we can use the querying string { *"Entity":{ "$all":["Xi Jinping", "Shanghai"]}}* to find any news article that *Xi Jinping* and *Shanghai* are recognized as entity, and use { *"Keyword":{ "$all":["earthquake", "death"]}}* to find any news article with keyword *earthquake* and *death*.

5   Conclusion and Future Work

The key idea of our method lies on the fact that news analysis work needs a meta-data extraction system to simplify the analyzing processes. In this article, we developed a Chinese news meta-data extraction system to meet such demand. The 28 kinds of meta-data we defined can describe the news from all aspect, laying the basis for any further news analysis work. Our methods are either innovative or improved from existing ones, thus ensuring the performance of the system. We haven't fully exploited the advantages of such system yet, because it transforms unstructured news text to structured object, the management of the latter should provides more vivid view about the news and innovative ways to query the news.

Acknowledgement. The work is supported by 973 Program (No. 2014CB340504), NSFC-ANR (No. 61261130588), Tsinghua University Initiative Scientific Research Program (No. 20131089256), THU-NUS NExT Co-Lab and National Natural Science Foundation of China (No. 61303075).

References

1. Arasu, A., Garcia-Molina, H.: Extracting structured data from web pages. In: Proceedings of the 2003 ACM SIGMOD International Conference on Management of Data. pp. 337–348. ACM (2003)
2. Garrido, A.L., Gómez, O., Ilarri, S., Mena, E.: An experience developing a semantic annotation system in a media group. In: Bouma, G., Ittoo, A., Métais, E., Wortmann, H. (eds.) NLDB 2012. LNCS, vol. 7337, pp. 333–338. Springer, Heidelberg (2012)
3. Hou, L., Li, J., Wang, Z., Tang, J., Zhang, P., Yang, R., Zheng, Q.: Newsminer: multifaceted news analysis for event search. Knowl.-Based Syst. **76**, 17–29 (2015)
4. Johnson, D.E., Oles, F.J., Zhang, T., Goetz, T.: A decision-tree-based symbolic rule induction system for text categorization. IBM Syst. J. **41**(3), 428–437 (2002)
5. Krishnalal, G., Rengarajan, S.B., Srinivasagan, K.: A new text mining approach based on HMM-SVM for web news classification. Int. J. Comput. Appl. **1**(19), 98–104 (2010)
6. Lee, L.H., Wan, C.H., Rajkumar, R., Isa, D.: An enhanced support vector machine classification framework by using euclidean distance function for text document categorization. Appl. Intell. **37**(1), 80–99 (2012)
7. Li, J., Zhang, K., et al.: Keyword extraction based on tf/idf for Chinese news document. Wuhan Univ. J. Nat. Sci. **12**(5), 917–921 (2007)
8. McCallum, A., Rosenfeld, R., Mitchell, T.M., Ng, A.Y.: Improving text classification by shrinkage in a hierarchy of classes. In: ICML, vol. 98, pp. 359–367 (1998)
9. Pawar, P.Y., Gawande, S.: A comparative study on different types of approaches to text categorization. Int. J. Mach. Learn. Comput. **2**(4), 423–426 (2012)
10. Piskorski, J., Yangarber, R.: Information extraction: past, present and future. In: Poibeau, T., Saggion, H., Piskorski, J., Yangarber, R. (eds.) Multi-source, Multilingual Information Extraction and Summarization. TANLP, pp. 23–49. Springer, Heidelberg (2013)

11. Shan, D., Zhao, W.X., Chen, R., Shu, B., Wang, Z., Yao, J., Yan, H., Li, X.: Eventsearch: a system for event discovery and retrieval on multi-type historical data. In: Proceedings of the 18th ACM SIGKDD International Conference on Knowledge Discovery and Data Mining, pp. 1564–1567. ACM (2012)
12. Trampuš, M., Novak, B.: Internals of an aggregated web news feed. In: Proceedings of the 15th International Information Science Conference IS SiKDD 2012, pp. 431–434 (2012)
13. Vadrevu, S., Nagarajan, S., Gelgi, F., Davulcu, H.: Automated metadata and instance extraction from news web sites. In: Proceedings of the 2005 IEEE/WIC/ACM International Conference on Web Intelligence, 2005, pp. 38–41. IEEE (2005)
14. Wang, W., Zhao, D., Zou, L., Wang, D., Zheng, W.: Extracting 5W1H event semantic elements from Chinese online news. In: Chen, L., Tang, C., Yang, J., Gao, Y. (eds.) WAIM 2010. LNCS, vol. 6184, pp. 644–655. Springer, Heidelberg (2010)
15. Zheng, Q., Li, J., Wang, Z., Hou, L.: Co-mention and context-based entity linking. In: Li, J., Qi, G., Zhao, D., Nejdl, W., Zheng, H.-T. (eds.) Semantic Web and Web Science. SPC, pp. 117–129. Springer, Heidelberg (2013)
16. Zhou, Y., Li, Y., Xia, S.: An improved KNN text classification algorithm based on clustering. J. Comput. 4(3), 230–237 (2009)

Bootstrapping Yahoo! Finance by Wikipedia for Competitor Mining

Tong Ruan[1(✉)], Lijuan Xue[1], Haofen Wang[1], and Jeff Z. Pan[2]

[1] East China University of Science and Technology, Shanghai 200237, China
{ruantong,whfcarter}@ecust.edu.cn, xuelijuanjsj@163.com
[2] The University of Aberdeen, Aberdeen, Scotland
jeff.z.pan@abdn.ac.uk

Abstract. Competitive intelligence, one of the key factors of enterprise risk management and decision support, depends on knowledge bases that contain a large amount of competitive information. A variety of finance websites have collected competitive information manually, which can be used as knowledge bases. Yahoo! Finance is one of the largest and most successful finance websites among them. However, they have problems of incompleteness, lack of competitive domain, and not-in-time updating. Wikipedia, which was built with collective wisdom and contains plenty of useful information in various forms, can solve the above-mentioned problems effectively, thus helping build a more comprehensive knowledge base. In this paper, we propose a novel semi-supervised approach to identify competitor information and competitive domain from Wikipedia based on a multi-strategy learning algorithm. More precisely, we leverage seeds of competition between companies and competition between products to distantly supervise the learning process to find text patterns in free texts. Considering that competitive information can be inferred from events, we design a learning-based method to determine event description sentences. The whole process is iteratively performed. The experimental results show the effectiveness of our approach. Moreover, the results extracted from Wikipedia supplement 14,000 competitor pairs and 8,000 competitive domains between rival companies to Yahoo! Finance.

Keywords: Competitor mining · Multi-strategy learning · Distant supervision · Relation reasoning

1 Introduction

Competitive intelligence analysis, one of the key factors of enterprise risk management and decision support, depends on knowledge bases that contain a large amount of competitive information. A variety of finance websites have collected competitive information manually, which can be used as knowledge bases. Yahoo!

This work was partially supported by the Fundamental Research Funds for the Central Universities (Grant No: 22A201514045) and the Project funded by China-Postdoctoral Science Foundation (project No: 137763).

G. Qi et al. (Eds.): JIST 2015, LNCS 9544, pp. 108–126, 2016.
DOI: 10.1007/978-3-319-31676-5_8

Finance, one of the largest and most successful finance websites among them, provides plenty of competitor information for listed companies contained in lists and tables. However, competitive information provided by them have problems of incompleteness, lack of competitive domain, and not-in-time updating. Wikipedia, which was built with collective wisdom and contains plenty of useful information in various forms, can solve the above-mentioned problems effectively, thus helping build a more comprehensive knowledge base. Hence, we can extract competitor information and competitive domain from Wikipedia to make a supplement to Yahoo! Finance.

Due to large numbers of companies on the market and the emergence of new companies, it is labor-intensive and time-consuming to manually collect competitor information and competitive domain. Therefore, automatic approaches have been proposed for this purpose. Ma et al. [1], Bao et al. [2], and Xu et al. [3] studied how to extract competitors from the web. Lappas et al. [4] and Wan et al. [5] focused on finding top-k competitive products. These approaches suffer from the following challenges: First, competitor information might be not only contained in semi-structured sources like lists or tables provided by Yahoo! Finance but mentioned in free texts of Wikipedia. How to fully utilize different kinds of data sources is a great challenge. Second, the sentence that contains company names might not describe the competitive relation between companies and instead describe an event between companies. For example, the sentence "Oracle sued SAP, accusing them of fraud and unfair competition" describes a litigation event between Oracle and SAP. Also, there are sentences that describe acquisition event, merged event and ranking event. The above-mentioned events widely exist in free texts. While the above-mentioned events cannot be regarded as competitive relation directly, competitor information can be derived from these events by rule-based reasoning. How to determine these events and reason more competitor information from them is another challenge. Finally, competitive domain extraction is as important as competitor information identification. How to determine competitive domain between rival companies is a hard problem.

In this paper, we present the effort to work on identifying competitor information and competitive domain from Wikipedia to enrich Yahoo! Finance. In order to solve the above mentioned challenges, we propose a semi-supervised approach to extract competitor information and competitive domain based on a multi-strategy learning algorithm. More precisely, we have the following contributions.

1. We use *complementary methods* to extract *different kinds of seeds* from *multiple sources*. Seeds of competition between companies are extracted from lists and tables by wrappers. Seeds of competition between companies and seeds of competition between products are extracted from free texts by Hearst patterns. Seeds of competition between products are inferred from seeds of competition between companies. In this way, the number of seeds increases significantly.

2. We integrate *extraction* with *reasoning*. Competitor information can be inferred from events between companies. Our approach uses a *learning-based method* to identify event description sentences and then extract events between companies from event description sentences. More competitor information can be inferred from events according to the defined rules.
3. We *enrich competitor information* of Yahoo! Finance. Our method also *provides competitive domain* that Yahoo! Finance does not offer. The results of our experiment supplement a large amount of competitor information, and provide a certain number of competitive domains.

The rest of the paper is organized as follows. Section 2 lists the related work. Section 3 gives a brief overview of our approach. Section 4 introduces the approach details. Section 5 shows the experiment results, and we conclude the paper in Sect. 6.

2   Related Work

There are two lines of research related to the problem we solve. They are competitor mining and information extraction, which will be discussed in the following subsection respectively.

2.1   Competitor Mining

Competitor mining has attracted more and more attention from both academia and industry. Recently, Ma et al. [1] proposed a method to infer competitor relationships on the basis of structure of an intercompany network derived from online news. In particular, they presented an algorithm that uses graph-theoretic measures and machine learning techniques. Bao et al. [2] developed an algorithm called *CoMiner*, which uses NLP approach to extract competitors, competitive domain, and competitive evidence from web search results given a query. Xu et al. [3] presented a graphical model to mine and visualize comparative relations between products from Amazon customer reviews. All their works mainly focus on mining competitors from Web pages. There is no existing research working on extracting competitor information from Wikipedia.

Lappas et al. [4] presented an algorithm to find top-k competitive items for a given product and evaluated competitiveness in different datasets ranging from Amazon.com and Booking.com to TripAdvisor.com. Wan et al. [5] proposed a method to find top-k profitable products. Their research explored competitiveness in the context of product. While relevant, the above work has a focus completely different from our own. Our approach extracts competitive products and treats them as competitive domain of companies.

2.2   Information Extraction

Competitor identification is an application of information extraction that combines relation extraction with named entity recognition. Information extraction has been studied for a considerable amount of time. Consequently, a multitude of tools and algorithms have been proposed for this purpose over the past few years. Previous works [6–9] focus on learning all relations from free texts, thus their works are called "Open Information Extraction". Different from their research, target relation is specified in advance in our work. NELL [10], SOFIE [11], and PROSPERA [12] are information extraction tools that need to give target categories and relations. The input data for NELL consisted of an initial ontology that contained hundreds of categories and relations as well as a small number of instances for each category and relation. SOFIE extracts ontological facts from natural language texts and links the facts into an ontology. PROSPERA relies on the iterative harvesting of n-gram-itemset patterns to generalize natural language patterns found in texts. All of the above works focus on extracting relations from a single data source. On the other hand, we propose a method to extract relations from multi-sources data.

One trend of information extraction is leveraging data published on the web, including Web pages, Linked Open Data, and lists and tables on dynamic Web sites. Gentile et al. [13] presented a methodology called multi-strategy learning, which combines text mining with wrapper induction to extract knowledge from lists, tables, and Web pages. While the method seems promising, there are no clear evaluation results in their paper. Ruan et al. [14] utilized the multi-strategy learning method to identify competitor information from prospectuses. On the other hand, distant supervision is an effective method to leverage redundancies among different sources, which has been used in [15,16].

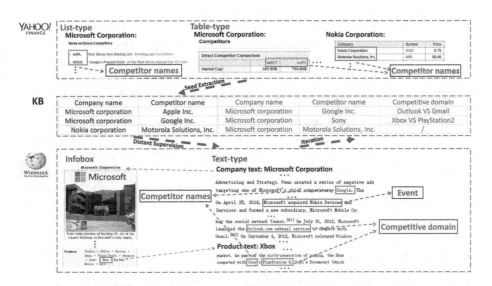

Fig. 1. Competitor information and competitive domain mining of Microsoft

3   Approach Overview

In this section, we start with a brief introduction of problem definition and analysis and then provide the overall workflow of our proposed approach.

3.1   Problem Definition

Wikipedia is a free, multilingual, and collaboratively edited Internet encyclopedia. It contains plenty of useful information in various forms. In Wikipedia, competitors are mentioned in free texts. Also, there are other competitive relations and events mentioned in free texts of Wikipedia. Yahoo! Finance, one of the largest and most successful finance websites, contains a large amount of competitive information. We find that Yahoo! Finance has a complete industry list, and each industry contains a number of companies. The category of the company can be obtained based on the industry list. We also discover that each company has a specific Web page to describe the company's competitors contained in lists and tables. The objective of our approach is to find a way to extract competitor information and competitive domain from Wikipedia, which can make a supplement to the competitive information of Yahoo! Finance. The input of our task is a large number of free texts from Wikipedia and a number of competitor pairs extracted from Yahoo! Finance. The output is a set of competitor pairs with competitive domain information. We denote it as $R = \{<C_i, C_j, D>|C_i \in C, C_j \in C\}$, where C is all companies in corpus, C_j is a competitor of C_i, and D is competitive domain between C_i and C_j. In our work, we use competition between products to denote competitive domain D.

Taking "Microsoft" as an example, Fig. 1 shows the competitor names of Microsoft contained in lists and tables provided by Yahoo! Finance. It also shows competitor names, events, and competitive domains of Microsoft mentioned in free texts from Wikipedia. The figure illustrates that competitor information can be extracted from semi-structured data first, and then the extracted competitor names are fed as seeds to distantly supervise the learning process to extract competitor names from free texts. Moreover, competition between products and events among companies can be extracted from free texts. Leveraging competition between products can identify competitive domain, and more competitor information can be inferred from events between companies. The whole process is iteratively performed. As a result, the competitors of Microsoft and the competitive domains where Microsoft's rivals compete with itself are obtained through multiple iterations.

3.2   Overall Workflow of Our Approach

We provide a workflow to describe the process of our approach in this section. In our task, there are three kinds of competitive relations. We use $CC=\{< C_i, C_j> |C_i \in C, C_j \in C\}$ to denote competition between companies, $CP=\{< C_i, P_j> |C_i \in C, P_j \in P\}$ to denote competition between company and product, $PP=\{< P_i,P_j> |P_i \in P, P_j \in P\}$ to denote competition between products.

Here, C is all companies, and P is all products in corpus. As shown in Fig. 2, there are three main components, namely, *Seed Extractor*, *Competitive Relation and Event Extractor*, and *Competitive Relation Reasoner*.

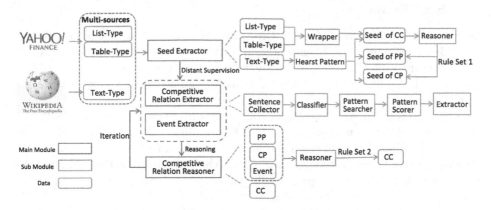

Fig. 2. Overall workflow of our approach

The *Seed Extractor* extracts competitive pairs in three different ways to obtain as many seeds as possible. First, seeds are extracted from semi-structured sources like lists or tables provided by Yahoo! Finance using specific HTML wrappers. Then a set of Hearst patterns is defined to extract seeds from free texts of Wikipedia. Finally, a set of inference rules is applied on the seeds obtained from previous steps to derive more seeds of competitive relation. *Competitive Relation and Event Extractor* consists of *Competitive Relation Extractor* and *Event Extractor*. The Competitive Relation Extractor leverages the seeds extracted by Seed Extractor to help competitive relation identification from free texts, which can be modeled as a distant supervision process. Competitive sentences containing seeds are collected from free texts. Then competitive lexical patterns are learned from contexts of these seeds in sentences, and competitive information is extracted by applying the above patterns to new sentences in other free texts. The Event Extractor first identifies event description sentences from sentences labeled by the seeds of competition between companies. Event description sentences identification is modeled as a classification problem. Then the extraction process is the same in extracting competitive relation from free texts. The *Competitive Relation Reasoner* makes use of the results of the Competitive Relation and the Event Extractor to infer more competitive relations based on the defined rules. The whole process is iteratively performed. In each iteration, the newly extracted competitive relations are treated as new seeds for bootstrapping.

4   Multi-strategy Learning Approach

In the following parts, we will describe the multi-strategy learning process in detail. We start with an introduction of seed extraction, and then present the process of competitive relation and event extraction, and finally provide the inference rules to reason more competitive information.

4.1 Seed Extraction

Seeds Extraction from Yahoo! Finance. We first extract seeds for competition between companies. Competitive information between companies is mentioned in lists, tables, and free texts. Since semi-structured sources are the easiest to deal with, we first extract competitors from semi-structured sources like lists or tables.

Lists are parallel structures, and such structure whose precedent words are "News on Direct Competitors". First, we find all lists of a web page. Then, the first string of each item is extracted as a competitor name. Because the extracted competitor names are initialisms, we obtain the full names of these companies from the title of the web pages. Finally, the extracted competitors are used as seeds for table processing and added to the seed set of CC.

While the formats of tables are diverse in real word, tables in our corpus are much simpler, and the competitors are always contained on the first row or first column. We need to determine whether the contents on the first row or first column are competitors. After observing the tables, we find that the contents of a table row or column have the same sort of data. If more than two cells of a row or column in a table contain competitor names, then all the cells of that row or column correspond to competitors. First, we extract all the tables of a web page. After that, we find competitive table rows and columns. For a table row or column, if it has more than two cells and the contents of these cells are recognized as competitor names according to results extracted from lists as seeds, the row or column is a competitive table row or column. Finally, we extract contents of the rows or columns returned in the previous step as competitor names and add these competitor names to the seed set of CC. All the competitors in the tables are obtained through multiple iterations.

Seeds Extraction by Hearst Patterns. To extract seeds of the other two kinds of competitive relations, we define a set of Hearst patterns. We describe the Hearst patterns that we exploit and give the corresponding examples from our data in Table 1. Here, we use CN to denote competitor name and EN to denote entity name.

Exploiting the defined Hearst patterns can extract competition between companies, between company and product, and between products from free texts, and add these competitive relations to the seed set of CC, CP, and PP, respectively.

Seeds Enrichment by Reasoning. We extract every company's products from Wikipedia's infobox, and the types of products are identified at the same time. Then we leverage Rule 1 and Rule 2 to enrich the seed set of CP and PP.

Table 1. Hearst patterns for seed extraction

Hearst pattern	Example
H1:EN's competitors (CN)* and CN	Microsoft's chief competitor Google.
H2:EN's competitors such as (CN)* and CN	Sina Weibo's competitors such as Tencent Weibo and Baidu's services.
H3:EN's competitors like (CN)* and CN	Ford's competitor like Toyota Prius.
H4:EN's competitors include (CN)* and CN	Dell's major competitors include Hewlett-Packard(HP), Acer, Fujitsu, Toshiba, Gateway, Sony, Asus, Lenova, IBM, MSI, Samsung, and Apple.
H5:EN competes with (CN)* and CN	The Xbox competes with Sony's PlayStation 2, Sega's Dreamcast, and Nintendo's GameCube.
H6:EN competes against CN	The Oracle database competes against PostgreSQL, Firebird, and MySQL

Rule 1 (competitive relation is independent of product release time):
compete (C_i, C_j) $\wedge$ owner (P_i, C_i) $\wedge$ owner (P_j, C_j) $\wedge$ type (P_i, T) $\wedge$ type (P_j, T) =>compete (P_i, P_j) $\wedge$ compete (P_i, C_j) $\wedge$ compete (P_j, C_i).

Rule 2 (competitive relation depends on product release time):
compete (C_i, C_j) $\wedge$ owner (P_i, C_i) $\wedge$ owner (P_j, C_j) $\wedge$ type (P_i, T) $\wedge$ type (P_j, T) $\wedge$ release time (P_i, t_1) $\wedge$ release time (P_j, t_2) $\wedge$ $t_1 \approx t_2$
=>compete (P_i, P_j) $\wedge$ compete (P_i, C_j) $\wedge$ compete(P_j, C_i).

E.g., compete (Samsung, Apple Inc.) $\wedge$ owner (Galaxy S4, Samsung) $\wedge$ owner (iPhone 5s, Apple Inc.) $\wedge$ type (Galaxy S4, Smartphone) $\wedge$ type (iPhone 5s, Smartphone) $\wedge$ release time (Galaxy S4, July 2013) $\wedge$ release time (iPhone 5s, September 2013) $\wedge$ September 2013 $\approx$ July 2013
=>compete (Galaxy S4, iPhone 5s) $\wedge$ compete (iPhone 5s, Samsung) $\wedge$ compete (Galaxy S4, Apple Inc.).

4.2 Competitive Relation and Event Extraction

Distant Supervision for Competitive Patterns in Free Texts. Competitive information identification on free texts requires seeds annotation in sentences to learn patterns. These patterns are further used in other sentences to extract competitive information. The quality of the extracted competitive information heavily depends on the number of annotated sentences, whereas manual annotation costs too much human effort. Here, we leverage seeds extracted from Sect. 4.1 to label free texts automatically. Such kind of distant supervision can save manual efforts of labeling sentences significantly. We first collect sentences that contain seeds and label these sentences. Then we generate extraction patterns from the annotated sentences. Finally, we use the generated patterns to extract new competitive information from other free texts. The whole process is iterative until no new competitive information can be extracted.

In our corpus, free texts are divided into two types: one is about company, and the other is about the company's product. The former is called *company text*, and the latter is called *product text*.

(1) Sentence Collecting

We leverage the seed set of CC, CP, and PP to collect sentences respectively. Here, we use S_{CC} to denote a sentence set collected by the seeds of CC, S_{CP} to denote a sentence set collected by the seeds of CP, and S_{PP} to denote a sentence set collected by the seeds of PP. According to the number of entities of the seed contained in the sentence, we collect sentences according to three kinds of situations:

- Situation 1: Sentences that contain both entities of seed.
- Situation 2: Sentences that contain one entity of seed and a pronoun like *it*, *the company*, or *the firm* at the same time. The sentences must appear in another entity's text.
- Situation 3: Sentences that only contain one entity of seed and appear in another entity's text.

For the second situation, the pronoun is replaced with the title of the company text or product text where the sentence appears. Thus, the second situation is merged into the first situation. We collect sentences according to the above three situations with the seed set of CC from company texts, seed set of PP from product texts, and seed set of CP from company texts and product texts. Then we add these sentences to sentence set S_{CC}, S_{PP}, and S_{CP}, respectively.

We observe that sentence set S_{CC} contains a number of special sentences that describe events between companies (e.g., acquisition event, merged event, and so on), and these events are useful for competitor extraction. A sentence that describes an event is called Event Description Sentence *(EDS)*. Thus, sentence set S_{CC} consists of EDSs and other sentences that describe competitive relation between companies. We use S_{EDS} and S_{OCC} to denote EDSs and other sentences in S_{CC}, respectively. So far, the main concern is how to identify EDSs from sentence set S_{CC}. We view EDS detection task as a classification problem and discuss it in the next part.

(2) Pattern Searching

For sentence set S_{OCC}, S_{CP}, and S_{PP}, we process them as follows: First, we leverage Stanford part-of-speech (POS) tagger[1] to tag each sentence in the sentence set. Second, we count the number of occurrences of each word marked with a POS tag and choose top-k as high-frequency words. We discover that

---

[1] http://nlp.stanford.edu/software/lex-parser.shtml.

competitive relation is always reflected by the verb. Thus, we restrict that high-frequency words must be a verb. Then, if a sentence contains both entities of seed, we delete all tokens that are not found between entities in the sentence. The entities are then replaced with the placeholders ?D? and ?R?, respectively. If a sentence only contains one entity of seed, we delete all tokens that are not found before this entity in the sentence. Then the entity is replaced with the placeholder ?R?, and we add the placeholder ?D? at the beginning of the string. Finally, we process the strings of the previous step. We replace the date, person, organization, and location in every string with the placeholder ?N?. If the processed string contains high-frequency words, we retain it as a pattern and denote it with p. Each p generated is used to create new relation instances.

Table 2. Example sentences for pattern search

Sentence contains two entities of seed	Sentence contains one entity of seed
On March 22, 2007, Oracle sued SAP, accusing them of fraud and unfair competition	It was being acquired by Wisconsin Energy Corporation for $ 9.1 billion

For example, consider the sentence that contains both entities of seed in Table 2. If the tuple <Oracle, SAP> becomes a seed, we replace "Oracle" with ?D?, as well as replace "SAP" with ?R?. These substitutions lead to the pattern ("?D? sued ?R?"). The sentence that only contains one entity of seed appears in the text of "Integrys Energy Group" in Table 2. If the tuple <Integrys Energy Group, Wisconsin Energy Corporation> becomes a seed, we replace "Wisconsin Energy Corporation" with ?R? and add ?D? at the beginning of the string. These operations lead to a pattern ("?D? It was being acquired by ?R?").

(3) Pattern Scoring

The pattern scoring part is actually to compute the support of the pattern. A good pattern p should be generated by several sentences. This characteristic is modeled by computing the support of the pattern. The support s(p) of the pattern p is the number of sentences that generate pattern p.

(4) Competitive Relation Extraction

Our approach generates a corresponding pattern set for each sentence set and gives a score to every pattern. Based on the pattern-generating process, different pattern sets are used for the extraction from different texts. The pattern set generated by sentence set S_{OCC} is used for extracting competition between companies from company text, generated by sentence set S_{PP} is used for extracting competition between products from product text, and generated by sentence set S_{CP} is used for extracting competition between company and product from company text and product text. The extraction process is carried out as follows: For each pattern p, we first retrieve sentences that contain p stripped from the placeholders "?D?" and "?R?". If the sentence comes from company text, it is subsequently processed by the StanfordNER[2] tool, which is able to detect entities. We extract all companies' product, and view them as a dictionary. If the sentence comes from product text, it is labeled according to the dictionary. If a sentence only has one entity detected, we add the title of company text or product text at the beginning of the sentence. The entities on the left and right of p are competitive relation instance that we need to extract.

Event Description Sentence Detection. In this section, we introduce the details of our learning-based method used to identify EDSs from S_{CC}. Events between companies mainly include acquisition event, merged event, litigation event, and ranking event in our work. We describe the events we need to extract and give the corresponding examples in Table 3.

Table 3. Examples of event

Event	Example
Acquisition event	IBM acquired Informix Software in 2001.
Merged event	In2001, FirstEnergy merged with GPU, Inc.
Litigation event	On March 22, 2007, Oracle sued SAP, accusing them of fraud and unfair competition.
Ranking event	Panasonic had around a 10 percent share of the consumer electronics market in Europe, ranking third behind Samsung Electronics

The whole process of identifying EDSs from sentence set S_{CC} is as follows:

1. If p contains the word "acquire" and the length of p is less than 4, the two entities replaced by the placeholders in p form the seed of acquisition event. Since the heuristic rule can find seeds with high precision, we use sentences labeled by these seeds as positive examples for training a learning-based model. If p contains the word "compete" and the length of p is less than 4, the two entities replaced by the placeholders in p form an entity pair. We use sentences labeled by these entity pairs as negative examples.
2. We count high-frequency words of training data as features. Words that can distinguish EDSs of an acquisition event from other sentences are selected using the feature selection method for performance improvements. The selected words are regarded as final features to constitute the learning model. Other aspects such as feature selection method and model selection are discussed in Sect. 5.2.
3. According to p whether contains the word "merge", "lawsuit", and "followed", we can obtain the seeds of merged event, litigation event, and ranking event, respectively. Thus, we use the approach described above to train a classification model for every event.

---

[2] http://nlp.stanford.edu/software/CRF-NER.shtml.

For the acquisition event, we need to distinguish active ones from passive ones. When the EDSs of every event are identified from sentence set S_{CC} by using the classification model, we generate patterns for every event. The pattern-generating method and the extraction process are the same as the competitive relation described in previous steps.

4.3   Competitive Relation Reasoning

Event between companies, competitive relation between companies, competitive relation between company and product, and competitive relation between products are obtained from previous steps. The competitive domain between two companies is determined by the competition between company and product or between products.

We will use some rules to reason more competitive relations from the above events and competitive relations. The rules that we will utilize are defined in the following part. We reason more competitor information from the event between companies according to the first four rules and leverage Rule 5 to reason more competition between products from the competitive relation between products. If the competition between the owners of products has not been extracted, we use Rule 6 and Rule 7 to infer the competitive relation between the owners of products. The whole process is iteratively performed. In each iteration, the newly inferred competitive relations are added to the seed set.

Rule 1: $\text{acquired}(C_i, C_j) \wedge \text{compete}(C_k, C_i) \Rightarrow \text{compete}(C_k, C_j)$
Rule 2: $\text{merged}(C_i, C_j) \wedge \text{compete}(C_m, C_i) \wedge \text{compete}(C_n, C_j)$
        $\Rightarrow \text{compete}(C_m, C_j) \wedge \text{compete}(C_n, C_i)$
Rule 3: $\text{litigated}(C_i, C_j) \wedge \text{domain}(C_i, D) \wedge \text{domain}(C_j, D) \Rightarrow \text{compete}(C_i, C_j)$
Rule 4: $\text{ranked}(C_i, C_j) \wedge \text{domain}(C_i, D) \wedge \text{domain}(C_j, D) \Rightarrow \text{compete}(C_i, C_j)$
Rule 5: $\text{compete}(P_i, P_j) \wedge \text{compete}(P_j, P_k) \Rightarrow \text{compete}(P_i, P_k)$
Rule 6: $\text{compete}(P_i, P_j) \wedge \text{owner}(P_i, C_i) \wedge \text{owner}(P_j, C_j) \Rightarrow \text{compete}(C_i, C_j)$
Rule 7: $\text{compete}(C_i, P_j) \wedge \text{owner}(P_j, C_j) \Rightarrow \text{compete}(C_i, C_j)$

5   Experiments

5.1   Experiment Setup

All corpora used in the experiment were crawled from Yahoo! Finance-en[3] and Wikipedia-en[4]. There are 8,000 companies in Yahoo! Finance. Among them, 5,000 companies have a web page to describe the companies' competitors. We crawled these 5,000 companies' web pages from Yahoo! Finance and obtained 3,184 products of these companies from the infobox of Wikipedia. We retrieved text data of these companies and products from Wikipedia.

In our work, we need to label data manually to assess the quality and coverage of the extracted results. We took 40 % samples in the results of the extracted competitive relations and events and labeled these data manually. Precision and the number of extracted instances were used as the evaluation metrics.

---

[3] http://finance.yahoo.com/.
[4] https://www.wikipedia.org/.

5.2   Results Evaluation

Seeds Statistics on Different Sources. As mentioned in Sect. 4, competitive relation extracted from free texts includes competition between companies, competition between company and product, and competition between products. In our work, seeds are obtained by three ways: HTML wrappers, Hearst patterns, and inference method. Figure 3(a) shows the source distribution of every seed set. We find that 96 percent of the seeds are extracted from semi-structure data, and only 4 % are extracted by Hearst patterns for the seed set of CC. For the seed sets of PP and CP, 95 percent are obtained by reasoning, and 5 % are extracted by Hearst patterns. Using the inference method can increase the seeds of PP and CP greatly. Then we study the contribution of every Hearst pattern for seed extraction. Seeds extracted by Hearst patterns contain CC, CP, and PP type seed. The number of each type of competitive relation extracted by every Hearst pattern is shown in Fig. 3(b). Pattern H1 and H6 have the most contributions to seed extraction.

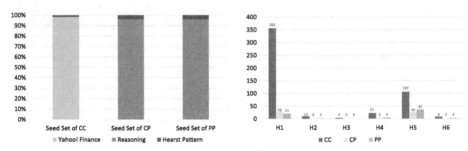

(a) Seeds Extracted from Different Sources   (b) Seeds Extracted by Hearst Patterns

Fig. 3. Seeds statistics on different sources

Extraction Performance. First, we study the contributions to the competitive relation extraction of every seed extraction method. The competitive relation between companies extracted by Hearst patterns increases the number of seeds of CC, while the effect on the extraction results is insignificant. As shown in Table 4, when we use the reasoning method to add the number of seeds of CP and PP, the number of extraction results improves significantly, while the precision might drop a bit.

Table 4. Contributions of seed reasoning

	Not use seed reasoning		Use seed reasoning	
	CP	PP	CP	PP
#Extracted	1,571	1,832	2,880	7,245
precision	0.9115	0.8889	0.875	0.8923

Second, we need to determine a threshold for the pattern scores of every pattern set. The extraction precision and the number of extracted instances will be impacted when we adjust the threshold. We try to analyze the precision distribution and the number of extracted instances by setting different thresholds. We pick the one achieving higher precision and getting more instances. The precision curves under different pattern score thresholds of different events are shown in Fig. 4(a). We also label the number of extracted instances on the curves. Figure 4(b) shows the precision distribution under different pattern score thresholds of different kinds of competitive relations. The range of thresholds are wide; thus we set the step value as 2 for all events and competitive relations to get the most suitable threshold. With the increase of the threshold, the number of extracted instances declines, but the precision might grow. We should consider both precision and the number of extracted instances for threshold selection. For example, when the threshold is 5, it is an optimal choice for acquisition event from the point of precision and the number of extracted instances. Therefore, we set 5 as the empirical threshold to be used for further learning. Similarly, we set the thresholds for merged event, ranking event, and litigation event as 7, 9, and 7, respectively. For competitive relation, we set the thresholds for CC, CP, and PP as 11, 5, and 7, respectively.

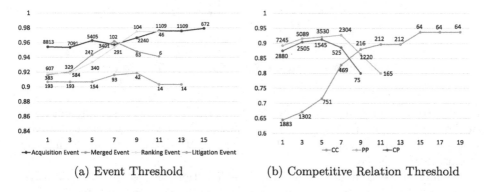

(a) Event Threshold   (b) Competitive Relation Threshold

Fig. 4. Threshold tuning

Then we study the influence on extraction under different numbers of seeds. We can obtain the category of the company from Yahoo! Finance. Thus, the seeds of CC are divided into different categories. We randomly select 20 %, 40 %, 60 %, 80 %, and 100 % of seeds from every category equally for pattern learning, and then use the above patterns to extract new instances. The precision distribution and the number of extracted instances under different numbers of seeds are shown in Fig. 5(a). We select all the seeds from 50, 100, 150, and 200 categories, respectively, for pattern learning. Figure 5(b) shows the extraction precision distribution and the number of extracted instances under different numbers of categories. For seeds of CP and PP, we randomly select 50, 100, 500,

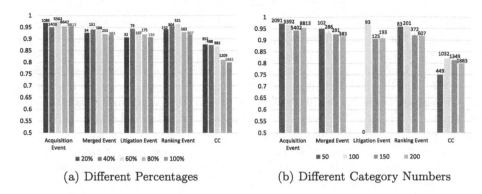

(a) Different Percentages   (b) Different Category Numbers

Fig. 5. Different numbers of CC-Type seed

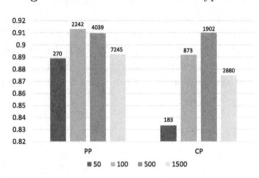

Fig. 6. Different numbers of PP-Type and CP-Type seed

and 1,500 seeds for pattern learning. The extraction results under different numbers of seeds of CP and PP are shown in Fig. 6. The seed size selection standard is the same as the threshold that should consider the precision and the number of extracted instances.

Finally, since competitive relation identification on free texts is iterative, we show the results after each iteration in Table 5. We use extracted relations and events to reason more competitive relation between companies, and then add these relations to the seed set of CC for iteration. In each iteration, event extraction performs iteration at the same time. The number of extracted candidates of CC is the number of reasoning from events plus the number of extracted CC. Since the precision of extracted events is very high, the confidence of competitive relation between companies inferred from events is also very high. While the precision of extracted competitive relation between companies is not very high, the precision of the final competitive relation between companies increases a lot because the competitive relation between companies inferred from events is added to it. For all types, after a small number of iterations, the whole process terminates. The number of extracted relation instances increases after each iteration, but the precisions might drop a bit.

Table 5. Iterations for competitive relation extraction

	1	2
# Extracted candidates of CP	1,515	2,293
# New CP	349	30
Precision	0.9108	0.8823

	1	2
# Extracted candidates of PP	2,837	3,015
# New PP	490	14
Precision	0.8997	0.9177

	1	2	3
# Extracted candidates of CC	3,359	3,445	3,498
# New CC	241	96	51
Precision	0.9461	0.9187	0.9033

Event Description Sentence (EDS) Detection Performance. In our work, EDS detection task is modeled as a classification problem. We learn a classification model for every event. In the following, we discuss which parameters are important to learn the best classification model.

First, we use different feature selection methods to select features that can distinguish EDSs of every event with other sentences for performance improvements. Here, Information Gain (IG), Chi Square (CS), and Information Gain Ratio (IGR) are chosen as three methods to compare. The identification performance after feature selection are compared under these methods. The identification performance is also compared under the condition that there is no feature selection (NFS). Then we choose the best one as the final feature selection method. NaiveBayes, one of the most widely used learning algorithms, is used to learn a model. As shown in Fig. 7(a), when Information Gain Ration is chosen as the feature selection method, we get the highest precision of acquisition event. Similarly, we select Information Gain Ration, Information Gain and Chi Square for merged event, litigation event, and ranking event, respectively.

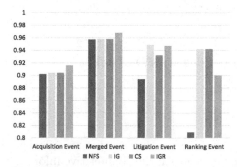

(a) Feature Selection Method Comparison

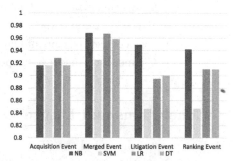

(b) Model Comparison

Fig. 7. Different feature selection methods and models comparison

Then we use different learning algorithms to train various models. The identification performance under these models is compared to select the best algorithm. Here, we choose Logistic Regression (LR), SVM, Decision Tree (DT), and NaveBayes (NB) as four algorithms to compare. 10-fold cross validation is used during model training. As shown in Fig. 7(b), the model learned by NaiveBayes achieves the best precision of all events except acquisition event, which is chosen as our learning model of these three events. Logistic Regression is chosen as the learning model of acquisition event.

Contributions to Yahoo! Finance. Yahoo! Finance has collected competitors of listed companies for users. Because the competitor information provided by Yahoo! Finance is incomplete, competitors extracted by our method make a supplement to Yahoo! Finance. More specifically, our approach determines competitive domains between companies that have competitive relation. Since we use events between companies to infer competitor information, the number of competitor pairs increases substantially. The final number of every event and the number of competitive relations inferred from each event are shown in Table 6. We show the contributions of our work to Yahoo! Finance in Table 7.

Table 6. Competitive relation inferred from event

Event	# Event	# Competitive relation
Acquisition event	8,902	14,726
Merged event	417	4,278
Litigation event	206	188
Ranking event	632	573

Table 7. Contributions to Yahoo! Finance

# Companies of Yahoo! Finance	5,000
# Competitor pairs of Yahoo! Finance	32,293
# Extracted Competitor pairs of Ours	19,765
# New competitor pairs of Ours	14,352
# Competitor pairs with competitive domain of Ours	8,145

6   Conclusions and Future Work

In this paper, we presented the effort to work on extracting competitor information and competitive domain from Wikipedia to enrich the competitive information of Yahoo! Finance. Competitive domain is identified by competition between company and product or competition between products. We proposed a multi-strategy learning method for competitive relation extracting. We leverage specific wrappers to extract competitors from semi-structured sources provided by

Yahoo! Finance, exploit Hearst patterns to extract competitive relations from free texts, and use reasoning method to infer competitive relations; and then the extracted competitive relations are fed as seeds to boost the extraction process from free texts. Distant supervision learning is used in these processes to avoid manual labeling efforts. We use classification model to determine event description sentences from sentences labeled by the seeds of competition between companies. Then we leverage the extracted events to infer more competitor information. Experimental results show that our method makes a good supplement to Yahoo! Finance. As for the future work, we plan to design ranking functions to rank competitors so that we can return the most competitive rivals. We also plan to extract temporal information so that we can analyze the change of competitors for a company.

References

1. Ma, Z., Pant, G., Sheng, O.R.: Mining competitor relationships from online news: a network-based approach. Electron. Commer. Res. Appl. **10**(4), 418–427 (2011)
2. Bao, S., Li, R., Yu, Y., Cao, Y.: Competitor mining with the web. IEEE Trans. Knowl. Data Eng. **20**(10), 1297–1310 (2008)
3. Xu, K., Liao, S.S., Li, J., Song, Y.: Mining comparative opinions from customer reviews for competitive intelligence. Decis. Support Syst. **50**(4), 743–754 (2011)
4. Lappas, T., Valkanas, G., Gunopulos, D.: Efficient and domain-invariant competitor mining. In: Proceedings of the 18th ACM SIGKDD International Conference on Knowledge Discovery and Data Mining, pp. 408–416. ACM (2012)
5. Wan, Q., Wong, R.C.W., Peng, Y.: Finding top-k profitable products. In: 2011 IEEE 27th International Conference on Data Engineering (ICDE), pp. 1055–1066. IEEE (2011)
6. Banko, M., Cafarella, M.J., Soderland, S., Broadhead, M., Etzioni, O.: Open information extraction for the web. IJCAI **7**, 2670–2676 (2007)
7. Wu, F., Weld, D.S.: Open information extraction using wikipedia. In: Proceedings of the 48th Annual Meeting of the Association for Computational Linguistics, pp. 118–127. Association for Computational Linguistics (2010)
8. Fader, A., Soderland, S., Etzioni, O.: Identifying relations for open information extraction. In: Proceedings of the Conference on Empirical Methods in Natural Language Processing, pp. 1535–1545. Association for Computational Linguistics (2011)
9. Schmitz, M., Bart, R., Soderland, S., Etzioni, O., et al.: Open language learning for information extraction. In: Proceedings of the 2012 Joint Conference on Empirical Methods in Natural Language Processing and Computational Natural Language Learning, pp. 523–534. Association for Computational Linguistics (2012)
10. Carlson, A., Betteridge, J., Kisiel, B., Settles, B., Hruschka Jr., E.R., Mitchell, T.M.: Toward an architecture for never-ending language learning. AAAI **5**, 3 (2010)
11. Suchanek, F.M., Sozio, M., Weikum, G.: SOFIE: a self-organizing framework for information extraction. In: Proceedings of the 18th International Conference on World Wide Web, pp. 631–640. ACM (2009)
12. Nakashole, N., Theobald, M., Weikum, G.: Scalable knowledge harvesting with high precision and high recall. In: Proceedings of the Fourth ACM International Conference on Web Search and Data Mining, pp. 227–236. ACM (2011)

13. Gentile, A.L., Zhang, Z., Ciravegna, F.: Web scale information extraction with lodie. In: 2013 AAAI Fall Symposium Series (2013)
14. Ruan, T., Lin, Y., Wang, H., Pan, J.Z.: A multi-strategy learning approach to competitor identification. In: Supnithi, T., Yamaguchi, T., Pan, J.Z., Wuwongse, V., Buranarach, M. (eds.) JIST 2014. LNCS, vol. 8943, pp. 197–212. Springer, Heidelberg (2015)
15. Mintz, M., Bills, S., Snow, R., Jurafsky, D.: Distant supervision for relation extraction without labeled data. In: Proceedings of the Joint Conference of the 47th Annual Meeting of the ACL and the 4th International Joint Conference on Natural Language Processing of the AFNLP: Volume 2, pp. 1003–1011. Association for Computational Linguistics (2009)
16. Roth, B., Barth, T., Wiegand, M., Singh, M., Klakow, D.: Effective slot filling based on shallow distant supervision methods (2014). arXiv preprint arXiv:1401.1158

Leveraging Chinese Encyclopedia for Weakly Supervised Relation Extraction

Xiyue Guo[1,3] and Tingting He[2(✉)]

[1] National Engineering Research Center for E-learning, Central China Normal University,
Wuhan 430079, China
`gxy@mails.ccnu.edu.cn`
[2] School of Computer, Central China Normal University, Wuhan 430079, China
`tthe@mail.ccnu.edu.cn`
[3] School of Information Technology, Xingyi Normal University for Nationalities,
Xingyi 562400, China

Abstract. In the research of named-entity relation extraction based on supervision, selecting relation features for traditional methods are usually finished by people, and it's hard to implement these methods for large-scale corpus. On the other hand, fixing relation types is the premise, so the practicabilities of these methods are not so ideal. This paper presents a weakly supervised method for Chinese named-entity relation extraction without man-made annotations, and the relation types in this method are not chosen artificially. The method collects entity relation types from the structured knowledge in encyclopedia pages, and then automatically annotates the relation instances existing in the texts based on these relation types. Simultaneously, the syntactic and semantic features of entity relations will be considered in this method, then the machine learning data will be completed, finally we use Support Vector Machine (SVM) model to train relation classifiers from training data, and these classifiers could try to extract entity relations from testing data. We carry out the experiment with the data from Chinese Baidu Encyclopedia pages, and the results show the effectiveness of this method, the overall F1 value reaches to 83.12 %. In order to probe the universality of this method, we also use the acquired relation classifiers to extract entity relations from news texts, and the results manifest that this method owns certain universality.

Keywords: Relation extraction · Weakly supervised · SVM · Baidu Encyclopedia

1 Introduction

Named-entity relation extraction aims to obtain the relationship between named-entities from natural language texts, which is a more advanced task based on named-entity recognition, and this task can supply basis for other researches, such as automatic question answering, event extraction, machine translation, etc. Nowadays the human have entered the age of big data, we have to deal with large scale natural language texts (or in other formats) every day, and extracting valuable information from these texts,

© Springer International Publishing Switzerland 2016
G. Qi et al. (Eds.): JIST 2015, LNCS 9544, pp. 127–140, 2016.
DOI: 10.1007/978-3-319-31676-5_9

for commercial or academic goals, is becoming more and more common in actual demand. However, the difficulty of this task will sharply increase with the expansion of data scale. One way to solve the problem is to recognize the named-entities and extract their relations from texts.

After analyzing the existing researches, we can learn that the existing supervised methods need vast annotated training data and stationary relation types, which could reduce the practicability of these methods. What's more, compared with the alphabetic languages, the peculiarity of language structure and the complexity of semantics for Chinese language result in the inefficiency of traditional relation extraction methods. For example, the position distributions of relation words in Chinese texts are very different from that in English texts. Actually the bigger obstacle is the lack of annotated corpus for Chinese information extraction. The researchers have to annotate a mass of original Chinese text data themselves, and it can cost much time, effort and financial resources. Besides, these approaches request the annotators to own professional knowledge of relative fields. All of these limit the progress of research on Chinese relation extraction.

Under this circumstance, this paper tries to present a weakly supervised relation extraction method for Chinese encyclopedia texts, looking forward to minimizing human intervention, trying to extract more relation types, and improving the relation extraction effects. This method utilizes a small quantity of structured information in encyclopedia pages to annotate the rest of texts, aiming to extract the entity relations in them.

The structure of this paper contains: Sect. 2 introduces the domestic and overseas research status of relation extraction, Baidu Encyclopedia, and the assistive toolkit LTP. Section 3 elaborates the main idea of this method. Section 4 exhibits the experimental design and results analysis. Finally, Sect. 5 gives the conclusion and future work.

2   Related Work

2.1   Named-Entity Relation Extraction

The traditional research approaches for relation extraction include the rule-based, dictionary-based, ontology-based, and machine learning-based. Recent researching trend shows that the idea of combining machine learning and other methods has become mainstream [1].

Fei Wu et al. proposed a novel method to increase recall from Wikipedia's long tail phenomenon of sparse classes, and it could improve the recall indeed, but the implementation procedure needed lots of manual annotation [2]. Mintz et al. (2009) presented a distant supervision method for relation extraction without labeled data, and they selected standard lexical and syntactic features, and adopted L-BFGS algorithm to train a relation classifier from a large scale of Wikipedia texts under the supervision of Freebase, their method didn't require labeled corpora, avoiding the domain dependence of ACE-style algorithms, and allowing the use of corpora of any size [3]. Fader et al. (2011) published an open information extraction system named ReVerb, and this system introduced two simple syntactic and lexical constraints on binary relations expressed by verbs, and it first identified relation phrases that satisfied the syntactic and lexical constraints, and then found a pair of NP arguments for each identified relation phrase, and the resulting extractions were then assigned a confidence score using a logistic regression classifier [4]. Based on ensemble learning, Thomas (2011) presented a method to extract the interaction between drugs, and based on the feature spaces of different languages, this method could select the best result between several machine learning methods [5]. Surdeanu (2012) thought out a new method by introducing multi-instances and multi-labels, this method integrated the entities and their annotations using the graph model with latent variables. To a degree, this method conquered the defect of distant supervision, and the experiment indicated the good performance for the texts from two different domains [6]. Mausam et al. (2012) also published their open language learning for information extraction (OLLIE), they alleged that they had conquered two important defects in state-of-the-art open IE systems, and the performance of the method was better than ReVerb and WOE [7]. Literature [8] (2013) announced a method based on positional semantic features. With the computability and operability of positional features and the comprehensibility and implementability of semantic features, this method integrated the information gain of words position and the computing result of semantics according to HowNet, the effect of relation extraction improved obviously.

The research on Chinese named-entity relation extraction has already made great achievements. He (2006) gave a relation extraction method based on seed self-expansion mechanism, it found out the named-entity pairs, and converted their contexts into vectors. It selected a few named-entity pair instances that had the relations wanted to be extracted and made them as initial relation seed set. The relation seed set was extended automatically in self-study process. It got the named-entity pairs by calculating the similarity of context vectors between named-entity pairs and relation seed set [9]. In 2007, Chen raised a method based on SVM and HowNet, this method collected the basic features of entity relation and the concept information in HowNet to form training data, and then gave the data to SVM to get the relation classifier. The F1 value of this method could reach 76.58 [10]. Chen (2013) issued their Chinese entity relation extraction approach based on convex combination kernel function. First, it chose lexical information, phrase syntactic information and dependent syntactic information as features, then got different high-dimensional matrixes though mapping by different convex combination kernel functions, finally the method got the optimal kernel by testing all classified models that trained all high-dimensional matrixes by SVM. The best F1 value of this method was 62.9 [11]. Literature [12] proposed a method for extracting attribute values from unstructured text in 2013, regarding attribute value extraction as a sequential data-labeling problem. In order to avoid labeling the corpus manually, the information in the basic information box of Baidu Encyclopedia was used to label the unstructured text as the training data. After the training data was generated, multidimensional features were used to train the sequential data-labeling model, and the performance was improved by using the context.

2.2 Brief Introduction to Baidu Encyclopedia

Similar to encyclopedias in other languages, Baidu Encyclopedia covers all areas of knowledge in Chinese, and its approval and authority are now increasing sharply.

Baidu Encyclopedia also contains complete, unified classification function (classifying the entries by hierarchical relationships, and allowing the editors to create open classification on the basis of entry relation graphs). Figure 1 shows this function of Baidu Encyclopedia, taking the homepage of education category as an example[1].

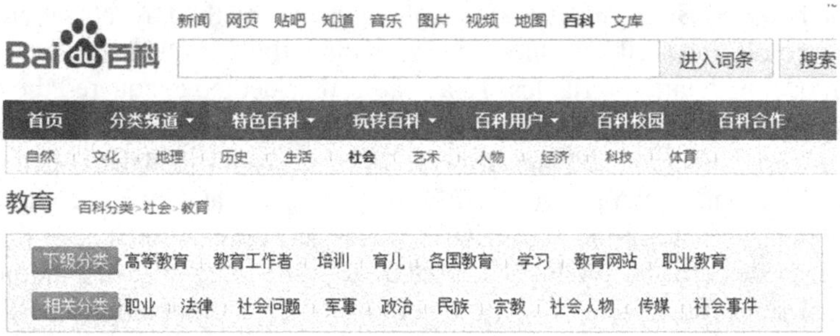

Fig. 1. Example for entry classification function in Baidu Encyclopedia.

The most obvious characteristic of Baidu Encyclopedia is allowing the editors to create a basic information box for the current entry in the front of the entry. The table provides much convenience for visitors to obtain important basic data, at the same time it also offers foundation and condition for relation extraction.

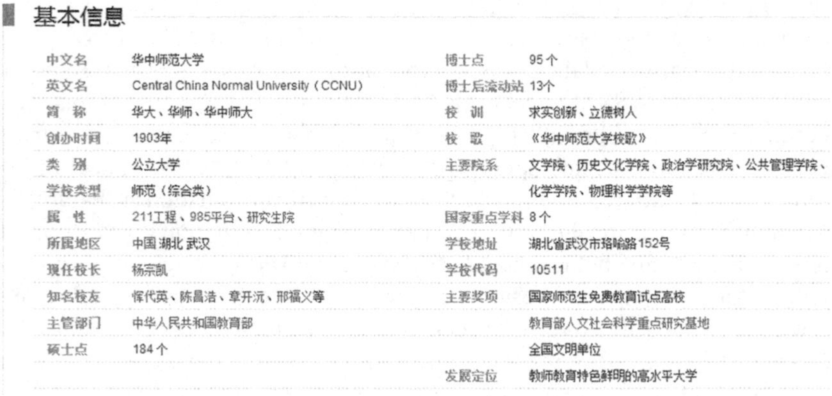

基本信息

中文名	华中师范大学	博士点	95个
英文名	Central China Normal University（CCNU）	博士后流动站	13个
简　称	华大、华师、华中师大	校　训	求实创新、立德树人
创办时间	1903年	校　歌	《华中师范大学校歌》
类　别	公立大学	主要院系	文学院、历史文化学院、政治学研究院、公共管理学院、
学校类型	师范（综合类）		化学学院、物理科学学院等
属　性	211工程、985平台、研究生院	国家重点学科	8个
所属地区	中国 湖北 武汉	学校地址	湖北省武汉市珞喻路152号
现任校长	杨宗凯	学校代码	10511
知名校友	恽代英、陈昌浩、章开沅、邢福义等	主要奖项	国家师范生免费教育试点高校
主管部门	中华人民共和国教育部		教育部人文社会科学重点研究基地
硕士点	184个		全国文明单位
		发展定位	教师教育特色鲜明的高水平大学

Fig. 2. Example of basic information box in Baidu Encyclopedia.

Taking the entry of *"Central China Normal University"* as an example[2], Fig. 2 shows the structure and content of basic information box in Baidu Encyclopedia. We can use current entry name and its basic information box to form plenty of tri-tuples, and use

[1] http://baike.baidu.com/fenlei/%E6%95%99%E8%82%B2.
[2] http://baike.baidu.com/view/4516.htm.

them to annotate and find out relations in the rest of texts. This procedure doesn't need too much human annotation.

2.3 Language Technology Platform Toolkit

Foundational text analysis is necessary during the procedure of relation extraction. Here we choose Language Technology Platform (LTP) as assistive tool to complete this task. LTP is developed by Harbin Institute of Technology and iFLYTE. The backend of LTP relies on the language technology platform which has been researched for over 10 years. LTP provides abundant text processing functions for users, including Chinese word segmentation, lexical analysis, syntactic analysis, sematic role labeling (SRL), especially named-entity recognition, and users can employ this tool locally or online with LTP-Cloud3[3]. The performance and efficiency of this tool are both fine. Figure 3. is an example of LTP processing result.

```
<xml4nlp>
<note sent="y" word="y" pos="y" ne="y" parser="y" wsd="n" srl="y"/>
<doc>
<para id="0">
<sent id="0" cont="百度百科于2006年4月20日上线。">
<word id="0" cont="百度" pos="n" ne="O" parent="1" relate="ATT"/>
<word id="1" cont="百科" pos="n" ne="O" parent="6" relate="SBV"/>
<word id="2" cont="于" pos="p" ne="O" parent="6" relate="ADV"/>
<word id="3" cont="2006年" pos="nt" ne="O" parent="4" relate="ATT"/>
<word id="4" cont="4月" pos="nt" ne="O" parent="5" relate="ATT"/>
<word id="5" cont="20日" pos="nt" ne="O" parent="2" relate="POB"/>
<word id="6" cont="上线" pos="v" ne="O" parent="-1" relate="HED">
<arg id="0" type="A0" beg="0" end="1"/>
<arg id="1" type="TMP" beg="2" end="5"/>
</word>
<word id="7" cont="。" pos="wp" ne="O" parent="6" relate="WP"/>
</sent>
</para>
</doc>
</xml4nlp>
```

Fig. 3. Example for LTP processing results

In Fig. 3, <word> tag stands for the word segmentation results, the attribute "id" in a <word> tag is the serial number of a word, "cont" means the content of a word, "pos" means POS tag, "ne" means named-entity recognition, "parent" means the parent word id of current word in dependency relation, and "relate" means the dependency type. And <arg> tag means SRL results; the attribute "id" in <tag> means serial number of current SRL block, "type" means which part it is in current SRL block, "beg" means the beginning word id of current SRL block, "end" the ending word id of current SRL block.

---

[3] http://www.ltp-cloud.com.

3 Weakly Supervised Relation Extraction from Baidu Encyclopedia

3.1 Flow Diagram of Method

According to the basic information box in Baidu Encyclopedia pages, our method treats the fieldnames in the box as the latent relation names, and uses the content in basic information box to build the relation tuples, then gets the annotations for entity relation features from these tuples, finally utilizes the results above to annotate the entry's text. We can consider the content in basic information box as small scale of labeled data, and the rest content of entry as un-labeled data, so the method should be weakly supervised. The core flow of this method is shown in Fig. 4.

It must be said, the LTP processing results have contained named-entity recognition annotations, and therefore our method will not contain other module for named-entity recognition.

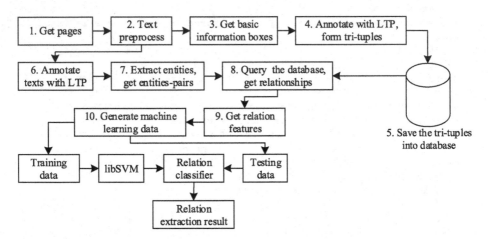

Fig. 4. Flow diagram of weakly supervised relation extraction from Baidu Encyclopedia.

Here are detailed explanations for the steps in Fig. 4.

Step 1: Write the web crawler program to download the Baidu Encyclopedia entry pages. We don't download the whole website of Baidu Encyclopedia, but just get several types of entries in practice.

Step 2: Parse the downloaded encyclopedia pages, and filter out the noise content contained in these files, and transfer them to plain text files.

Step 3: Read the content in basic information box, combine the current entry name, attribute names and attribute values in the box into tri-tuples. For example, Fig. 2 shows the basic information box of Central China Normal University, and we can at least build tri-tuples as follows:

(Central China Normal University, Wuhan, district),
(Central China Normal University, Zongkai Yang, incumbent president),

(Central China Normal University, Daiying Yun, famous alumni),
(Central China Normal University, Changhao Chen, famous alumni)
Step 4: Send the elements in tri-tuples above to LTP, and get the initiatory annotation results, then extract the contained entity types and drop the tuples without entities. We can get such results:
(Central China Normal University Ni, Wuhan Ns, district),
(Central China Normal University Ni, Zongkai Yang Nh, incumbent president),
The results above contain the entities and the relations between them, and *Ni* stands for organization entity, *Ns* stands for toponymy, and *Nh* means human name entity.
Step 5: Save the tri-tuples above into database, to prepare for being retrieved in step 8.
Up to this point, we have finished the work that getting entity data from basic information box. The steps below will elaborate how to use these data to annotate the rest of the text.
Step 6: Back to the plain text files in step 2, send these files to LTP sentence by sentence, and obtain their LTP annotation results.
Step 7: Traverse the results above, extract all entity information of current sentence, and build a set of entity-pairs which is formed by combining any two different entities from the entity set in each sentence.
Step 8: Query each tuple-pair above from tri-tuple database in step 5 to get the matched records, and the relation of the current matched record is the relation of two entities in current query bi-tuple.
For example, the current sentence contains two entities, "Central China Normal University" and "Zongkai Yang", then put these two entities as query term to query from tri-tuple database, and we will be able to get the query result "incumbent president", here we can be sure that this query result is the relation of the two current entities in current sentence.
Step 9: Get the relation features of current tuple-pair, including syntactic features, semantic features, lexical features, etc. For more details see Sect. 3.2.
Step 10: Considering the efficiency and ease of use, we choose libSVM as machine learning tools. So we have to make machine learning data according to the desire of libSVM, and then divide the data into two parts: training data and testing data (See Sect. 4.1). Give the training data to libSVM to learn, get a classifier eventually. Finally, we can use this classifier to extract entity relations from testing data.

3.2 Relation Features

Relying on basic information box in Baidu Encyclopedia pages, our method avoids manual annotation work. However, relation features still need to be chosen for machine learning. According to literature [11] which is our previous research achievement, we choose two types of relation features: basic relation features, syntactic and semantic features. The former includes length of entity, types of entity, content of entity, POS tags of words and context of entity. The latter contains dependency parsing, semantic role labeling distance between current entity and the core predicate.

3.3 Improve the Quality of Training Data

As we all know, there are much noisy data in training data, and these data may be caused by the procedure of automatic annotation, or may come from the original texts. In order to remove the effect of these noisy data, we adopt two measures, (1) filter out the useless HTML tags and their contents in original texts, only keep the <body> tag and its content, continue to delete useless tags in <body>, e.g. <img>. And (2) correct the wrong relation tri-tuples before saving them into database, for example, after being processed by LTP, the value of "School location" will be divided into several parts, while in fact they should be combined together. In this case, we will directly put the whole value as an element in relation tri-tuple. On the other hand, some values should be divided, for example, the value of "Famous alumni" usually has more than one person, and each of them should be considered as an independent element in relation tri-tuples, so we split them according to the existing separator, and generate a group of relation tri-tuples.

3.4 Model Training

Support Vector Machine (SVM) is a very effective machine learning model for relation extraction [12]. The SVM model constructs a hyperplane in a high- or infinite-dimensional space, which can be used for classification, regression, or other tasks. In our method, we use libSVM as the relation training model. The libSVM[4] is one of the most famous toolkits for SVM model, and it implements the SMO algorithm for kernelized SVMs, supporting classification and regression.

Here we will introduce the SVM model used in our method with the simple linearly classification which is a typical quadratic programming problem. The target of SVM is to find out the maximum margins among data. We assume the dataset is $(x_1, y_1), \ldots, (x_n, y_n)$, where x_i is an input vector and $y_i \in \{-1, +1\}$ is a binary label corresponding to it. Then the mathematic representation of the target is:

$$W(\alpha) = \sum_{i=1}^{N} \alpha_i - \frac{1}{2} \sum_{i=1}^{N} \sum_{j=1}^{N} \alpha_i \alpha_j y_i y_j K\left(x_i, x_j\right) \tag{1}$$

Where α_i are Lagrange multipliers, and $K(x_i, x_j)$ is the kernel function which is supplied by user. And the constraint conditions of Eq. (1) are:

$$\sum_{i=1}^{N} \alpha_i y_i = 0 \tag{2}$$

$$0 \leq \alpha_i \leq C, \forall i \tag{3}$$

Where C is an SVM hyperparameter, which is also provided by user.

SMO is one of optimization algorithms for SVM. SMO also decomposes a complex quadratic programming problem into a series of smallest possible sub-problems, which are then solved analytically. Every time SMO selects only two elements to optimize simultaneously under the condition of fixing other parameters, then calculates the

---

[4] http://ntucsu.csie.ntu.edu.tw/~cjlin/libsvm/.

optimal value, and updates α. Since each iteration takes less time, the overall efficiency of SMO can be improved tremendously [13].

We build the training data using the features mentioned in Sect. 3.2 according the requirements of libSVM, and use cross-validation to obtain the best training parameters.

4 Experiment and Analysis

4.1 Experimental Design

Prepare for corpus. We write program specially for downloading the Baidu Encyclopedia pages, crawl all sub-categories and entry pages in the categories of education, Chinese universities, Chinese geographic from Baidu Encyclopedia website. More than 50,000 entry pages are downloaded, and the size of these files is about 10 GB. What's more, in order to make the relations distribute averagely, the original files and sentences sequences are disrupted deliberately and rearranged randomly. Here we directly take out 80 % of the corpus as training data and the residual 20 % as testing data.

Determine the relation types. Traverse the basic information boxes in training data, and calculate the frequency of each fieldname. Most fieldnames appear less frequently, while the frequency of the fieldnames appeared in more than one entries is visibly high. From the perspective of academic research, we select top 10 fieldnames as the relation types, although theoretically the amounts of relation types are unlimited in our method.

Assessment criteria. Our method still treats relation extraction as classification problem, so this method should be assessed by precision, recall and F1 value.

4.2 Results and Analysis

After the experiment has finished, first of all we explore the distribution of relation types according to the different corpus size. As we referred above, the frequency of most relation types are very low. In this experiment, some relation types appear for over 10,000 times (e.g. District), while about 20,000 relation types appear less than 5 times (e.g. Main achievement, Founder), so the gap between these frequencies is huge. In order to see the distribution feature of these relation types, we have to normalize these frequencies. We try several normalization functions, and find that the logarithmic transformation method has the best visual effect. The equation is:

$$Y = \log_{max} X \tag{4}$$

In this equation, max stands for the *maximum frequency* among all the relation type frequencies, X stands for the *real frequency* of each relation type. This equation can compress the frequencies between 0 and 1. Then we can draw the distribution diagram for new relation types changing with different corpus size (Fig. 5).

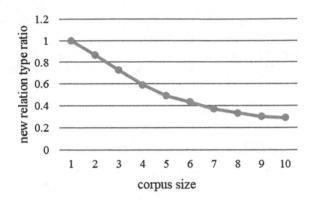

Fig. 5. Quantity of new relation types changing with the expansion of corpus.

The quantity of new relation types is very large at the beginning, but then it reduces quickly with the growth of corpus until levels off at a low ratio before the end of corpus. This figure meets the regularity that the new relation types will be fewer and fewer, and this conclusion accords with what's mentioned in literature [6]. This figure is very useful when we are thinking about what's the suitable corpus size. Our conclusion is that, for Baidu Encyclopedia, corpus size between 8 GB and 9 GB is suitable, because the scales of new relation types here are approximately stable.

Secondly, we calculate the frequency of each fieldnames, and select the top 10 field-names as the relation types, as following: district, Chinese name, government location, birthday, incumbent president, school location, famous alumni, nationality, birthplace, grad-uate institutions. The relation extraction results for these 10 relation types are shown in Table 1.

Table 1. Statistical results of relation extraction.

Relation name	Precision	Recall	F1 value
District	87.22 %	53.25 %	66.13 %
Chinese name	75.70 %	95.78 %	84.56 %
Government location	59.71 %	69.94 %	64.42 %
Birthday	94.04 %	74.72 %	83.27 %
Incumbent president	63.64 %	75.33 %	68.99 %
School location	65.45 %	51.47 %	57.62 %
Famous alumni	88.71 %	84.29 %	86.44 %
Nationality	95.60 %	94.09 %	94.84 %
Birthplace	74.29 %	72.22 %	73.24 %
Graduate institutions	71.05 %	79.41 %	75.00 %
Entirety	77.54 %	75.05 %	76.28 %

From Table 1, we can find that extraction results of most relation types are fine, without extensive manual intervention. There are 7 kinds of relation types whose precision are larger than 70 %, 7 kinds of relation types whose recall are larger than 70 %, and 6 kinds of relation types whose F1 value are larger than 70 %. However, we still realize that the results of few kinds of relation types are not so well, including *"District"*, *"Government location"* and *"School location"*. After deeply analyzing the original data, we find out at least 3 reasons which lead to this disadvantageous result.

1. The 3 kinds of relation types all belong to location, and they have the similar relation features, so the classifier is hard to distinguish each relation type from the other one decidedly.
2. Some of the entities in the different relational tuples are repeated, so some entities are ambiguous.
3. The scale of experimental data of some relation types is relatively less, which results in insufficient learning for machine learning, so the classifier is difficult to deal with. For example, there are only about 1,000 relation instances of *"District"* in training data and testing data.

In order to solve those problems above, we adjust the experiment strategy by decomposing multi-classification into multiple binary classifications, and choosing the closest classification results. Concretely, every time we extract training data which just contains two different relations, and get a relation classifier for the two relation types. After that we put these classifiers to extract relations from testing data, get results for the two relation types. Each relation type will be processed together with all the other relation types one by one, finally we get the best result as the result for current relation type.

Table 2. Statistical results of relation extraction after adjusted.

Relation name	Precision	Recall	F1 value
District	92.23 %	85.20 %	88.58 %
Chinese name	75.70 %	95.78 %	84.56 %
Government location	80.87 %	75.49 %	78.09 %
Birthday	94.04 %	74.72 %	83.27 %
Incumbent president	63.64 %	75.33 %	68.99 %
School location	100.00 %	90.00 %	94.74 %
Famous alumni	88.71 %	84.29 %	86.44 %
Nationality	95.60 %	94.09 %	94.84 %
Birthplace	74.29 %	72.00 %	73.13 %
Graduate institutions	71.05 %	79.41 %	75.00 %
Entirety	83.61 %	82.63 %	83.12 %

What's more, We adopt cross-validation to solve the problem of uneven distribution of the instances without expanding the size of corpus any more. The final results of experiment after being adjusted are shown in Table 2.

It's obvious that the innovated result is better than before, the precision and recall of *"District"* greatly improves, F1 value of this relation type reaches 88.58 %; Precision of *"School location"* increases to 100 %, and recall of it is 90 %. Also, the result of *"Government location"* boosts clearly. And the performance of entirety exceeds 80 %.

We compared our method with recent relative research achievements [14, 15, 16], which are all focusing on Chinese relation extraction published. See the details in Table 3.

Table 3. Comparison among several methods.

Methods	Amount of relation types	Precision	Recall	F1
Literature [14] (2012)	5	89.04 %	83.16 %	85.90 %
Literature [15] (2014)	5	–	–	78.53 %
Literature [16] (2015)	Infinite theoretically 4 types as examples	87.50 %	54.82 %	67.41 %
Our method	Infinite theoretically 10 types as examples	83.61 %	82.63 %	83.12 %

Table 3 indicates that every assessing criteria of outstanding methods has exceeded 80 %, and that's a great progress for Chinese relation extraction. Although our method doesn't produce the best F1 value, it does extract much more relation types. Moreover, this method doesn't need to annotate the corpus manually during the whole process. The results of some relation types are very excellent, while it's true that some of them still need to be improved. If we just select the results of the best top 5 relation types, the scores of entirety can exceed 90 %.

4.3   Universality Testing

Now that we have got the named-entity relation classifiers, can it be used in other texts? We conduct another experiment based on news texts choosing part of *The People's Daily* (1998) as the testing data, and the results of relation extraction are in Table 4.

Comparing Table 4 with Table 2, it's clear that the relation extraction result from news text is not as well as that from Baidu Encyclopedia. Results of most relation types go down, such as *"School location"*, *"Birthday"*, *"Incumbent president"*, etc. We guess the reasons for this phenomenon are:

1. The relation instances in news texts are more dispersive than that in Baidu Encyclopedia, so the sparsity of data is stronger relatively.
2. Our method maybe pays attention to the words in entities excessively, and the words in relation tri-tuples are stationary, so the classifier doesn't have flexible ability to identify relations from the entity instances with different words.
3. Objectively, the language structures and expressive styles are different from various corpus, this could lead to the unfitness when the classifier is processing news texts.

Table 4. Relation extraction result from *The People' Daily* (1998).

Relation name	Precision	Recall	F1 value
District	88.34 %	49.77 %	63.67 %
Chinese name	65.50 %	95.50 %	77.70 %
Government location	55.53 %	40.01 %	46.51 %
Birthday	91.76 %	31.61 %	47.02 %
Incumbent president	51.54 %	43.38 %	47.11 %
School location	47.81 %	51.20 %	49.45 %
Famous alumni	85.71 %	64.29 %	73.47 %
Nationality	90.74 %	85.69 %	88.14 %
Birthplace	54.66 %	69.29 %	61.11 %
Graduate institutions	60.63 %	75.00 %	67.05 %
Entirety	69.22 %	60.57 %	64.61 %

In a word, although this method can get satisfying relation extraction result from Baidu Encyclopedia, the result reflects that its universality still need to be studied unceasingly according to the reasons above.

5 Conclusion

This paper proposes a weakly supervised Chinese named-entity relation extraction method from Baidu Encyclopedia, which not only reduces manually annotation but also expands the extracted relation types. The experimental results indicate that the overall F1 value of this method is 83.12 %, and it's better than most previous methods. Besides, we try to use the classifier to extract relations from news texts, and the results show that we should do more work on its universality.

The relation types of this method are all from the basic information box in Baidu Encyclopedia entries, so they are impossible to cover all the relation types appeared in entry texts, and this problem will result in the phenomenon that the degree of coverage won't be so well. On the other side, we find that the effects of few relation types are bad than expectation, and the universality of this method is not so well. We will focus on solving these issues in the future.

Acknowledgments. We are very indebted to the reviewers who reviewed the papers very carefully. This work was supported by the major project of national social science fund (No. 12 & 2D223), the international cooperation project of Hubei Province (No. 2014BHE0017) and the self-determined research funds of CCNU from the colleges' basic research and operation of MOE

(CCNU15ZD003). The authors wish to thank Guangyou Zhou, Xinhui Tu for improving the research idea, Fanghong Jian, Jie Yuan and Peng Mo for providing help in experiments and text-proofing.

References

1. Zelenko, D., Aone, C., et al.: Kernel methods for relation extraction. J. Mach. Learn. Res. **3**, 1083–1106 (2003)
2. Mintz, M., Bills, S., Snow, R., Jurafsky, D.: Distant supervision for relation extraction without labeled data. In: Proceedings of the Joint Conference of the 47th Annual Meeting of the AC, pp. 1003–1011 (2009)
3. Fader, A., Soderland, S., Etzioni, O.: Identifying relations for open information extraction. In: Proceedings of the Conference on Empirical Methods in Natural Language Processing. Association for Computational Linguistics, pp. 1535–1545 (2011)
4. Thomas, P., Neves, M., et al.: Relation extraction for drug-drug interactions using ensemble learning. In: Drug-Drug Interaction Extraction, Huelva, Spain, pp. 11–18 (2011)
5. Surdeanu, M., Tibshirani, J., et al.: Multi-instance multi-label learning for relation extraction. In: Conference on Empirical Methods in Natural Language Processing and Natural Language Learning, Jeju Island, Korea, pp. 455–465 (2012)
6. Li, H., Wu, X., et al.: A relation extraction method of Chinese named entities based on location and semantic features. Appl. Intell. **38**(1), 1–15 (2013)
7. He, T., Xu, C., et al.: Named-entity relation extraction method based on seed self-expansion. Comput. Eng. **32**(21), 183–184, 193 (2006)
8. Xu, F., Wang, T., Chen, H., et al.: SVM-based Chinese entity relation extraction. In: 9th Chinese National Conference on Computational Linguistics, Dalian, China, pp. 497–502 (2007)
9. Chen, P., Guo, J., et al.: Chinese field entity relation extraction based on convex combination kernel function. J. Chin. Inf. Process. **27**(5), 144–148 (2013)
10. Zeng, D., Zhao, J., et al.: Open entity attribute-value extraction from unstructured text. J. Jiangxi Norm. Univ. (Nat. Sci. Ed.) **37**(3), 279–283 (2013)
11. Guo, X., He, T., et al.: Chinese named-entity relation extraction based on the syntactic and semantic features. J. Chin. Inf. Process. **28**(6), 183–189 (2014)
12. Zhang, H., Huang, M., Zhu, X.: A unified active learning framework for biomedical relation extraction. J. Comput. Sci. Technol. **27**(6), 1302–1313 (2012)
13. Platt, J.C.: Fast training of support vector machines using sequential minimal optimization. In: Advances in Kernel Methods. MIT Press, pp. 185–208 (1999)
14. Xianyi, C., Qian, Z.: A study of relation extraction of undefined relation type based on semi-supervised learning framework. J. Nanjing Univ. Nat. Sci. **48**(4), 466–474 (2012)
15. Wang, H., Qi, Z., Hao, H., Xu, B.: A hybrid method for chinese entity relation extraction. In: Zong, C., Nie, J.-Y., Zhao, D., Feng, Y. (eds.) NLPCC 2014. CCIS, vol. 496, pp. 357–367. Springer, Heidelberg (2014)
16. Zhen, J., Dake, H.E., et al.: Relation extraction from Chinese online encyclopedia based on weakly supervised learning. CAAI Trans. Intell. Syst. **10**(1), 113–119 (2015)

Improving Knowledge Base Completion by Incorporating Implicit Information

Wenqiang He[(✉)], Yansong Feng, and Dongyan Zhao

Peking University, Beijing, China
{hewenqiang,fengyansong,zhaodongyan}@pku.edu.cn

Abstract. Over the past few years, many large Knowledge Bases (KBs) have been constructed through relation extraction technology but they are still often incomplete. As a supplement to training a more powerful extractor, Knowledge Base Completion which aims at learning new facts based on existing ones has recently attracted much attention. Most of the existing methods, however, are only utilizing the explicit facts in a single KB. By analyzing the data, we find that some implicit information should also been captured for a more comprehensive consideration during completion process. These information include the intrinsic properties of KBs (e.g. relational constraints) and potential synergies between various KBs (i.e. semantic similarity). For the former, we distinguish the missing data by using relational constraints to reduce the data sparsity. For the later, we incorporate two semantical regularizations into the learning model to encode the semantic similarity. Experimental results show that our approach is better than the methods that consider only explicit facts or only a single knowledge base, and achieves significant accuracy improvements in binary relation prediction.

Keywords: Knowledge base completion · Implicit information · Relational constraints · Semantical regularizations

1 Introduction

With the development of Semantic Web, much work has gone into the automatic construction of large knowledge bases with using the relation extraction technology. Notable endeavors include YAGO [24], DBpedia [2], Freebase [4], and NELL [7]. These KBs, containing millions of facts, have played an important role in multiple AI related applications, such as Question Answering [27,28], Information Extraction [10,26] and etc.

However, these knowledge bases are far from complete. For example, 93.8 % of persons from Freebase have no place of birth, and 78.5 % have no nationality [16]. There is a necessary demand for increasing their coverage of facts to make them more useful in practical applications. This phenomenon has given rise to a new research direction: **knowledge base completion**.

The goal of knowledge base completion is to predict missing relations between entities by utilizing the existing facts in knowledge base. For example, since

© Springer International Publishing Switzerland 2016
G. Qi et al. (Eds.): JIST 2015, LNCS 9544, pp. 141–153, 2016.
DOI: 10.1007/978-3-319-31676-5_10

we have known the fact *birthPlace(X, Y)* in DBpedia, we can infer the missing fact *nationality(X, Y)* to a great degree. It can be viewed as an important supplement to train a more powerful extractor for finding more new facts in web or news corpus. In previous methods, the existing facts were either to used as features in a logistic regression classifier [13, 14], or treated as the optimization objective in the learning model [5, 6, 18]. The common characteristic of these methods is that they inferred new facts with only the explicit resources (i.e. existing facts). Recently, surface-level textual patterns [9] or first-order logic rules [19, 21] are combined into the completion process. Nevertheless, acquiring of these external resources is time-consuming and brought other drawbacks (e.g. sparsity problem).

By analyzing the data, we find that there are many implicit information existing in knowledge bases themselves. For example, the arguments of relation *birthPlace* need to be *person* and *location* entities respectively. We call this feature as *relational constraints*. Besides, after combining various knowledge bases, we can find that some relations from different knowledge bases (e.g. *birthPlace* in DBpedia, *wasBornIn* in Yago) represent the same relationship even though they are different on the lexical-level. We call this feature as *semantic similarity*. This kind of feature also exists in entities (e.g. entities in the same category should be more similar than those in different categories).

Based on these observations, we integrate different knowledge bases and make full use of the explicit and implicit information together to infer missing facts. Specifically, we utilize the explicit facts as optimization objective in learning model similar to previous methods. For the implicit information, we distinguish the missing data into true ones and false ones by using *relational constraints* feature, which restricts that the candidate entities' type should meet the corresponding relation's requirements. The false ones can be added into the training dataset as negative samples. Moreover, we incorporate two semantical regularizations into the learning model to encode the *semantic similarity* feature.

The advantages of our approach can be summarized as follows: (1) The use of *relational constraints* translates some unqualified missing facts to training dataset and greatly reduces the data sparsity. (2) The semantical regularizations smooth the learning model and take advantage of potential synergies between various knowledge bases reasonably. (3) It gives a more comprehensive consideration of knowledge base's intrinsical properties during completion process.

The remainder of this paper is organized as follows. In Sect. 2, we provide an overview of several major approaches for knowledge base completion and related work. The technical details of using two implicit features are presented in Sects. 3 and 4 respectively. The results of an empirical analysis are reported in Sect. 5, followed by the conclusion in Sect. 6.

2   Related Work

The literatures about knowledge base completion can be categorized into three major groups: (1) methods based on rules; (2) methods based on path ranking algorithms; (3) methods based on representation learning.

The first group inferred new facts by applying first order Horn clause rules [3,7]. They cannot scale to make complex inferences from large knowledge bases, for the reason that computing cost of learning first-order Horn clauses is expensive. The second group viewed the knowledge base as a graph and found paths which connect the source and target nodes of relation instance through random walk [9,13,14]. These works improved significantly over traditional Horn-clause learning and inference method. However, they were limited by the connectivity of the knowledge base graph. The third group aimed at representing entities and relations in a low-dimensional vector space and predicting new facts through operating on these latent factors [17,22]. These methods were able to capture some unobservable but intrinsic properties of entities and relations [11].

Furthermore, the third group methods can be further classified into three fine-grained classes. (1) methods based on neural networks to capture the intrinsic geometric structure of the vector space [5,6,15,22]; (2) methods based on matrix/tensor factorization that modeled the knowledge base as a tensor or matrix and then performed factorization techniques on it [8,17,20]; (3) methods based on Bayesian clustering that embedded the latent factors of entities and relations into a nonparametric Bayesian clustering framework [12,25].

Our method is based on representation learning and using matrix factorization techniques, but differs in that it not only utilize the explicit facts but also consider implicit information, which exists inherently but have not been taken advantage of by previous work. Furthermore, the idea of incorporating implicit information can be easily extended to other methods. For instance, it can provide additional features in path ranking algorithm and constraints in neural network architectures.

3   Relational Constraints for Constructing Matrix

Following the W3C Resource Description Framework (RDF) standard, facts of knowledge base are represented in the form of binary relationships. The formalized structures, in particular, are numerous *(subject, relation, object)* (SRO) triples, where *subject* and *object* are entities and *relation* indicates the inter-relationship between entities. This structure can be translated into a matrix naturally and then the knowledge base completion problem can be regarded as matrix completion.

Specifically, we set E to the set of entity pairs *(subject, object)*, R to the set of binary relations, and F to all the observed triples in various KBs. Assuming that we index an entity pair with $e \in E$, a relation with $r \in R$, and $r(e)$ is true when $(e_{subject}, r, e_{object}) \in F$, false otherwise. Producing a matrix X from various knowledge bases, which consist of m relations and n entity pairs, is straightforward: each row in the matrix corresponds to a relation r and each column an entity pair e. Each matrix cell is denoted as X_{er} and the size of the matrix is $m \times n$.

Note that our method integrates various knowledge bases into a unified matrix, and faces the entity linking problem between different knowledge bases at the same time. Fortunately, many of these public knowledge bases have been connected in the Linked Open Data Cloud (LOD cloud) [1]. The LOD cloud provides *sameAs* links between equivalent entity identifiers across different knowledge bases. We can use this resource to solve the entity linking problem simply. Certainly, we can also utilize more complex methods [23], but this is not the core problem we consider in this paper.

We define the value of X_{er} means the probability that a relation r holds an entity pair e. Then we can initialize the matrix as follows:

$$X_{er} = \begin{cases} 1, & \text{if } r(e) \text{ is true} \\ ?, & \text{otherwise} \end{cases} \tag{1}$$

where $X_{er} = ?$ means that the corresponding position in matrix is a missing data and needs to predict a value for completing the matrix. If the value is greater than a threshold, it means that we have found a new relationship between the corresponding entity pair. Hence, the knowledge base completion problem can be translated into completing missing values in the matrix.

However, the knowledge base matrix is very spare in this way. In RDF standard, the relation of knowledge base has a domain and a range. The domain limits the scope of subjects where such relation can be applied and the range is the set of possible objects of the relation. For example, *Birth-Place* is a relation using the class *Person* as domain and its range is *Place*. we call this implicit feature as *relational constraints*. Following this characteristic, we distinguish the true-missing and false-missing data in knowledge base matrix as follows.

Definition 1. *We say an entity pair e(subject, object) is consistent with relation r iff the type of subject is equivalent to or the subset of the domain of r and the type of object meets the same requirement with the range of r. For any triples $(e_{subject}, r, e_{object}) \notin F$, which are missing data in knowledge base, if e(subject, object) is consistent with relation r, it is **true-missing data** in knowledge base matrix, otherwise it is **false-missing data**.*

Based on this definition, the triple *BirthPlace(Barack-Obama, Hawaii)* can be treated as true-missing data, but the triple *BirthPlace(BarackObama, Michelle Obama)* cannot be, even though they are all missing data in knowledge base matrix. In other words, true-missing data are just the potential target we need to predict.

Then we modify the Eq. 1 as follows:

$$X_{er} = \begin{cases} 1, & \text{if } r(e) \text{ is true} \\ 0, & \text{if } r(e) \text{ is false missing data} \\ ?, & \text{otherwise} \end{cases} \tag{2}$$

where $X_{er} = 0$ is also consistent with the intuition that X_{er} means the probability of a relation r holds an entity pair e as mentioned above.

This strategy gets rid of these false-missing data for all the missing data and increases the density obviously. Note that the final performance is sensitive to the number of negative data and we will describe it in more detail in Sect. 5.

4   Semantically Smooth Factorization Model

In this section, we discuss how to complete the knowledge base using matrix factorization with semantical regularization.

4.1   Basic Matrix Factorization Model

After constructing the knowledge bases matrix, we need to approximate the $m \times n$ matrix X by a multiplication of L-rank factors:

$$X \approx U^T V \tag{3}$$

where $U \in R^{l \times m}$ and $V \in R^{l \times n}$ with $l < min(m, n)$, respect to the relation-specific and entity pair-specific matrices, respectively.

We can get these low-dimensional vector spaces by utilizing matrix factorization technology. Specifically, we can minimize the reconstruction error to be compatible within the existing facts in knowledge bases. The mathematical form of this optimization problem is defined as follows:

$$\min_{U,V} \frac{1}{2} \sum_{r=1}^{m} \sum_{e=1}^{n} I_{er}(X_{er} - g(U_r^T V_e))^2 + \frac{\lambda_1}{2} \|U\|_F^2 + \frac{\lambda_2}{2} \|V\|_F^2 \tag{4}$$

where $\|\cdot\|_F^2$ denotes the Frobenius norm, I_{er} is the indicator function that is equal to 1 if $X_{er} = 1$ or $X_{er} = 0$, which means $(e_{subject}, r, e_{object})$ is an existing or negative fact, and equal to 0 otherwise. The nonlinear logistic function $g(x) = 1/(1 + \exp(-x))$ bounds the range of the inner product value into $[0, 1]$ for consisting with the definition of X_{er}. In order to avoid overfitting, the norms of U and V are added as regularization terms and $\lambda_1, \lambda_2 > 0$, as the regular parameter.

Gradient based approaches can be applied to find a local minimum of the Eq. 4. After the low-dimensional matrices U and V are learned, the next step is to predict whether the entity pair e_j hold a relationship r_i for the missing triples. Given a threshold, we can predict $r_i(e_j)$ is true if the probability value $\widehat{X}_{ij} = g(U_i^T V_j)$ is greater than threshold, and false otherwise.

In the following sections, we will reveal the potential synergies between various knowledge bases and encode the semantic similarity into the objective function to smooth the factorization model.

4.2   Semantic Similarity Measure

After combining various knowledge bases, we find some interesting phenomenons. On the one hand, some of the relations present the same relationship even though they are different on the lexical-level. On the other hand, these knowledge bases share a large part of their entities. Here we take two popular knowledge bases, YAGO and DBpedia, as an example. The relation *graduatedFrom* in YAGO has 30,389 entity pairs and the relation *almaMater* in DBpedia has 64,928 entity pairs. They are semantic equivalence and the former has 27,523 (90 %) entity pairs that appear together with the latter. Note that, these semantic similar relations can not be found in a single knowledge base because of the uniqueness condition. This is an important reason that we integrate various knowledge bases into one unified matrix to utilizing potential synergies.

In consideration of the structure of matrix, we can represent a relation r using a row vector $\boldsymbol{r_i}$ and the item value 0 or 1 means whether the relation holds an entity pair. Hence, we can define the *semantic similarity* of relations using Vector Space Similarity as follows:

$$Sim(r_i, r_j) = \cos(\boldsymbol{r_i}, \boldsymbol{r_j}) = \frac{\sum\limits_{k \in I(i) \cap I(j)} r_{ik} \cdot r_{jk}}{\sqrt{\sum\limits_{k \in I(i) \cap I(j)} r_{ik}^2 \cdot \sum\limits_{j \in I(i) \cap I(j)} r_{jk}^2}} \tag{5}$$

where k belongs to the subset of entity pair which are both held by relation r_i and relation r_j. From the above definition, we can see that if two relations share more entities pairs, they are more similar with each other.

Further more, the semantic similarity of entities can be defined similarly. The difference is that the entity is represented as a column vector which indicates whether the entity pair is selected by the relation.

4.3   Semantic Regularizations

Based on the interpretation above, we introduce two semantic regularizations to encode the similarity and smooth the matrix factorization model semantically .

Relation Similarity Regularization: *If relation r_i is similar to relation r_j, we can assume that the low-dimensional vectors U_i and U_j of these two relations are close in the geometric structure.*

Hence, we minimize the following relation similarity regularization:

$$R_1 = \min_{U} \sum_{i=1}^{m} \sum_{j \in S(i)} Sim(i, j) \|U_i - U_j\|_F^2 \tag{6}$$

where $S(i)$ is the set of relations that are similar to the relation r_j, $Sim(i, j) \in [0, 1]$ is a weight item that indicates the smaller difference between two relation's latent vectors, the greater influence in regularization function.

Considering the large number of entities in knowledge bases, we can not define the entity similarity regularization similar to the above. We find that each entity has a *type property* which denotes which *category* it belongs to. Along with the *relation constraints* as mentioned above, we can define entity similarity in the same category.

Entity Similarity Regularization: *If entity e_i belongs to one category, we can assume that the low-dimensional vectors V_i of the entity and the representation of the category are close in the geometric structure. The representation of category can be viewed as an a combination of the entities which are in the same category.*

Hence, we minimize the following entity similarity regularization:

$$R_2 = \min_V \sum_{i=1}^{n} \|V_i - \frac{1}{|C^+(i)|} \sum_{j \in C^+(i)} V_j\|_F^2 \tag{7}$$

where $C^+(i)$ is the set of entities which are in the same category with the entity e_i, and $|C^+(i)|$ denotes the number of entities in the set $C^+(i)$.

In the above objective function, the representation of one category is equivalent to the average of all entities belongs to it. This assumption can not incorporate the interaction between entities. Therefore, we select the most similar k entities to perform a weighted average. We thus change the regularization term in Eq. 7 to

$$R_2 = \min_V \sum_{i=1}^{n} \|V_i - \frac{\sum_{j \in C_k^+(i)} (Sim(i,j) \times V_j)}{\sum_{j \in C_k^+(i)} Sim(i,j)}\|_F^2 \tag{8}$$

where $Sim(i,j) \in [0,1]$ is the similarity function to indicate the similarity between entity e_i and entity e_j. $C_k^+(i)$ is the set of top-k entities.

Hence we have the overall objective function:

$$
\begin{aligned}
\min_{U,V} L(X, U, V) = {}& \frac{1}{2} \sum_{r=1}^{m} \sum_{e=1}^{n} I_{er}(X_{er} - g(U_r^T V_e))^2 \\
& + \frac{\alpha}{2} \sum_{i=1}^{m} \sum_{j \in S(i)} Sim(i,j) \|U_i - U_j\|_F^2 \\
& + \frac{\beta}{2} \sum_{i=1}^{n} \|V_i - \frac{\sum_{j \in C_k^+(i)} (Sim(i,j) \times V_j)}{\sum_{j \in C_k^+(i)} Sim(i,j)}\|_F^2 \\
& + \frac{\lambda_1}{2} \|U\|_F^2 + \frac{\lambda_2}{2} \|V\|_F^2
\end{aligned}
\tag{9}
$$

where $\alpha, \beta > 0$, which controls the weights of relation and entity similarity in the factorization model, respectively. A local minimum of the objective function given by Eq. (9) can be found by performing gradient descent in feature vectors U_r and V_e:

$$
\begin{aligned}
\frac{\partial L}{\partial U_r} = {}& \sum_{e=1}^{n} I_{er} g'(U_r^T V_e)(g(U_r^T V_e) - X_{er})V_e + \lambda_1 U_r \\
& + \alpha \sum_{j \in Sim(i)} Sim(r,j)(U_r - U_j)
\end{aligned}
\tag{10}
$$

$$
\begin{aligned}
\frac{\partial L}{\partial V_e} = {}& \sum_{r=1}^{m} I_{er} g'(U_r^T V_e)(g(U_r^T V_e) - X_{er})U_i + \lambda_2 V_e \\
& + \beta(V_i - \frac{\sum_{j \in C_k^+(i)} (Sim(i,j) \times V_j)}{\sum_{j \in C_k^+(i)} Sim(i,j)})
\end{aligned}
\tag{11}
$$

5   Experiment

5.1   Data Sets

In this paper, we evaluate our methods based on the *PERSON* subset of two typical knowledge bases, YAGO 2s [24] and DBpedia 3.9 [2]. Two criteria are considered to filter some triples during the construction of knowledge base matrix: the object of the triple can not be a literal and the number of each relation instances must be greater than 400. Entity linking resource is downloaded from LOD Cloud [1], Relational constraints information is obtained from *yagoSchema* and *dbpedia:owl*. Considering the category taxonomy of the two knowledge bases is not unified, we use the *Category:Wikipedia* as a pivot to normalize entity category information and keep non-overlapping categories.

After the preprocessing, we list statistics of these data sets in Table 1. Where Rel., Ent., Fac. denotes the number of relations, entities, facts in each knowledge base, and #Rel., #Ent., #Fac. is the number we used, respectively. #Cat. is the number of non-overlapping categories. We split the known entity pairs of these relations (include existing and negative triples) into 80 % training and 20 % testing parts. We tune the parameters for our methods using a coarse, manual grid search with cross validation on the training data.

Table 1. Statistics of the data used in our experiments.

	Rel.	Ent.	Fac.	#Rel.	#Ent.	#Fac.	#Cat.
YAGO	75	5.7M	7.8M	37	2.5M	2.6M	14
DBPedia	1358	19.3M	21.3M	409	11.4M	11.6M	14

5.2   Experiment Setting

In our experiments, we seek to answer the following questions:

1. Whether the strategy of using relational constraints for constructing matrix has a positive effect on the results.
2. Whether encoding similarity regularizations into the matrix factorization model help to improve the performance.

We design several groups of compared methods to address these questions. As shown in Table 2, **Basic** means using only existing facts, **Full** means incorporating implicit information as mentioned above. **-RelCon.** means removing the relational constraints during the construction of knowledge base matrix, **-SemReg.** means removing the semantic regularizations in the matrix factorization model.

Knowledge base completion, whose goal is predicting new relationships between entities, can also be viewed as a binary classification task. Given the missing triple (*subject, relation, object*) and threshold, we need to jude whether the missing triple (*subject, relation, object*) is correct or not. So we use the classic classification evaluation metrics *Precision* and *Recall* to measure the predication quality of our method.

5.3   Results

From Table 2, we can observe that **-RelCon.**, **-SemReg** method performs all better than **Basic** method, and **Full** method generally achieves best performance of all. The results suggest that incorporating implicit information, including relational constraints and semantic similarity, into the factorization model improves knowledge base completion.

Actually, these implicit information promote the performance in two directions. First, relational constraints feature helping filter out the false-missing data and reducing the solution space in a direct way. Translating these data into negative data can also reduce the data sparsity significantly. Secondly, semantic similarity feature indicates the potential synergies between various knowledge bases and captures the intrinsic geometric structure of low-dimension vectors. It can be viewed as further constraints on the solution space in an indirect way.

Table 2. Comparison of performance of different methods.

	Precision	Recall	F1
Basic	0.424	0.240	0.306
Full	0.712	0.424	0.531
-RelCon.	0.627	0.326	0.429
-SemReg.	0.601	0.343	0.437

5.4   Impact of Parameters α and β

The adjustable parameters α, β control how much our methods should incorporate the implicit information into the completion process. A very small value of α, β means that we ignore these implicit information, while a very large value of α, β may lead the completion process to be dominated by these implicit information. In many cases, we do not want to set α, β to these extreme values since they will potentially hurt the prediction performance.

The impact of parameters α and β on F-measure are shown Fig. 1. We set α log values ranging from -6.0 to 0.0 and β ranging from -4.0 to 1, with setting the iteration at 120 and the threshold at 0.6. Given a fixed parameter β, We observe that the value of α impacts the prediction results significantly. When α increases, the F-measure value also increases at first, but after α exceeds a certain threshold like 0.01, the F-measure value decreases with further increments of the value α. Besides, when given a fixed parameter α, the interrelation between β and F-measure is similar to that. Interestingly, the changing of β has a less impacts than α. The possible reason is that the number of relations and entities in knowledge bases is incoordinate. The former is much less than the later and the semantical similarity between relations may have a strong constraints on the factorization model.

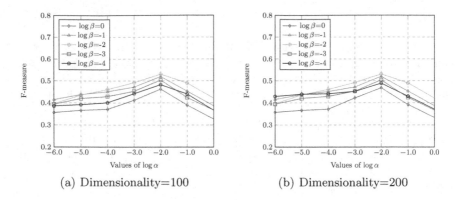

Fig. 1. Impact of parameters α and β

5.5 Impact of Negative Data and Threshold

As mentioned in Sect. 3, we use the relational constraints to initialize the negative data. The number of these data also has heavy impacts on the performance. From Fig. 2(a), we observe that a balanced ratio of negative and positive data (i.e. existing facts) helps to improve the performance, while an excess of negative data is worthless. The reason is obvious, if the number of negative data is much larger than positive data, the reconstruction error in the factorization model mainly depends on the former and the influence of existing facts has been hidden.

In addition to this, the parameter threshold determines the final prediction results. Figure 2(b) illustrates the impact of threshold on precision. It turns out that the precision increases simultaneously with the threshold. This phenomenon coincides with the intuition that a high threshold will result in more reliable prediction.

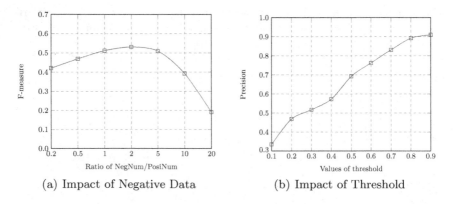

Fig. 2. Impact of Negative Data and Threshold

6   Conclusion

In this paper, we first tease out some implicit information which should be captured in knowledge base completion, and then design two strategies to incorporate these information. The method we propose not only takes the intrinsic properties of knowledge bases into consideration, but also makes a good use of the synergies between various knowledge bases. Experiments show that both the two strategies can help improve the performance of predicting new relations between existing entity pairs.

Acknowledgments. This work was supported by the National High Technology R&D Program of China (Grant No.2014AA015102, 2015AA015403), National Natural Science Foundation of China (Grant No.61272344, 61202233, 61370055) and the joint project with IBM Research. Any correspondence please refer to Wenqiang He.

References

1. http://linkeddata.org/
2. Auer, S., Bizer, C., Kobilarov, G., Lehmann, J., Cyganiak, R., Ives, Z.G.: DBpedia: a nucleus for a web of open data. In: Aberer, K., et al. (eds.) ASWC 2007 and ISWC 2007. LNCS, vol. 4825, pp. 722–735. Springer, Heidelberg (2007)
3. Berant, J., Dagan, I., Goldberger, J.: Global learning of typed entailment rules. In: Proceedings of the 49th Annual Meeting of the Association for Computational Linguistics: Human Language Technologies, vol. 1, pp. 610–619. Association for Computational Linguistics (2011)
4. Bollacker, K., Evans, C., Paritosh, P., Sturge, T., Taylor, J.: Freebase: a collaboratively created graph database for structuring human knowledge. In: Proceedings of the 2008 ACM SIGMOD international conference on Management of data, pp. 1247–1250. ACM (2008)
5. Bordes, A., Glorot, X., Weston, J., Bengio, Y.: A semantic matching energy function for learning with multi-relational data. Mach. Learn. **94**(2), 233–259 (2014)
6. Bordes, A., Usunier, N., Garcia-Duran, A., Weston, J., Yakhnenko, O.: Translating embeddings for modeling multi-relational data. In: Advances in Neural Information Processing Systems, pp. 2787–2795 (2013)
7. Carlson, A., Betteridge, J., Wang, R.C., Hruschka Jr., E.R., Mitchell, T.M.: Coupled semi-supervised learning for information extraction. In: Proceedings of the third ACM International Conference on Web Search and Data Mining, pp. 101–110. ACM (2010)
8. Chang, K.W., Yih, W.t., Yang, B., Meek, C.: Typed tensor decomposition of knowledge bases for relation extraction. In: Proceedings of the 2014 Conference on Empirical Methods in Natural Language Processing (EMNLP), pp. 1568–1579 (2014)
9. Gardner, M., Talukdar, P.P., Kisiel, B., Mitchell, T.M.: Improving learning and inference in a large knowledge-base using latent syntactic cues. In: EMNLP, pp. 833–838 (2013)

10. Hoffmann, R., Zhang, C., Ling, X., Zettlemoyer, L., Weld, D.S.: Knowledge-based weak supervision for information extraction of overlapping relations. In: Proceedings of the 49th Annual Meeting of the Association for Computational Linguistics: Human Language Technologies, vol.1, pp. 541–550. Association for Computational Linguistics (2011)

11. Jenatton, R., Roux, N.L., Bordes, A., Obozinski, G.R.: A latent factor model for highly multi-relational data. In: Advances in Neural Information Processing Systems. pp. 3167–3175 (2012)

12. Kemp, C., Tenenbaum, J.B., Griffiths, T.L., Yamada, T., Ueda, N.: Learning systems of concepts with an infinite relational model. In: AAAI, vol. 3, p. 5 (2006)

13. Lao, N., Mitchell, T., Cohen, W.W.: Random walk inference and learning in a large scale knowledge base. In: Proceedings of the Conference on Empirical Methods in Natural Language Processing, pp. 529–539. Association for Computational Linguistics (2011)

14. Lao, N., Subramanya, A., Pereira, F., Cohen, W.W.: Reading the web with learned syntactic-semantic inference rules. In: Proceedings of the 2012 Joint Conference on Empirical Methods in Natural Language Processing and Computational Natural Language Learning, pp. 1017–1026. Association for Computational Linguistics (2012)

15. Lin, Y., Liu, Z., Sun, M., Liu, Y., Zhu, X.: Learning entity and relation embeddings for knowledge graph completion. In: Proceedings of AAAI (2015)

16. Min, B., Grishman, R., Wan, L., Wang, C., Gondek, D.: Distant supervision for relation extraction with an incomplete knowledge base. In: HLT-NAACL, pp. 777–782 (2013)

17. Nickel, M., Tresp, V., Kriegel, H.P.: A three-way model for collective learning on multi-relational data. In: Proceedings of the 28th International Conference on Machine Learning (ICML 2011), pp. 809–816 (2011)

18. Nickel, M., Tresp, V., Kriegel, H.P.: Factorizing yago: scalable machine learning for linked data. In: Proceedings of the 21st International Conference on World Wide Web, pp. 271–280. ACM (2012)

19. Wang, Q., Bin Wang, L.G.: Knowledge base completion using embeddings and rules. In: AAAI (2015, pages to appear)

20. Riedel, S., Yao, L., Marlin, B.M., McCallum, A.: Relation extraction with matrix factorization and universal schemas. In: Joint Human Language Technology Conference/Annual Meeting of the North American Chapter of the Association for Computational Linguistics (HLT-NAACL 2013), June 2013

21. Rocktäschel, T., Singh, S., Riedel, S.: Injecting logical background knowledge into embeddings for relation extraction. In: Proceedings of the 2015 Human Language Technology Conference of the North American Chapter of the Association of Computational Linguistics (2015)

22. Socher, R., Chen, D., Manning, C.D., Ng, A.: Reasoning with neural tensor networks for knowledge base completion. In: Advances in Neural Information Processing Systems, pp. 926–934 (2013)

23. Suchanek, F.M., Abiteboul, S., Senellart, P.: Paris: Probabilistic alignment of relations, instances, and schema. Proc. VLDB Endow. 5(3), 157–168 (2011)

24. Suchanek, F.M., Kasneci, G., Weikum, G.: Yago: a core of semantic knowledge. In: Proceedings of the 16th International Conference on World Wide Web, pp. 697–706. ACM (2007)

25. Sutskever, I., Tenenbaum, J.B., Salakhutdinov, R.R.: Modelling relational data using bayesian clustered tensor factorization. In: Advances in Neural Information Processing Systems, pp. 1821–1828 (2009)

26. Weston, J., Bordes, A., Yakhnenko, O., Usunier, N.: Connecting language and knowledge bases with embedding models for relation extraction (2013). arXiv preprint arXiv:1307.7973
27. Yao, X., Van Durme, B.: Information extraction over structured data: question answering with freebase. In: Proceedings of ACL (2014)
28. Zou, L., Huang, R., Wang, H., Yu, J.X., He, W., Zhao, D.: Natural language question answering over RDF: a graph data driven approach. In: Proceedings of the 2014 ACM SIGMOD International Conference on Management of Data, pp. 313–324. ACM (2014)

Automatic Generation of Semantic Data for Event-Related Medical Guidelines

Yuling Fan[1,2], Rui Qiao[1,2(✉)], Jinguang Gu[1,2], and Zhisheng Huang[3]

[1] College of Computer Science and Technology,
Wuhan University of Science and Technology, Wuhan 430065, China
1023293818@99.com
[2] Hubei Province Key Laboratory of Intelligent Information Processing
and Real-Time Industrial System, Wuhan 430065, China
[3] Departments of Computer Science, VU University Amsterdam,
1081HV Amsterdam, The Netherlands

Abstract. Medical Guidelines pay an important role in medical decision making systems. Medical guidelines are usually involved with event-related actions or procedure. However, little research has been done on how event-related medical guidelines can be converted into its semantic representation such as RDF/OWL data. This paper proposes an approach of automatic generation of semantic data for event-related medical guidelines. This generation is achieved by using the logic programming language Prolog with the support of medical ontologies such as SNOMED CT. We will report the experiments with the automatic generation of the semantic data for event-related Chinese medical guidelines, and show the relevant results.

Keywords: Medical events · Medical guidelines · Semantic data · RDF · Prolog

1 Introduction

Clinical practice guidelines (CPGs) are varieties of clinical guidance developed by a system, which is beneficial for doctors and patients to make appropriate treatment, choice, decision when focus on a specific clinical problem [1]. CPGs aim to improve the quality of care, limit unjustified practice variations and reduce healthcare costs. Clinical actions will produce different results or states, we call diagnosis and treatment status which is worthy of observing as medical event, or an event. Definitely, what is presented here is more specific than people's common understanding of events, and can be regarded as an event in a narrow sense.

The National Natural Science Foundation of China under Grant Nos. 60803160, 61100133, 61272110; The major program of the National Social Science Foundation under Grant No. 11&ZD189; Provincial Key Laboratory Open Fund (znss2013B011); Hubei Provincial Education Department under Grant (Q20151111); Wuhan University of Science and Technology under grants (2013xz012,2014xz019); Rui Qiao is the corresponding author of this paper.

G. Qi et al. (Eds.): JIST 2015, LNCS 9544, pp. 154–163, 2016.
DOI: 10.1007/978-3-319-31676-5_11

In the famous medical ontology SNOMED CT, medical event is defined as all kinds of accidents, such as "falling at home", "being hit by a car", a totally of thirty thousand different medical events list. This paper adopts the definition of events made by SNOMED CT, defining the medical events as a specific status that should be fully avoided, which is an accident. And then, this paper will deal with events of Chinese medical guidelines, convert the event knowledge of natural language into computer-interpretable representations.

Medical guidelines cover a wide amount of data, and contain a large number of raw data from different sources. The characteristics of medical guidelines also determine its particular and high complexity of data, information and knowledge. These information is readable for user, while is difficult for machine to understand, and that's the reason why we have to automatically generate the semantic data for event-related Chinese medical guideline.

In this paper, we propose a method of Prolog to transfer the guidelines knowledge involving events of natural language description and generate their corresponding semantic data. This method analyzes the patterns of event sentences of guidelines, and summarizes the types of sentences. Guideline events are expressed in rule expression, and DCG is used to parse the sentence patterns. Since Prolog is a regular expression language, it is easy to generate and describe the rules based on knowledge representation. Therefore, compared with other methods, this method is easy to read and grasp, and the corresponding results are more accurate.

The rest of this paper is organized as follows: Sect. 2 gives a brief description of Prolog and semantic data representation of medical guidelines. In Sect. 3, it specifically introduces the semantic data generation of the Chinese guidelines based on Prolog. The Sect. 4 describes the semantic processing. And Sect. 5 presents the result of experimental data, and the corresponding analysis. The last section is the discussion and conclusion, the work of this paper and the following work is discussed.

2 Prolog and Medical Guidelines Semantic Representation

2.1 Introduction of Prolog

Prolog [2] is a logic programming language, the most notable feature of which is that it is a descriptive language. Prolog is used to solve problems by declarative method. It is based on known facts and rules, and utilizes deductive reasoning to solve problems automatically [3]. In the paper [4], it uses Java method based on Pattern, however, excessive use of string processing and regular expression makes the precision rate and recall rate very low.

Prolog is a rule-based language. The main advantages of the rule-based language are:

- Declaration. A rule-based language is a declarative one that expresses the logic of a computation without the need of exactly describing its control flow. That is significantly different from the traditional programming languages, like Java, which use a procedural approach for the specification of control flow in the computation. A declarative approach of formalization is more suitable for knowledge representation and reasoning because it needs no carefully design its computation (or reasoning) procedure. Thus, a rule-based language would provide a more convenient and efficient way for the automatic generation of medical guidelines, compared with other procedural approaches, like the script-based language, which relies procedural scripts, and the pattern-based approach, which is based on SPARQL queries with regular expressions.

- Easy Maintenance. A rule-based language provides an approach in which specified knowledge is easy to be understood for human users, because they are very close to human knowledge. It would not be too hard for human users to check the correctness of the knowledge representation if they are formalized as a set of rules. Furthermore, changing or revising a single rule would not make an effect on other part of the formalization significantly, because the meaning of the specification is usually represented in the specific rule. Thus, it is much easier for maintenance of knowledge, compared with procedural/scripting approaches of medical guideline processing.

- Reusability. In a rule-based language, a single rule (or a set of rules) is usually considered to be independent from other part of knowledge. Thus, it is much more convenient to re-use some rules of a processing. The common processing knowledge can be designed to be a common library, which can be re-used for other processing.

- Expressiveness. Automatic knowledge processing usually involves comprehensive scenarios of deliberation and decision-making procedures. To facilitate those capabilities, it may require sophisticated data processing in workflows. An expressive rule-based language can support various functionalities of data processing. Thus, it provides the possibility to build workflows for various scenarios of medical applications.

2.2 Medical Guidelines Semantic Representation

This paper chooses the clinical application guiding principles of antimicrobial drugs [5] (hereinafter referred to as antimicrobial guidelines) as experimental data. In this paper, the definition of events increases the importance of dimensions that events are accidents should be fully avoided. Mapping the definition to the antimicrobial guidelines, events will be the considerations described for antimicrobial agents. The description of penicillin in the guidelines is as follows: "You have to stop using drugs immediately once allergic reaction occurs. (应用本类药物时如发生过敏反应, 须立即停药)". If A is using penicillin (应用青霉素抗生素), and B is allergic reaction occurs (发生过敏反应), hence A +B is a state that should be avoided, at the same time, "stopping the medication immediately (立即停药)" will be an event.

This paper uses the representation of event in paper [6] to express an event as a statement (陈述), containing three parts of eventPrecondition (事件前提), eventOperator (事件操作符), event (事件). Based on the definition and characteristics of Chinese guidelines, this paper summaries 9 kinds of event operations, which are "Must (必须)", "May (可能)", "Likely (容易)", "CanNot (不可)", "Can (可以)", "Forbidden(禁止)", "Need (需要)", "Should (应该)", "ShouldNot (不宜)" separately.

Semantic technology provides a common technical framework for data interoperability. And semantic data is easy to be processed by machine or even reasoning [7]. The paper uses resource description framework (RDF) to express semantic representation of guideline events. We introduce specific predicates such as "haseventPrecondition", "haseventOperator", "hasEvent" to describe the precondition, the event operator and the event name for a single event.. For example, for the description: "Potassium penicillin cannot rapidly intravenous inject. (青霉素钾盐不可快速静脉注射)", the RDF triples are shown as follows:

```
<http://wasp.cs.vu.nl/sct/id#g1001zsh140331><http://www.w3.org/1999/02/22-rdf-syntax-ns#type><http://wasp.cs.vu.nl/sct/sct#Guidelines>.
<http://wasp.cs.vu.nl/sct/id#g1001zsh140331><http://wasp.cs.vu.nl/sct/sct#About>"Rational Use of Anitibiocs"
<http://wasp.cs.vu.nl/sct/id#g1001zsh140331><http://wasp.cs.vu.nl/sct/sct#Title>"抗菌药物临床应用指导原则"
<http://wasp.cs.vu.nl/sct/id#g1001zsh140331><http://wasp.cs.vu.nl/sct/sct#hasStatements><http://wasp.cs.vu.nl/sct/id#g1001zsh140331_1> .
<http://wasp.cs.vu.nl/sct/id#g1001zsh140331_1><http://wasp.cs.vu.nl/sct/sct#Text>"青霉素钾盐不可快速静脉注射".
<http://wasp.cs.vu.nl/sct/id#g1001zsh140331_1><http://wasp.cs.vu.nl/sct/sct#hasStatement><http://wasp.cs.vu.nl/sct/id#g1001zsh140331_11>
<http://wasp.cs.vu.nl/sct/id#g1001zsh140331_11><http://wasp.cs.vu.nl/sct/sct#haseventPrecondition>"青霉素钾盐"
<http://wasp.cs.vu.nl/sct/id#g1001zsh140331_11><http://wasp.cs.vu.nl/sct/sct#haseventOperator><http://wasp.cs.vu.nl/sct/guideline#CanNot>
    <http://wasp.cs.vu.nl/sct/guideline#CanNot><http://www.w3.org/2000/01/rdf-schema#label>"不可"@cn,
    <http://wasp.cs.vu.nl/sct/guideline#CanNot><http://www.w3.org/2000/01/rdf-schema#label>"CanNot"@en.
    <http://wasp.cs.vu.nl/sct/id#g1001zsh140331_11><http://wasp.cs.vu.nl/sct/sct#hasevent >"快速静脉注射"
```

Fig. 1. Semantic data representation of events

From Fig. 1, we can see that the RDF Ntriples describe the bilinguistic expressions (both Chinese and English) to express the description.

The event operator "CanNot (不可)" is represented in English, like this:

<http://wasp.cs.vu.nl/sct/guideline#CanNot><http://www.w3.org/2000/01/rdf-schema#label>"CanNot"@en;

The "CanNot" will be linked to "http://wasp.cs.vu.nl/sct/guideline", which can establish their relevance with the medical ontology.

3   Automatic Medical Events Processing by Using Prolog

Automatically generation of the semantic data of events in the medical guidelines is also the process of event extraction. So far, there are various methods of event extraction like method of machine learning [8], method based on pattern [9], and the rule-based method [10]. Generally speaking, event extraction needs to preprocess the input text. Firstly, it will divide the text into simple sentences using sentence detection procedures, and punctuate simple sentences through white spaces into a word (tokens), then mark the text and do the corresponding analysis. But Prolog is a rule-based method that does not require the above process.

3.1   Guidelines Preprocessing

According to the structure of the Chinese guidelines and the characteristics of Chinese language, it's necessary to preprocess the antimicrobial guidelines before processing the medical events. After the analysis of antimicrobial guidelines, it is obvious to find that many sentences containing "this kind of medicine (本类药物)", "the medicine (本药)", "using the drug (给药)" and other pronouns which affect the correct interpretation of the sentence. In this paper, these pronouns of specific drugs will be replaced with string processing.

Some special sentences are found after observation of the antimicrobial guidelines. As the description of Carbapenems said: "Contraindicated in patients who are allergic to this drug as well as its compatibility ingredient. (禁用于对本类药物及其配伍成分过敏的患者)". Obviously, it is a flip-sentence, which means patients who are sensitive to this drug and its compatibility ingredient should be forbidden to use this drug. After processing, the sentence will be: patients who are sensitive to Carbapenems and its compatibility ingredient should be forbidden to use Carbapenems. (对碳青霉烯类抗生素及其配伍成分过敏的患者禁用碳青霉烯类抗生素).

对任何一种头孢菌素类抗生素有过敏史及有青霉素过敏性休克史的患者禁用头孢菌素类抗生素
服用头孢菌素类抗生素前必须详细询问患者先前有否对头孢菌素类、青霉素类或其他药物的过敏史
有青霉素类、其他β内酰胺类及其他药物过敏史的患者，有明确应用指征时应谨慎使用头孢菌素类抗生素
在服用头孢菌素类抗生素过程中一旦发生过敏反应，须立即停药
如发生过敏性休克，须立即就地抢救并予以肾上腺素等相关治疗
头孢菌素类抗生素多数主要经肾脏排泄，中度以上肾功能不全患者应根据肾功能适当调整剂量
中度以上肝功能减退时，头孢哌酮、头孢曲松可能需要调整剂量
氨基糖苷类和第一代头孢菌素注射剂合用可能加重前者的肾毒性，应注意监测肾功能
头孢哌酮可导致低凝血酶原血症或出血，合用维生素K可预防出血
头孢菌素类抗生素亦可引起戒酒硫样反应
服用头孢菌素类抗生素期间及治疗结束后72小时内应避免摄入含酒精饮料
对碳青霉烯类抗生素及其配伍成分过敏的患者禁用碳青霉烯类抗生素

Fig. 2. Result of antimicrobial guidelines reprocessing

3.2 Specific Rule Generation

We can summarize the following types of considerations after detailed analysis of antimicrobial guidelines. (1) Simple sentences, containing only one event operation, directly consist of eventPrecondition, eventOperator, and event. For instance, "Penicillin is not used for intrathecal injection. (青霉素不用于鞘内注射)". (2) Sentences include two different event operations while they share the same event precondition. For example, "Aminoglycosides and first-generation cephalosporins injection combination may aggravate the former renal toxicity so as to pay attention to monitor renal function. (氨基糖苷类和第一代头孢菌素注射剂合用可能加重前者的肾毒性, 应注意监测肾功能)". (3) Sentences contain the same event precondition and event operation, but different events. Like "Once anaphylactic shock, he/she should be rescued instantly, supplied with oxygen, and injected with adrenaline, adrenal corticosteroids as well as other anti-shock therapy. (一旦发生过敏性休克, 应就地抢救, 并给予吸氧及注射肾上腺素、肾上腺皮质激素等抗休克治疗)". (4) Sentences contain two different event operations and different event precondition such as "Cefoperazone may cause hypoprothrombinemia or bleeding, combined with vitamin K can prevent bleeding. (头孢哌酮可导致低凝血酶原血症或出血, 合用维生素K可预防出血)". (5) Complex sentences contain three or more event operations. Take an example, "Quinolones may cause skin photosensitivity reactions, joint disease, tendon rupture, etc., and even cause QT interval prolongation, etc., which should be observed during treatment. (喹诺酮类抗菌药可能引起皮肤光敏反应、关节病变、肌腱断裂等, 并偶可引起心电图QT间期延长等, 用药期间应注意观察)".

Sentences in antimicrobial guidelines contain the programmatic guidance rules related to the use of antimicrobial agents. We need to summarize a general rule to indicate the above sentences. The following rules are defined in the experiment.

Rule (Type1, Type2, Type3, Drug1, People, Situation1, Situation2, Situation3, Drug2, EventOperator, Operator1, Operator2, Danger, Action1, Action2, Action3)– > precondition (Type1, Type2, Type3, Drug1, People, Situation1, Situation2, Situation3, Drug2), eventoperator (EventOperator), event (Operator1, Operator2, Danger, Action1, Action2, Action3).

Experiments show that the rule above can express most of the event description of antimicrobial guidelines, but not all.

3.3 Automatic Generation of Semantic Data

We propose a method of Prolog to automatically generate semantic data of the events knowledge of antimicrobial guidelines. There are 2 mainly advantages: on the one hand, Prolog is a regular expression language, which is easy to generate and describe the rules based on knowledge representation; on the other hand, Prolog provides DCG support, which can be used as a more convenient syntactic parsing tool.

We have implemented event-based semantic data generation automatically of Chinese medical guidelines using Prolog. The text file after preprocessing is loaded into the Prolog system. The system will utilize DCG to parse the rules, and then generate its corresponding semantic triples. Predicate "rule_processing" introduced corresponds to the main flow of DCG processing, which is to use the above rule to call DCG rule through appropriate information in order to obtain the corresponding parameter information such as "eventPrecondition", "eventOperator", "event". In the meanwhile, generating its corresponding semantic triples by "write_ntriple".

DCG rules are designed to parse the rules. It obtains the corresponding parameters by matching precondition, eventoperator as well as event in rules. Based on this, the following DCG rules are used to parse different rules of precondition, eventoperator, event. The following is one of examples:

Precondition ([], [], [], Drug1, [], [], [], [], Drug2)– > drug (Drug1), "和", drug (Drug2), "合用". This rule can match the sentence: "Aminoglycosides and first-generation cephalosporins injection combination may aggravate the former renal toxicity. (氨基糖苷类和第一代头孢菌素注射剂合用可能加重前者的肾毒性)". We can find that Drug1 is Aminoglycosides (氨基糖苷类), Drug2 is first-generation cephalosporins injection (第一代头孢菌素注射剂). Therefore, this DCG rule can match "Aminoglycosides and first-generation cephalosporins injection combination. (氨基糖苷类和第一代头孢菌素注射剂合用)".

Eventoperator ("可能")–>"可能". This rule match the eventoperator of "May (可能)". And event ([], [], Danger, [], [], [])– > danger (Danger). The Danger stands for the event "aggravate the former renal toxicity. (加重前者的肾毒性)".

4   Semantic Processing

Medical guidelines are usually represented as natural language text. In order to convert those textual data in the structured semantic data, we have to use the medical ontologies such as SNOMED CT and Drugbank to detect those medical concepts. There are the following different approaches to embed the medical domain knowledge into the processing system:

- Encoded Medical Knowledge in the rule-based language Prolog. Namely we formalize the medical knowledge/ontologies in the form of the Prolog DCG rules. Here is an example of the DCG rule which states the label of a SNOMED CT concept which is a drug is a drug name:

  drug (Drug)– > Drug, {rdf (ConceptID, rdf: isA, snomedct: drug),
  rdf (ConceptID, snomedct: hasLabel, Drug)}.

- Medical Knowledge via a SPARQL endpoint. Medical domain knowledge can be obtained via a semantic server like a SPARQL endpoint of a triple store. That can be achieved by using the Prolog's semantic web library such as:

  user: file_search_path (rdfql, './rdfql').
  :-use_module (rdfql ('sparql_json_result')).
  The following is a Prolog code example which posts a SPARQL query at the SPARQL endpoint at the localhost with the port 8183.
  Findall (Row, (sparql_query (Query, Row, [host ('localhost'), port(8183), path ('/ sparql/')])), Rows).

5   Experiments and Data Tests

According to the definition of events in this paper, what we focus on is to study the considerations described for antimicrobial agents. Through statistics, there are a total of 157 notes in the guidelines. After preprocessing of sentence segmentation and pronoun replacement, the results shown in Fig. 2 should be 217. The results of event extraction should be 273 as Fig. 1 showed, while the actual results of the experiment are 218, in which proved to be correct are 177. In terms of simple sentence, the number of corresponding data is 153, 147 and 138.

In this paper, the precision rate, recall rate and F1-measure are used to measure the results of event extraction. Specific methods of calculation are shown as (1) (2) (3).

$$\Pr ecision = \frac{\sum_{i=1}^{n} a_i}{\sum_{j=1}^{n} b_j} \tag{1}$$

$$Recall = \frac{\sum\limits_{i=1}^{n} a_i}{\sum\limits_{k=1}^{n} c_k} \tag{2}$$

$$F1 - measure = \frac{2 * \Pr ecision * \text{Re} call}{\Pr ecision + \text{Re} call} \tag{3}$$

a_i in the above formula shows the results that meet the definition without errors, b_j represents the actual results of the experiment, and c_k stands for the results that should be obtained. When evaluating event extraction, if the final result contains eventPre-conditon, eventOperator and event, the event extraction is considered to be correct.

After calculated, the precision rate of event extraction is 81.19 %, recall rate is 64.83 %, and F1-measure is 72.09 %. The precision of simple sentence extraction is 93.87 %, recall rate is 90.19 %, F1-measure is 91.99 %. The following is the bar chart.

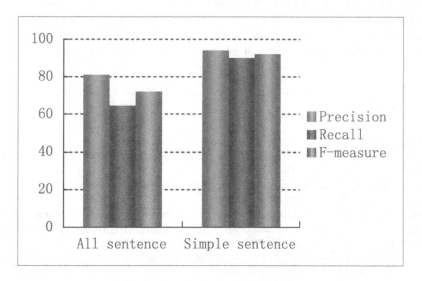

Fig. 3. Bar chart of event extraction

The complexity and variability of the Chinese guidelines, polysemous or synony-mous words in the document make it more difficult to extract events. If a sentence representing the reasons contains the same words with event operations, the event extraction of this sentence will be incorrect. For example, "During taking pyrazinamide avoid exposure to sunlight, because sunlight can cause photosensitive reactions or der-matitis. (服用吡嗪酰胺期间避免曝晒日光, 因可引起光敏反应或日光皮炎)". The eventOperator in this sentence is "avoid(避免)", but "can(可能)" will also be extracted.

As shown in Fig. 3, the recall rate of simple sentence extraction is 90.19 %. However, the rules proposed by this paper are not good when dealing with complex sentences, such as cases (4) and (5) in Sect. 3.2. Because complex sentence is composed of simple sentences, it contains different event preconditions, event operations, events between every comma. Therefore, it puts forward higher requirements for the rules. In the paper [4], we propose a Java-based method which uses the pattern processing. The precision is only 61.78 %, and the recall is 65.77 %. Compared to the Java-based method, the improvement on the precision and recall for the Prolog-rule-based method is still rather small. However, the use of Prolog rule-based method is more simple and convenient, the code repetition is also low. We will continue to develop more rules to deal with complex sentences.

6   Discussion and Conclusion

It is of great significance to present guidelines events knowledge of the antimicrobial guidelines as its corresponding semantic data. We can establish a basic set of events for the use of antimicrobial drugs. These event sets constitute a collection of states that diagnosis and clinical decision system should pay special attention to. We can use this particular set of states to monitor the patients' status and the actions in a diagnosis and treatment system. If a patient or a system enters into a state of such an event, early earning information can be provided timely, which is conducive to intelligent monitoring and knowledge management of rational use of antimicrobial agents [11].

References

1. Yali, Z., Shuqi, C.: The research of international clinical practice guidelines. Clin. Educ. Gen. Pract. **3**, 176–178 (2004). (赵亚利, 崔树起. 国际临床实践指南的研究进展[J]. 全科医学临床与教育, 2004, 2 (3): 176-178)
2. Wielemarks, J., Huang, Z., van der Lourens, M.: SWI-Prolog and the web. J. Theor. Pract. Logic Program. **8**(1), 1–30 (2008)
3. Xiaohong, B., Weihu, Z., Shiqin, W.: Overview of the characteristics of Prolog. J. Yanan Univ. (Nat. Sci. Ed.) **18**(1), 23–26 (1999). (白晓虹,张威虎, 王世勤. Prolog语言特点综述[J] 延安大学学报(自然科学版), 1999, 18 (1): 23-26)
4. Fan, Y., Gu, J., Huang, Z.: Event processing and automatic generation of its semantic data in Chinese medical guidelines. China Digital Medicine. In press. (范玉玲, 顾进广, 黄智生. 中文医学指南的事件处理及其语义数据自动生成[J]. 中国数字医学)
5. Chinese Medical Association, Chinese Hospital Pharmacy Specialized Committee Management Institute, China Pharmaceutical Specialty Hospital Committee. clinical application guiding principles of antimicrobial drugs[EB] (中华医学会, 中华医院管理学会药事管理专业委员会, 中国药学会医院学专业委员. 抗菌药物应用指导原则[EB])
6. Fan, Y., Gu, J., Huang, Z.: Event-oriented semantic data generation for medical guidelines. In: Zhao, D., Du, J., Wang, H., Wang, P., Ji, D., Pan, J.Z. (eds.) CSWS 2014. CCIS, vol. 480, pp. 123–133. Springer, Heidelberg (2014)
7. Jinguang, G., Qing, H., Zhisheng, H.: Semantic transformation and generation of the knowledge of antimicrobial guidelines. China Digital Med. **8**(4), 20–24 (2013). (顾进广, 胡青, 黄智生. 抗菌药物指南知识的语义转换与生成[J]. 中国数字医学, 2013, 8 (4) : 20-24)

8. Bjorne, J., Heimonen, J., Ginter, F., et al.: Extracting complex biological events with rich graph-based feature sets. In: BioNLP 2009 Proceedings of the Workshop on BioNLP: Shared Task, Stroudsburg, PA, USA: Association for Computational Linguistics, pp. 10–18 (2009)

9. Hakenberg, J., Solt, I., TIKK, D., et al.: Molecular event extraction from Link Grammar parse trees. In: BioNLP 2009 Proceedings of the Workshop on BioNLP: Shared Task, Stroudsburg, PA, USA: Association for Computational Linguistics, pp. 86–94 (2009)

10. Sarafraz, F., Eales, J., Mohammadi, R., et al.: Biomedical event detection using rules, conditional random fields and parse tree distances. In: BioNLP 2009 Proceedings of the Workshop on BioNLP: Shared Task, Stroudsburg, PA, USA: Association for Computational Linguistics, pp. 115–118 (2009)

11. Jingsong, L., Zhisheng, H.: Biomedical semantic technology. Zhejiang University Press, Hangzhou (2012). (李劲松, 黄智生. 生物医学语义技术[M] 杭州 : 浙江大学出版社, 2012)

Knowledge Engineering
and Management

Evaluating and Comparing Web-Scale Extracted Knowledge Bases in Chinese and English

Tong Ruan$^{(\boxtimes)}$, Xu Dong, Haofen Wang, and Yang Li

Department of Computer Science and Engineering,
East China University of Science and Technology, Shanghai 200237, China
{ruantong,whfcarter}@ecust.edu.cn,
dongxu0220@qq.com, marine11y@163.com

Abstract. DBpedia and YAGO are the two main data sources serving as the hub of Linking Open Data (LOD), and they both contain Chinese data. Zhishi.me and SSCO extract Chinese knowledge from Wikipedia and other Chinese Encyclopedic Web sites like Baidu-Baike and Hudong-Baike. The quality of these Knowledge Bases (KBs) are not well investigated while their qualities are key to smart applications. In this paper, we evaluate three large Chinese KBs including DBpedia Chinese, zhishi.me and SSCO, and further compare them with English KBs. Since traditional methods on evaluating Web ontology can not be easily adapted to web-scale extracted KBs, we design two metric sets considering Richness and Correctness based on a quasi-formal conceptual representation to measure and compare these KBs. We also design a novel metric set on overlapped instances of different KBs to make the metric results comparable. Finally, we employ random sampling to reduce human efforts for assessing the correctness. The findings in these KBs give a detailed status report of the current situation of extracted KBs in both Chinese and English.

1 Introduction

In recent years, an increasing number of semantic data sources are published on the Web. These sources are further interlinked to form Linking Open Data (LOD). Among LOD, DBpedia[1] and YAGO[2] are the two main data sources serving as the hub. The DBpedia project [1] extracts structured information from Wikipedia and publishes this information on the Web. DBpedia is currently one of the largest hubs of LOD. YAGO [2] is another well-known huge semantic knowledge base (KB), derived from Wikipedia, WordNet and GeoNames.

This work was partially supported by the National Science Foundation of China (project No: 61402173), and Software and Integrated Circuit Industry Development Special Funds of Shanghai Economic and Information Commission (project No: 140304).

[1] http://wiki.dbpedia.org/.
[2] http://www.mpi-inf.mpg.de/yago/.

G. Qi et al. (Eds.): JIST 2015, LNCS 9544, pp. 167–184, 2016.
DOI: 10.1007/978-3-319-31676-5_12

Due to the multilingual nature of Wikipedia, both DBpedia and YAGO contain semantic data in Chinese. While Wikipedia is one of the largest encyclopedias on the Web, the portion of Chinese articles are much fewer than that of articles in English or German. Thus DBpedia or YAGO does not contain adequate Chinese knowledge compared with the size of knowledge expressed in English. On the other hand, in China, the articles in Hudong and Baidu are 10 times as many as the Chinese version of Wikipedia. There emerge projects such as zhishi.me [3], SSCO [4] and XLore [5] trying to extract structured Chinese information from a combination of Chinese encyclopedia Web sites including Hudong-Baike, Baidu-Baike and Chinese Wikipedia. Both zhishi.me[3] and SSCO[4] have Web sites with user friendly GUI for users to access.

Since there are so many KBs in different languages which are extracted from different sources via different methods, it is natural to ask questions such as: Are the quality of Chinese KBs better or worse than their English correspondences? How the qualities of extracted KBs are impacted by multiple data sources? Will these KBs share similar errors or not?

To address the assessment requirements of comparing web-scale extracted KBs, we focus on two quality dimensions namely *Richness* and *Correctness*. The reason is, *whether a KB is web-scale depends on the richness of the data, and extracted data is prone to errors.* To find suitable metric sets to measure the above quality dimensions, we survey the researches of metrics and methodologies on LOD evaluation, as all the above KBs are aspired by the design principles of LOD. The author of [6] summarized 68 metrics and categorized them into 4 dimensions namely *Availability*, *Intrinsic*, *Contextual* and *Representational*. The sub dimensions of *Intrinsic* includes *Syntactic validity*, *Semantic accuracy*, *Consistency* and *Completeness* etc. Our *Correctness* dimension relates to *Syntactic validity*, *Semantic accuracy* and *Consistency*, and our *Richness* dimension relates to *Completeness* sub dimensions in [6]. However, the metrics in a metric set of a sub dimension from [6] are collected from different research works and they logically overlap and interweave. In the meanwhile, they do not share a unified representation. Glenn and Dave[5] listed 15 metrics to assess the quality of a data set. The metrics includes *Accuracy*, *Completeness*, *Typing* and *Currency* etc. However, they do not give any formulas on how to calculate these metrics.

We provide a graph-based conceptual representation and define metric sets of the two dimensions in a quasi-formal way. The goal of the conceptual representation is to achieve data set independency. The approach is different from TripleCheckMate [7] which is solely based on DBpedia. The conceptual representation consists of schema graph and data graph, and so do the metrics we define. However, we reenforce the metrics on data graph because our Chinese KBs have little schema information.

Furthermore, the quality of a data set could be better understood by comparing its values of evaluation metrics with those of other data sets. For example,

---

users may have no idea about the meaning of instances size 100,000, while they could easily understand that a data set is larger than another data set. The comparisons become more meaningful if they are carried out under the same or similar condition. In that case, to calculate the metrics on the overlapped instances or the overlapped domains is more fair and more reasonable. Therefore, we also design a novel comparative metric set to facilitate quality comparison and to investigate complementarity between KBs.

We evaluate and compare KBs with the three metric sets. While we employ *Simple Random Sampling* to lessen human efforts and design a tool KBMetrics to calculate metrics such as richness automatically and to streamline human evaluating processes, it still has taken about 12 people 30 days to assess the correct ratio of the data sets because the evaluation process of correctness is time-consuming. We find many interesting results during evaluation:

- While DBpedia English is much *richer* than DBpedia Chinese, DBpedia Chinese and English are *complementary* to some extent. We find the *person* domain are more *language-specific* than *place* with the help of our comparative metrics.
- The correctness ratio of DBpedia Chinese and DBpedia English are almost the *same* statistically.
- Zhishi.me and SSCO have *more instances* than DBpedia Chinese, since they consolidate data from three Encyclopedic Web sites.
- The correctness ratio of YAGO from our assessment is *not as high* as declared in [8], and we give a detail analysis of the reason for the phenomenon.
- The metrics on *Overlapped Instances* are not only useful in investigating the real quality of each KB, but also helpful to trace the characteristic of each KB.
- Multiple data sources may increase richness of data. However, whether it could improve the correctness of the KB depends on how they utilize the data sources.

The sampling data sets and the results of the evaluation are published on our web site[6]. The paper is organized as follows. Section 2 introduces related work. Section 3 proposes major steps in our evaluation process including metric definition, random sampling and instance matching algorithms. Section 4 shows the results of the metrics on zhishi.me, SSCO, DBpedia and YAGO and analyzes the results. Section 5 gives a conclusion and points out the future direction of our work.

2   Related Work

Our work focuses on how to evaluate and compare extracted KBs following design principles of LOD to some extent. Therefore we survey literatures on *metrics on LOD* and *Quality Evaluation and Improvements on KBs*. Since we have developed KBMetrics tool to help calculate the metrics, we describe related works about quality assessment tools as well.

---

[6] http://kbeval.nlp-bigdatalab.com/results201505.rar.

2.1   Metrics on LOD

A systematic review of different approaches for assessing the data quality of LOD can be referred to [6]. The author of [6] surveyed 68 metrics and categorized them into 4 dimensions, Availability, Intrinsic, Contextual and Representational. A lot of researches focus on evaluating special aspects of LOD quality. Labels were considered as an important quality factor of LOD in [9], and the authors introduced a number of related metrics to measure completeness, accessibility, and other quality aspects of labels. Zhang Hongyu et al. [10] designed a few complexity metrics on Web ontologies. Gueret et al. [11] focused on assessing the quality of links in Linked Data. They assumed that unsuitable network structures were related to low quality of links. Hogan et al. [12] believed RDF publishers were prone to making errors and discussed common errors in RDF publishing. Besides, Michael Farber et al. [13] gave a survey on major cross-domain data sets of LOD cloud. They compared DBpedia, Freebase, OpenCyc, Wikidata and YAGO from 35 aspects including schema constraints, data types, LOD linkages and so on. However, they used natural languages and checklist instead of quantitative metrics to describe special characteristics of data sets.

While there are a lot of metrics to assess the data quality of LOD, they do not target at web-scale exacted KBs. In this paper,we carefully design three metric sets dedicated for evaluating and comparing web-scale exacted KBs.

2.2   Quality Evaluation and Improvements on Extracted KBs

Quality evaluation and quality improvements on extracted data becomes a hot research topic recently. Amrapali Zaveri et al. [14] classified the errors in DBpedia into four dimensions including Accuracy, Relevancy, Representational-Consistency and Interlinking, and in such a way they defined the notion "Correct" indirectly. However, error classification in [14] is too specific to DBpedia so that it cannot be used in the context of multiple KBs. Dominik Wienand et al. [15] detect incorrect numerical data in DBpedia by unsupervised numerical outlier detection methods. Heiko Paulheim and Christian Bizer [16] added missing type statements and removed faulty statements in DBpedia and NELL by using statistical distributions. Kontokostas et al. [17] presented a methodology for test-driven quality assessment which automatically generated test cases based on pre-defined test patterns. While all these works evaluated DBpedia and found particular errors of DBpedia, they did not give any idea on how "correct" DBpedia is. The purposes of these papers are different from ours.

Besides, major data set providers evaluate the correctness of their data sets before publishing them. For example, YAGO2 (a version of YAGO) [8] used statistic sampling with *Wilson score interval* to reduce human efforts when evaluating the correctness of the YAGO2 manually. However, the evaluation does not give a formal definition of "Correctness". The evaluation of Google Knowledge Vault [18] was based on "local closed world assumption" proposed by the author, and the correctness value is determined by checking the relation between training set and testing set. To show the effect of "local closed world assumption", the author also evaluated the assumption via comparisons with human evaluation.

Till now, all these evaluation works are executed on extracted KBs in English. To the best of our knowledge, there are no evaluation work on KBs on Chinese, especially in a comparative way.

2.3   Quality Assessment Tools

Flemming[7] was a data quality assessment tool in German. Users can get the ultimate quality value of a data set after a few interactive steps with Web UI. Sieve [19] was a framework which fused data from different sources with metrics designed for the fusion purpose. WIQA [20] presented a framework to filter out low quality information. The article aimed at filtering high quality information from web, and defined a policy language for expressing information filtering policies. EvoPat [21] defined several patterns to capture evolution and refactoring operations on data and schema. TripleCheckMate [7] is a crowdsourcing quality assessment tool focusing on correctness evaluation of DBpedia. Except for TripleCheckMate, none of the above tools support human evaluation process mentioned in Sect. 3, while TripleCheckMate only supports DBpedia.

3   Methods of Evaluation

3.1   The Target to Evaluate

Generally the KB can be viewed as a graph G, as shown in Fig. 1. The graph G consists of schema graph G_s, data graph G_d and the relations R between G_s and G_d, denoted as $G = <G_s, G_d, R>$.

The schema graph $G_s = <N_s, P_s, E_s>$, where N_s is a set of nodes representing classes; P_s is a set of nodes representing properties, and P_s contains

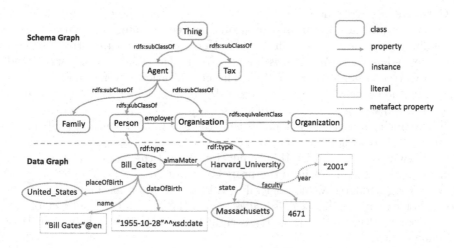

Fig. 1. Conceptual representation of an extracted KB

---

[7] http://linkeddata.informatik.hu-berlin.de/LDSrcAss/datenquelle.php.

rdfs:subClassOf, rdfs:equivalentClass and other user defined properties such as *employer*; and E_s is a set of edges representing the relationships between classes in the graph G_s. $E_s \subseteq N_s \times P_s \times N_s$. e.g. The domain of the property *employer* in Fig. 1 is *Person* and the range of the property is *Organisation*.

The data graph $G_d = <N_d, P_d, E_d>$, where N_d is a set of nodes representing instances and literals; P_d is a set of nodes representing properties; and E_d is a set of edges representing the relationships between nodes in the graph G_d. Actually each edge (Subject, Predicate, Object) stands for a fact. N_d includes two disjoint parts, namely instances (N_i) and literals.

The relations R between schema graph G_s and data graph G_d are the relations which links the instances in data graph to classes in schema graph by the *rdf:type* property. R = {(instance, *rdf:type*, class)|instance $\in N_i$, class $\in N_s$}.

3.2   Metrics

The metrics used in this paper are listed in Table 1. We defined these metrics on the whole KB, including data graph, schema graph and the relations between the two. However, our evaluation focuses on data graph, since many Chinese KBs have no schema as shown in Table 3.

The *Data Size* and *Schema Size* metrics give us a general idea on how large the target KBs are. We further extend data size metrics to different domain as *Data Size over Domain* metrics. Besides, we use *Degree of Network* to measure connections between instances and classes. *Instantiation Ratio* is used to measure linkages between schema graph and data graph. We use *Average Fact Number (AFN)* metric to calculate average outgoing degrees of instances. Since the data graph of our KBs is fairly sparse, we use metrics *Zero-out Instances* and *Zero-in Instances* to find ratio of nodes with zero outgoing degree and zero ingoing degree.

Correctness metrics describe ratio of errors in KBs. The errors may appear in schema graph, data graph and relations between schema and data, as corresponding correctness metrics in Table 1. Here we focus on the errors in data graph, defined as *Data Correctness* Metric. We further classify the errors related to Data Correctness Metric in Table 2. The classification is very easy to understand. For the edge (S, P, O) in data graph, the errors in P or O both lead to the edge (fact) wrong. The details of the six types of errors are as below.

(a) No Such Property (NSP) means that the P is wrong, which indicates that the instances should not have this property. For example, (李义伟, width, 150^^ <http://www.w3.org/2001/XMLSchema#int>), and the subject of the fact is a person, which should not have the property "width".

(b) Wrong Property Value Logically (WPVL) means that the O is logically wrong. For example, (沈从文别集, 页数, 全20 册@zh), and the subject of the fact is a book. Although the property is right for the instance, it is obvious that the value is wrong.

(c) Wrong Object Type (WOT) means that the O is correct, but its type (instance or literal) is wrong. For example, (任永顺, countryofbirth, 中国@zh). The object of the fact should be a instance instead of a literal.

Table 1. Metrics definitions

Metric Set	Category	Metrics	Description
Richness	Data Size	Number of Instances (NOI)	Total number of instances in KB $\quad |N_i|$
		Number of Facts (NOF)	Total number of facts in KB $\quad |E_s|$
	Schema Size	Number of Classes (NOCL)	Total number of classes in KB $\quad |N_s|$
		Number of Categories (NOC)	Total number of categories in KB $noc = \{C \mid (i, dct:subject,\ C) \in E_d\} \quad |noc|$
		Number of Properties (NOP)	Total number of properties in KB $\quad |P_d|$
	Degree of Network	Average Fact Numbers (AFN)	Average fact number per instance $\quad |E_s| / |N_i|$
		Zero-out Instances (ZOI)	Percentage of instances having no facts $zoi = \{i \mid out-degree(i) = 0,\ i \in N_i\} \quad |zoi| / |N_i|$
		Zero-in Instances (ZII)	Percentage of instances that are not property values of any instances. $zii = \{i \mid in-degree(i) = 0,\ i \in N_i\} \quad |zii| / |N_i|$
		Instantiation Ratio	Percentage of instances which have rdf:type relations to classes $ir = \{i \mid (i, rdf:type,\ C),\ C \in N_s,\ i \in N_i\} \quad |ir| / |N_i|$
	Data Size over Domain	Number of Instances over Domain (NOID)	Total number of instances in particular domain C, $C \in N_s$ $i(C) = \{i \mid (i, rdf:type, c), c \in C_s, C_s$ is the set of descendant classes of C$\} \quad |i(C)|$
		Number of Facts over Domain (NOPD)	Total number of facts in particular domain C $f(C) = \{(i, p, n) \mid (i, p, n) \in E_d,\ i \in i(C)\} \quad |f(C)|$
Correctness	Data Correctness	Correctness Ratio of Instances (CROI)	The percentage of correct facts of an instance, by average. $\frac{1}{n}\sum_{i=1}^{n}\left(1 - \frac{The\ number\ of\ wrong\ facts\ of\ instance\ i}{The\ number\ of\ facts\ of\ instance\ i}\right)$
		Correctness Ratio of Facts (CROF)	The percentage of correct facts. $1 - \frac{The\ number\ of\ wrong\ facts}{The\ number\ of\ all\ sampled\ facts}$
	Schema Correctness	Correctness Ratio of subClassOf Property (CROCP)	The percentage of correct subClassOf properties between classes $1 - \frac{The\ number\ of\ wrong\ subClassOf\ properties\ between\ classes}{The\ number\ of\ all\ sampled\ subClassOf\ properties\ between\ classes}$
	Relation Correctness	Correctness Ratio of rdf:type Property (CROTP)	The percentage of correct rdf:type properties between classes and instances $1 - \frac{The\ number\ of\ wrong\ rdf:type\ properties\ between\ classes\ \&\ instances}{The\ number\ of\ all\ sampled\ rdf:type\ properties\ between\ classes\ \&\ instances}$
Comparative metrics	Richness Comparison (Data)	Overlapped Instances (OI)	Number of overlapped instances for some KBs $\quad |I_k|$
		Overlapped Instances over Domain (OIOD)	Number of overlapped instances in KB K for particular domain C $oiod(K,C) = \{i \mid i \in I_k, (i, rdf:type,\ C) \in E_d\} \quad |oiod(K,C)|$
		Facts of Overlapped Instances (FOI)	Number of facts of overlapped instances for KB K $foi(K,C) = \{(i, p, n) \mid (i, p, n) \in E_d,\ i \in I_k\} \quad |foi(K,C)|$
	Data Correctness Comparison	Correctness Ratio of Facts over Overlapped Instances (CRFO)	The percentage of correct facts of overlapped instances. $1 - \frac{The\ number\ of\ wrong\ facts}{The\ number\ of\ all\ sampled\ facts}$
		Correctness Divergence (CD)	Distribution divergence on correctness of facts over overlapped instance of different KBs.

(d) Wrong Literal Format (WLF) means that the value of the datatype property has wrong literal format. For example, (洗车河, length, 86^^ <http://www.w3.org/2001/XMLSchema#integer>). The subject of the fact is a river, and its length maybe 86 km, so the value of its length has a wrong format.

(e) Lack of Context (LOC) means that the property value is ambiguous if there is no context. For example, (华盛顿·斯泰卡奈罗·塞凯拉, isAffiliatedTo, 富明尼斯足球), and it means a player is affiliated to a sport team. But he could not serve for the team forever, and it could be a time slot. So some annotated information are needed to supplement the fact, namely *metafact*.

(f) Out of Date or Data Source Error (ODDS) means the fact is wrong because it is not synchronized with the newest extract source or the source itself is wrong. As to the Out of Date error, for example, (亚马维尔省, hasNumberOf-People, 282600^^xsd:nonNegativeInteger), and the value is out of date.

The error type classification can be helpful for error understanding and tracing. It could be used to find possible ways to improve KBs. For example, *Wrong Literal Format (WLF)* can be discovered automatically when all properties are restricted by domain and range.

Table 2. Error definitions

No.	Error codes	Description
1	No Such Property (NSP)	The property itself is meaningless and has no semantic relation with particular instance.
2	Wrong Property Value Logically (WPVL)	The value of a property is logically wrong.
3	Wrong Object Type (WOT)	An object property has Literal Facts and vice versa.
4	Wrong Literal Format (WLF)	A datatype property has facts with wrong literal format.
5	Lack of Context (LOC)	The property value is ambiguous if no context is given.
6	Out of Date or Data Source Error (ODDS)	The data is wrong because it is not synchronized with the newest Wikipedia or the Wikipedia itself is wrong.

For *Metrics on Overlapped Instances*, overlapped instances of different KBs are collected by instance matching. The overlapped instances in one KB is denoted as set I_1, and another as I_2. $|I_1| = |I_2|$. $\forall i_1 \in I_1$, $\exists i_2 \in I_2$ and (i_1, *owl:sameAs*, i_2), and vice versa.

To measure whether KBs are complementary on correctness, we design a metric which measures the divergence on correctness, named *Correctness Divergence* (CD), as formula (1) shows.

$$CD(m,n) = \frac{\sum_{f \in facts(m,n))} Diff(correct(f,m), correct(f,n))}{|facts(m,n)|} \tag{1}$$

$facts(m,n)$ is the set of all facts belonging to the overlapped instances of KBs m and n. $correct(f,m) \in \{0,1\}$ represents whether the fact f is evaluated as correct or not. The value of $diff(s,t)$ is 0 if $s = t$, else is 1. The metric checks whether correctly extracted facts in one KB are also extracted correctly in another, and vice versa. The range of CD is $[0,1]$, where 0 means the two KBs extract data from Encyclopedic web sites in the same way, and 1 means the two KBs are fully complementary.

3.3 Evaluation Process

Quality evaluation process consists of six steps, *Import KBs, Select Metrics, Instance Matching, Calculate Metrics, Visualize Metrics Results, Compare/Analyze results*, as shown in Fig. 2. For the *Import KBs* step, serialized text data is imported into Jena Graph database, as most LOD data sets are in text format such as *ttl* and *nt*. Then we can use SparQL on Jena to calculate *Machine-Computable* metrics such as Richness. If metrics on *Overlapped Data* are selected, additional instance matching step is required as presented in Sect. 3.5.

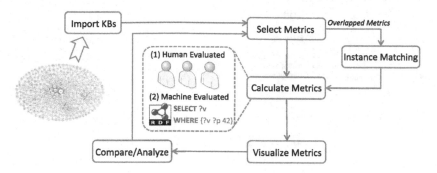

Fig. 2. The process of KB evaluation

3.4 The Process of Calculating Human Evaluated Metrics

The general process of calculating human evaluated metrics is shown in Fig. 3. Here we use human evaluated metrics to denote the metrics which can not be easily calculated by machine and have to be evaluated by human. It's beyond the scope of this paper to define which metrics should be evaluated by human. However, metrics evaluated by human may require too much human efforts and suffer from individual subjectivity from evaluators. In this case, to lessen human efforts and to reduce individual subjectivity are two major challenges in calculating human evaluated metrics. If we use sampling to lessen human efforts, we have to determine the number of samples required and sample targets such as fact-based, instance-based, property-based or other else. In this paper, it is assumed all possible M samples of objects are equally likely to occur, as we do not prefer one triple to the other. In that case, random sampling is used[8]. The formula is:

$$n = \left[\left(z^2 * p * q \right) + ME^2 \right] / \left[ME^2 + z^2 * p * q/N \right] \qquad (2)$$

In this formula, N is size of the data set, ME is the marginal of error. It is assumed we do not have prior knowledge of the correctness ratio, and p is set to 0.5 in default. User could input Alphas equals (1-confidence level) for an estimation problem, and z can be derived from Alpha. Here Normal distribution is assumed, as the size of LOD is always very large. As to the sampling objects, we take two approaches, random sample on facts of each KB, and random sample on overlapped instances of KBs. YAGO took another approach, it sampled on the facts of the same properties.

Since human evaluator may give opposite results on the same data item, we have to employ more than one person to reduce individual subjectivity. In our experiments, each data item is evaluated by three persons, and if inconsistency exists among these persons, another domain expert does the final judgement. Furthermore, to make evaluation fairness on different KBs, we require each evaluator evaluate equal portion of instances/facts of each KB. For example, if we have 6 evaluators, each data item should be evaluated by three evaluators, and

---

[8] http://stattrek.com/sample-size/simple-random-sample.aspx.

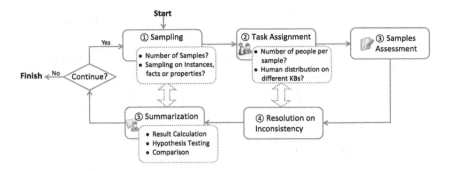

Fig. 3. The process of human evaluated metrics

the size of target KBs are A, B and C. Then each evaluator should evaluate data size of A/2, B/2 and C/2.

3.5 Instance Matching

We use two methods to collect the overlapped instances. One is directly using the *owl:sameAs* relation in KBs, and we obtain the overlapped instances between DBpedia English and DBpedia Chinese with the *owl:sameAs* links in the two KBs. For KBs containing no *owl:sameAs* links, we use a heuristic rule. If labels of the instance in different KBs are same, we regard the instances as same. However, the formats of Chinese label of instances are different in KBs. Take the woman 网球运动员 (tennis player)" 李娜(LINA)" as an example, DBpedia Chinese is labeled as "李娜 (网球运动员)" YAGO is "李娜(网球运动员)" zhishi.me is "李娜 (网球运动员)" and SSCO is " 李娜(网球运动员)". We use a pre-processing step to replace different separators with a unified one. We take the second approach in finding overlapped instances of DBpedia Chinese, zhishi.me and SSCO. The precision of this approach is more than 98 %. While the recall may be low, the approach has successfully collected 49,333 instances, which is adequate for our assessment purpose.

4   Evaluation Results

4.1 General Comparisons on the KBs

We mainly evaluate and compare four KBs, DBpedia English (DBpedia-EN), DBpedia Chinese (DBpedia-CN), SSCO and zhishi.me. We use YAGO2s and DBpedia 2014 version. We also take YAGO at a glance. While YAGO is basically in English and does not have separate Chinese dumps, Chinese alias of instances and concepts are given in separate files via *rdfs:label* and language tags in YAGO. We are interested in the quality of these labels, so that we filtered out the instances and classes in YAGO containing Chinese label and name the KB as YAGO-CN.

Availability. All these KBs can be assessed from their Web sites by http URIs. DBpedia-EN, DBpedia-CN, zhishi.me and YAGO have SparQL endpoints to execute SparQL queries, while SSCO does not provide any SparQL service. The RDF dumps of DBpedia-EN, DBpedia-CN can be downloaded, but the download versions of zhishi.me and SSCO are not available due to the license problem mentioned below.

License. The DBpedia data sets are licensed under the terms of the Creative Commons Attribution-ShareAlike License and the GNU Free Documentation License. YAGO is licensed under a Creative Commons Attribution 3.0 License by the YAGO team of the Max-Planck Institute for Information. SSCO and zhishi.me are extracted from Baidu-Baike which is a commercial encyclopedia so that SSCO and zhishi.me can not be used without Baidu's permission.

Chinese Data Representation. The linked localized versions of DBpedia are available in 125 languages. The same instances of different language versions are linked by *owl:sameAs*, such as <http://zh.dbpedia.org/resource/北爱尔兰> owl:sameAs <http://dbpedia.org/resource/Northern_Ireland>. Besides, there exist localized textual descriptions such as *rdfs:label*, *rdfs:comment* and *dbpedia-owl:abstract* in DBpedia English. YAGO adopts another approach. All names of entities and classes in YAGO are in English, and some of them have *rdfs:label* values in localized languages. Therefore YAGO has no separate versions in Chinese. SSCO and zhishi.me are solely Chinese KBs, and they have no multilingual descriptions such as *rdfs:label*. Data in zhishi.me are mostly in Chinese. Since Chinese Wikipedia contains many vocabularies in English, the two KBs also are mixed with English vocabularies.

Schema. The comparisons of KBs regarding to schema are shown in Table 3. Generally speaking, English KBs contain much more schema information than Chinese KBs. DBpedia Ontology, which could be considered as part of DBpedia-EN, contains altogether 683 classes and 2795 properties. However, DBpedia contains more than 50,000 distinct properties in its fact files. The schema graph and data graph are more consistent in YAGO. YAGO defines 75 properties and its fact files contain no other properties. There are no direct *rdf:type* relations between DBpedia-CN and DBpedia ontology. While part of instances from DBpedia-CN have *owl:sameAs* relationship with DBpedia-EN, so instances in DBpedia-CN link to DBpedia ontology indirectly, and we take DBpedia-CN as schemaless as a whole. DBpedia-CN, zhishi.me and SSCO only have "category" instead of "class". The categories are derived from categories of articles in Wikipedia. Furthermore, properties of Chinese KBs are not defined, which is different from English KBs.

InterLinks. SSCO and zhishi.me are extracted from three Encyclopedic Web sites, while DBpedia-CN is only from Wikipedia. Zhishi.me and SSCO consolidate the three sources in different ways. Zhishi.me use *owl:sameAs* relationship to link the same instances from different sources, while SSCO directly consolidates them into one.

Table 3. Comparing of the KBs regarding the schema

	DBpedia-EN	DBpedia-CN	SSCO	zhishi.me	YAGO
classes	yes	no	no	no	yes
rdfs:subClassOf relation	yes	no	no	no	yes
rdfs:equivalentClass relation	yes	no	no	no	no
rdfs:subPropertyOf relation	yes	no	no	no	yes
property domain	yes	no	no	no	yes
property range	yes	no	no	no	yes
relation between instance and class	yes	no	no	no	yes
rdf:sameAs relation	yes	yes	yes	yes	yes

4.2   Evaluation Results of Metrics

4.2.1   Comparison of Richness Between DBpedia-CN and DBpedia-EN

DBpedia-EN is richer than DBpedia-CN, as shown in Fig. 4(a). The numbers of instances, properties, facts and categories of DBpedia-EN are much larger than those of DBpedia-CN. Instances in DBpedia-EN have more facts too. DBpedia-EN also covers most of the instances and facts in DBpedia-CN. The number of overlapped instances of two KBs is 432,135, accounting for 63 % instances of DBpedia-CN, accounting for 10 % of DBpedia-EN. Furthermore, on the overlapped instances, DBpedia-EN has about two times the number of facts as much as DBpedia has.

DBpedia-CN has its Own Data. For example, the instance <http://zh.dbpedia.org/resource/秦汉三国历史年表> only appears in DBpedia-CN, and this instance is related to Chinese history. Furthermore, property values in Chinese sometimes may be of higher quality than those in English. Take the instance of "Albert_Einstein" as an example, the property value of "almaMater" in English is "* ETH Zurich \n * University of Zurich" in literal, while its value in Chinese is <http://zh.dbpedia.org/resource/ETH> and

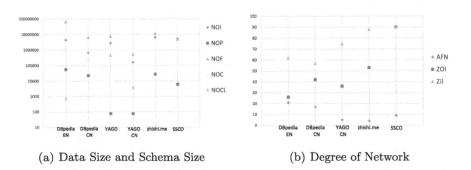

(a) Data Size and Schema Size   (b) Degree of Network

Fig. 4. Results of data size, schema size and degree of network

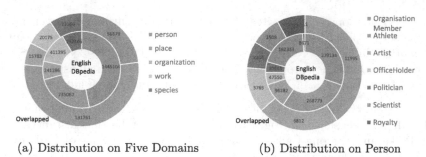

(a) Distribution on Five Domains (b) Distribution on Person

Fig. 5. Instance distribution on domain

<http://zh.dbpedia.org/resource/University_of_Zurich> links to a resource. In this case, DBpedia-CN exemplifies the design principle of Linked Open Data.

DBpedia-CN and DBpedia-EN are fairly different, referring to the distribution of overlapped instances on major domains. We use *rdfs:type* relation in DBpedia-EN to calculate the *Data size on Domain* metrics. The inner circle of Fig. 5(a) shows the percentage of five major domains of DBpedia-EN, ranking from high to low. They are "person", "place", "organisation", "work" and "species". DBpedia-CN does not have *rdfs:type* relation and we infer the difference between the KBs from the overlapped instances. We found the percentage of "person" and "work" in DBpedia-EN are smaller than that in overlapped data sets, but the "place" is on the contrary. The phenomenon leads us to think that "person" and "work" are more language-specific, and "place" about geographic information is more internationalized. The distribution of instances on subclasses of "person" is shown in Fig. 5(b), and we further find "OrganisationMember", the subdomain of "person", is more locally values than other subdomains.

4.2.2 Comparison of Richness Between Chinese KBs

Data Size and Schema size Fig. 4(a) shows the number of facts(NOF), instances (NOI), categories(NOC) and properties(NOP) of four KBs. It's no surprising that zhishi.me has the largest number on the four metrics, since it collected data

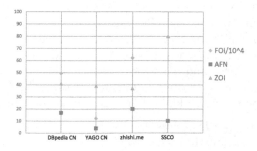

Fig. 6. The results of data richness on overlapped instances

from three Encyclopedic Web sites. The NOI value of zhishi.me is even larger than that of DBpedia-EN. However, while the NOI of SSCO is almost as much as that of zhishi.me, the NOF of SSCO is less than DBpedia-CN, since SSCO filtered out low frequency properties. The size of YAGO-CN in all aspects is very small, we found only 5 % of YAGO's instances contain Chinese label.

Degree of Network. The AFN metric of zhishi.me and SSCO are smaller than that of DBpedia-CN, as shown in Fig. 4(b). The reason may be that instances from Baidu-Baike and Hudong-Baike contain little facts. The values of ZOI metric show that 53 % of the instances of zhishi.me have no facts and only 10 % instances of SSCO have facts. In general, the instances in DBpedia-CN are linked to each other tighter than those in SSCO and zhishi.me.

Data Richness on Overlapped Instances. We find 49,333 matched instances with our heuristic instance matching algorithm proposed in Sect. 3.5. Figure 6 shows the Data Richness metrics on these overlapped instances. The AFN of zhishi.me on overlapped instances is larger than that of DBpedia-CN, which is opposite to the situation on the whole KBs, the reason is that zhishi.me has a large amount of instances with no facts in the whole KB. Take the overlapped instance "利耶帕亚国际机场" for example. This instance has 15 facts in DBpedia-CN, 1 in YAGO, 17 in zhishi.me and zero in SSCO. And the 17 facts in zhishi.me contain the 15 facts in DBpedia-CN.

4.2.3   Comparison of Correctness

Correctness on Sampled Facts We sampled 423 triples from each KB according to the random sampling, and the results of correctness are shown in Fig. 7. (a)DBpedia-CN and DBpedia-EN have the similar level of correctness statistically via hypothesis testing[9]. We find there is no significant difference between the two KBs when Significance Level is set to 0.1 using two-proportion z-test. (b) Many facts in DBpedia-CN have property names in English and property values in Chinese. We trace the reason, and find that the original property names in the infobox of Wikipedia Chinese are in English. Although we label them as correct, these facts are difficult to understand to native Chinese. (c) The correctness ratio of SSCO is much higher than that of zhishi.me and DBpedia-CN, and is comparable to that of YAGO. (d) YAGO's accuracy is 88.4 % in our evaluation but 95.02 % in its web site. One reason of the gap may be due to different evaluation criterion. For example, we take metafact into correctness consideration. e.g. An athlete belongs to a club during a certain period of time. YAGO provides meta property called "occursSince" and "occursUntil". However, some metafacts may only contain "occursSince" and miss "occursUntil", and in such cases we judge some of the facts as wrong. Since the evaluation taken by YAGO themselves sampled [8] on properties, such kind of errors may not be found.

Correctness on Overlapped Instances. We further sampled 100 instances on overlapped instances of the four KBs. In general, the correctness ratios are similar

---

[9] http://stattrek.com/hypothesis-test/difference-in-proportions.aspx.

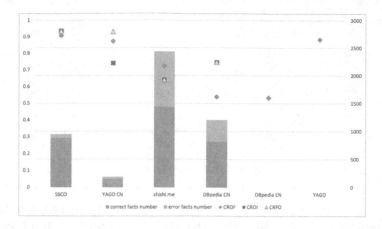

Fig. 7. The result of correctness. The bar in the figure is corresponded to the right Y-axis, and the point in the figure is corresponded to the left.

to the results of random sampling on facts. Zhishi.me has the largest number of correct facts, while SSCO has the highest correct ratio. With the comparison of similar instances, we can find the reason why SSCO achieves high in correctness. Since SSCO has very small number of facts, we infer that SSCO compares data from different sources and filters out the low frequency properties. With data from Baidu-Baike and Hudong-Baike, SSCO also successfully replaced property named in English with property named in Chinese for some instances. We guess that zhishi.me and DBpedia do not have such filtering mechanism so that they have more extracted data with less accuracy.

Correctness Divergence (CD). The values of the metric between zhishi.me and DBpedia, SSCO and zhishi.me, DBpedia and SSCO are 0.41, 0.62 and 0.76. It shows that these KBs are complementary, especially for DBpedia and SSCO.

Classification of Errors. Figure 8(a) and (b) show the detailed error types on different KBs. Figure 8(a) is the result of CROF (Correctness Ratio of Facts) which samples 423 triples for each KB randomly and Fig. 8(b) is the result of CRFO (Correctness Ratio of Facts over Overlapped Instances) which samples 100 overlapped instances randomly. With the help of error classification, we find that error distribution are different in different KBs. We can also infer that each KB has its own ways to extract and process data.

(a) No Such Property (NSP). NSP happens when the extractor extracts layout information hidden in Wikipedia's pages as property names by mistaken, such as "imagewidth" and "downloaded". Another reason may be that a triple should be a metafact instead of a fact. For example, we find "獴" has an property called *wasCreatedOnDate*, actually the property refers to the creation date of binomial nomenclature of "獴". These two situation frequently occur in DBpedia-CN and zhishi.me. We also find that Hudong-Baike contains many incorrect information, which leads errors in SSCO and zhishi.me. For instance, the infor-

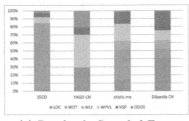
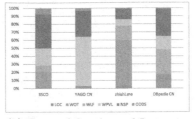

(a) Randomly Sampled Facts   (b) Facts of Overlapped Instances

Fig. 8. Errors classification

mation in the Hudong-Baike page of "越南人民军" (Vietnam People's Army)"
actually describes "越南" (Vietnam)". **(b) Wrong Property Value Logically
(WPVL).** The reason for this error is mainly due to the extracted methods.
For example, for the wrong triple (universe Jones, Duration, "USA"), the proper
value should be "USA: 128min". We also find YAGO has many incorrect Chinese labels. For example, Australi is labeled as "澳大利亚经济" (Economics of
Australia)" @zho. **(c) Wrong Object Type (WOT).** Zhishi.me and SSCO
have more instances with this type of error than DBpedia-CN, as many literal
data in Hudong-Baike are represented as links so that the KBs take them as
resources. **(d) Wrong Literal Format (WLF).** The error happens mostly in
DBpedia-CN and zhishi.me, since they neglect the unit of numerical data. SSCO
seldom has this type of error and we infer from the data that SSCO has special algorithm to extract the numerical data and its unit as well. **(e) Lack of
Context (LOC).** Only YAGO has metafact facility to encode such situation.
DBpedia and zhishi.me extracts the contextual information incorrectly. SSCO
just discards certain data sources (usually tables) which contain triples relying
on contexts.

5   Conclusion and Future Work

This paper focuses on evaluating KBs extracted from Encyclopedia sites. As the
experiment shows, the metric sets we designed can give us a quick survey of the
characteristic of extracted KBs. The sampling and task assignment techniques
employed greatly facilitate metrics calculating. Furthermore, metrics on overlapped instances not only help us to compare the differences on these KBs, but
also give many hints on the construction mechanisms of these KBs. In the future
we plan to evaluate more KBs and try to construct new KBs from existing ones
based on the comparative metrics in our paper.

References

1. Bizer, C., Lehmann, J., Kobilarov, G., Auer, S., Becker, C., Cyganiak, R., Hellmann, S.: Dbpedia-a crystallization point for the web of data. Web Semant.: Sci.
   Serv. Agents World Wide Web **7**(3), 154–165 (2009)

2. Suchanek, F.M., Kasneci, G., Weikum, G.: Yago: a core of semantic knowledge. In: Proceedings of the 16th international conference on World Wide Web, pp. 697–706. ACM (2007)
3. Niu, X., Sun, X., Wang, H., Rong, S., Qi, G., Yu, Y.: Zhishi.me - Weaving chinese linking open data. In: Aroyo, L., Welty, C., Alani, H., Taylor, J., Bernstein, A., Kagal, L., Noy, N., Blomqvist, E. (eds.) ISWC 2011, Part II. LNCS, vol. 7032, pp. 205–220. Springer, Heidelberg (2011)
4. Hu, F., Shao, Z., Ruan, T.: Self-supervised chinese ontology learning from online encyclopedias. The Scientific World Journal, Accepted
5. Wang, Z., Li, J., Wang, Z., Li, S., Li, M., Zhang, D., Shi, Y., Liu, Y., Zhang, P., Tang, J.: Xlore: A large-scale english-chinese bilingual knowledge graph. In: Proceedings of the International Semantic Web Conference (2013)
6. Zaveri, A., Rula, A., Maurino, A., Pietrobon, R., Lehmann, J., Auer, S.: Quality assessment methodologies for linked open data. SWJ (2012)
7. Kontokostas, D., Zaveri, A., Auer, S., Lehmann, J.: TripleCheckMate: A tool for crowdsourcing the quality assessment of linked data. In: Klinov, P., Mouromtsev, D. (eds.) KESW 2013. CCIS, vol. 394, pp. 265–272. Springer, Heidelberg (2013)
8. Hoffart, J., Suchanek, F.M., Berberich, K., Weikum, G.: Yago2: A spatially and temporally enhanced knowledge base from wikipedia. In: Proceedings of the Twenty-Third International Joint Conference on Artificial Intelligence, pp. 3161–3165. AAAI Press (2013)
9. Ell, B., Vrandečić, D., Simperl, E.: Labels in the web of data. In: Aroyo, L., Welty, C., Alani, H., Taylor, J., Bernstein, A., Kagal, L., Noy, N., Blomqvist, E. (eds.) ISWC 2011, Part I. LNCS, vol. 7031, pp. 162–176. Springer, Heidelberg (2011)
10. Zhang, H., Li, Y.F., Tan, H.B.K.: Measuring design complexity of semantic web ontologies. J. Syst. Softw. **83**(5), 803–814 (2010)
11. Guéret, C., Groth, P., Stadler, C., Lehmann, J.: Assessing linked data mappings using network measures. In: Simperl, E., Cimiano, P., Polleres, A., Corcho, O., Presutti, V. (eds.) ESWC 2012. LNCS, vol. 7295, pp. 87–102. Springer, Heidelberg (2012)
12. Hogan, A., Harth, A., Passant, A., Decker, S., Polleres, A.: Weaving the pedantic web. In: Proceedings of the WWW Workshop on Linked Data on the Web (2010)
13. Färber, M., Ell, B., Menne, C., Rettinger, A.: A comparative survey of dbpedia, freebase, opencyc, wikidata, and yago
14. Zaveri, A., Kontokostas, D., Sherif, M.A., Bühmann, L., Morsey, M., Auer, S., Lehmann, J.: User-driven quality evaluation of dbpedia. In: Proceedings of the 9th International Conference on Semantic Systems, pp. 97–104. ACM (2013)
15. Wienand, D., Paulheim, H.: Detecting incorrect numerical data in DBpedia. In: Presutti, V., d'Amato, C., Gandon, F., d'Aquin, M., Staab, S., Tordai, A. (eds.) ESWC 2014. LNCS, vol. 8465, pp. 504–518. Springer, Heidelberg (2014)
16. Paulheim, H., Bizer, C.: Improving the quality of linked data using statistical distributions. Inter. J. Semant. Web Inf. Syst. (IJSWIS) **10**(2), 63–86 (2014)
17. Kontokostas, D., Westphal, P., Auer, S., Hellmann, S., Lehmann, J., Cornelissen, R., Zaveri, A.: Test-driven evaluation of linked data quality. In: Proceedings of the 23rd International Conference on World Wide Web, pp. 747–758. ACM (2014)
18. Dong, X., Gabrilovich, E., Heitz, G., Horn, W., Lao, N., Murphy, K., Strohmann, T., Sun, S., Zhang, W.: Knowledge vault: A web-scale approach to probabilistic knowledge fusion. In: Proceedings of the 20th ACM SIGKDD International Conference on Knowledge Discovery and Data Mining, pp. 601–610. ACM (2014)

19. Mendes, P.N., Mühleisen, H., Bizer, C.: Sieve: linked data quality assessment and fusion. In: Proceedings of the 2012 Joint EDBT/ICDT Workshops, pp. 116–123. ACM (2012)
20. Bizer, C., Cyganiak, R.: Quality-driven information filtering using the wiqa policy framework. Web Semant.: Sci. Serv. Agents World Wide Web 7(1), 1–10 (2009)
21. Rieß, C., Heino, N., Tramp, S., Auer, S.: EvoPat – Pattern-based evolution and refactoring of RDF knowledge bases. In: Patel-Schneider, P.F., Pan, Y., Hitzler, P., Mika, P., Zhang, L., Pan, J.Z., Horrocks, I., Glimm, B. (eds.) ISWC 2010, Part I. LNCS, vol. 6496, pp. 647–662. Springer, Heidelberg (2010)

Computing the Semantic Similarity of Resources in DBpedia for Recommendation Purposes

Guangyuan Piao[(✉)], Safina showkat Ara, and John G. Breslin

Insight Centre for Data Analytics, National University of Ireland Galway,
IDA Business Park, Lower Dangan, Galway, Ireland
{guangyuan.piao,safina.ara}@insight-centre.org,
john.breslin@nuigalway.ie

Abstract. The Linked Open Data cloud has been increasing in popularity, with DBpedia as a first-class citizen in this cloud that has been widely adopted across many applications. Measuring similarity between resources and identifying their relatedness could be used for various applications such as item-based recommender systems. To this end, several similarity measures such as *LDSD* (*Linked Data Semantic Distance*) were proposed. However, some fundamental axioms for similarity measures such as "equal self-similarity", "symmetry" or "minimality" are violated, and property similarities have been ignored. Moreover, none of the previous studies have provided a comparative study on other similarity measures. In this paper, we present a similarity measure, called *Resim* (*Resource Similarity*), based on top of a revised *LDSD* similarity measure. *Resim* aims to calculate the similarity of any resources in DBpedia by taking into account the similarity of the properties of these resources as well as satisfying the fundamental axioms. In addition, we evaluate our similarity measure with two state-of-the-art similarity measures (*LDSD* and *Shakti*) in terms of calculating the similarities for general resources (i.e., any resources without a domain restriction) in DBpedia and resources for music artist recommendations. Results show that our similarity measure can resolve some of the limitations of state-of-the-art similarity measures and performs better than them for calculating the similarities between general resources and music artist recommendations.

Keywords: Similarity measure · Recommender system · DBpedia

1 Introduction

The term Web of Data, often referred to as the Semantic Web, Web 3.0 or Linked Data, indicates a new generation of technologies responsible for the evolution of the current Web [10] from a Web of interlinked documents to a Web of interlinked data. The goal is to discover new knowledge and value from data, by publishing them using Web standards (primarily RDF [4]) and by enabling connections between heterogeneous datasets. In particular, the term Linked Open Data (LOD) denotes a set of best practices for publishing and linking structured data

© Springer International Publishing Switzerland 2016
G. Qi et al. (Eds.): JIST 2015, LNCS 9544, pp. 185–200, 2016.
DOI: 10.1007/978-3-319-31676-5_13

on the Web. The project includes a great amount of RDF datasets interlinked with each other to form a giant global graph, which has been called the Linked Open Data cloud[1]. DBpedia[2] is a first-class citizen in the LOD cloud since it represents the nucleus of the entire LOD initiative [3]. It is the semantic representation of Wikipedia[3] and it has become one of the most important and interlinked datasets on the Web of Data. Compared to traditional taxonomies or lexical databases (e.g., WordNet [17]), it provides a larger and "fresher" set of terms, continuously updated by the Wikipedia community and integrated into the Web of Data. A resource in DBpedia represents any term/concept (e.g., Justin Bieber) as a dereferenceable URI (e.g., http://dbpedia.org/resource/Justin_Bieber) and provides additional information related to the resource. We use the prefix *dbpedia* for the namespace http://dbpedia.org/resource/ in the rest of this paper. That is, *dbpedia:Justin_Bieber* denotes http://dbpedia.org/resource/Justin_Bieber.

On top of DBpedia, many approaches from different domains have been proposed by manipulating DBpedia resources and the relationships among them. For example, the resources can be used to represent a multi-domain user profile of interests across different Online Social Networks. In this case, an interest can be represented by a resource in DBpedia and the interest could be any topical resource that the user is interested in (e.g., *dbpedia:Justin_Bieber* or *dbpedia:Food*). Then, the user profile of interests can be used for personalization or recommendations [1,2,20]. It also has been widely adopted for improving the performance of recommender systems [6,11,18,21,22]. For instance, Heitmann et al. [11] proposed building open recommender systems which can utilize Linked Data to mitigate the *sparsity* problem of collaborative recommender systems [14].

Measuring similarity between resources and identifying their relatedness could be used for various applications, such as community detection in social networks or item-based recommender systems using Linked Data [23]. In this regard, several similarity measures were proposed for item-based recommendations [9,13,23,24]. However, none of these studies evaluated over one or some of other similarity measures. Instead, each study proposed its own evaluation method for its measure. Hence, the performance compared to other similarity measures were not proven.

Secondly, despite different aspects of relatedness were considered in different similarity measures, property similarity is not incorporated within these measures. The Merriam-Webster Dictionary[4] defines property as "a special quality or characteristic of something". Also, from the definition of an ontology, the properties of each concept describe various features and attributes of the concept [19]. Thus, property similarity is important when there is no direct relationship between resources. For example, the similarity of two resources *dbpedia:Food* and *dbpedia:Fruit* using the *LDSD* (*Linked Data Semantic Distance*) similarity measure [23] (we will discuss the similarity measure in detail in Sect. 3.1)

---

[1] http://lod-cloud.net/.

[2] http://wiki.dbpedia.org/.

[3] https://www.wikipedia.org/.

[4] http://www.merriam-webster.com/.

is 0 since there is no direct link or shared resource via any properties. Similarly, the similarity of *dbpedia:Food* and *dbpedia:Rooster* is 0 as well. As a result, the recommender system cannot recommend any items (e.g., adverts) on *dbpedia:Fruit* or *dbpedia:Rooster* to the user if he or she is interested in the topic of *dbpedia:Food*. This can be addressed by considering shared incoming/outgoing properties since *dbpedia:Fruit* and *dbpedia:Food* have incoming properties such as *dbpedia-owl:industry* and *dbpedia-owl:product* in common (the prefix *dbpedia-owl* denotes the namespace http://dbpedia.org/ontology/ in the rest of this paper).

Thirdly, some fundamental axioms [7] are violated for distance-based similarity measures such as:

- Equal self-similarity: $sim(A, A) = sim(B, B)$, for all stimuli A and B.
- Symmetry: $sim(A, B) = sim(B, A)$, for all stimuli A and B.
- Minimality: $sim(A, A) > sim(A, B)$, for all stimuli $A \neq B$.

These are also common axioms for all word similarity measures in WordNet [16] and popular similarity measures based on graphs such as *SimRank* [12]. However, we found that the state-of-the-art similarity measures such as *LDSD* or *Shakti* [13] for calculating resource similarity do not satisfy at least two of these axioms (we will discuss this in detail in Sect. 2).

In this paper, we propose a new similarity measure named *Resim* (*Resource Similarity*), which is built on top of a revised *LDSD* and incorporates the similarity of properties. In this regard, *Resim* has two major components, one is a revised *LDSD* similarity measure to satisfy the three axioms mentioned above, and the other is a newly proposed property similarity measure. We choose *LDSD* since it works well in single-domain recommendations (see Sect. 4) and also has comparable results to supervised learning approaches [5,6]. In addition, we compare and evaluate our similarity measure with *LDSD* and another state-of-the-art similarity measure named *Shakti*. These similarity measures were both devised for calculating the similarity between resources and recommendation purposes. To the best of our knowledge, this is the first comparative study of semantic similarity measures for DBpedia resources.

On top of that, we investigate if the performance of the item-based recommender system suffers from "*Linked Data sparsity*". Here, the *Linked Data sparsity problem* means that a lack of information on resources (e.g., small numbers of incoming/outgoing relationships from/to other resources) can decrease the performance of a recommender system.

The organization of the rest of the paper is as follows. In Sect. 2, we discuss related work on similarity measures of resources for recommendation purposes. In Sect. 3, we introduce our similarity measure - *Resim* - to calculate the similarity of resources in DBpedia. Section 4 elaborates on the experimental setup for the evaluation of our similarity measure with others, and Sect. 5 highlights the results. In Sect. 6, we study if Linked Data sparsity has an effect on the performance of the item-based recommender system that adopts the similarity measure for calculating the similarity between items (resources in DBpedia in this paper). Finally, Sect. 7 concludes the paper and gives some ideas for future work.

$$LDSD(r_a, r_b) = \cfrac{1}{1 + \sum_i \frac{C_d(l_i, r_a, r_b)}{1 + \log(C_d(l_i, r_a, n))} + \sum_i \frac{C_d(l_i, r_b, r_a)}{1 + \log(C_d(l_i, r_b, n))} \\ + \sum_i \frac{C_{ii}(l_i, r_a, r_b)}{1 + \log(C_{ii}(l_i, r_a, n))} + \sum_i \frac{C_{io}(l_i, r_a, r_b)}{1 + \log(C_{io}(l_i, r_a, n))}} \qquad (1)$$

$$LDSD_{sim}(r_a, r_b) = 1 - LDSD(r_a, r_b) \qquad (2)$$

2 Related Work

Maedche et. al [15] defined a set of similarity measures for comparing ontology-based metadata by considering different aspects of an ontology separately. They propose differentiating across three dimensions for comparing two resources: taxonomic, relational and attribute similarities. However, the similarity measures depend on some strong assumptions about the model such as "Ontologies are strictly hierarchical such that each concept is subsumed by only one concept", which is not the case in terms of DBpedia.

Passant [23] proposed a measure named $LDSD$ to calculate semantic distance on Linked Data. The distance measure (Eq. (1)) considers direct links from resource A to resource B and vice versa (C_d, C_{ii} and C_{io} functions are detailed in Sect. 3.1). In addition, it also considers the same incoming and outgoing nodes via the same properties of resources A and B in a graph (an example is given in Fig. 1). The distance measure has a scale from 0 to 1, where a larger value denotes less similarity between two resources. Thus, the similarity measure can be defined using (Eq. 2), and we will use $LDSD_{sim}$ to denote the similarity measure in the rest of the paper. In later work, the author used the $LDSD$ similarity measure in a recommender system based on DBpedia resources which recommends similar music artists based on the artists in a user's preference profile [22].

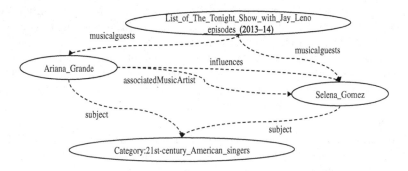

Fig. 1. Example of relationships of two resources in DBpedia

While $LDSD_{sim}$ works well in single-domain recommendations, there are several problems that need to be addressed. Since the measure is based on a count of direct/indirect links for resources, a higher number of these relationships can lead to higher similarity. However, the similarity would never be 1 even for the same resources, which leads to "different self-similarity". That is, $sim(r_a, r_a)$ and $sim(r_b, r_b)$ will be different even though both are close to 1. For instance, the similarity of *dbpedia:Doctor* with itself is 0.967 while the similarity of *dbpedia:Professor* with itself is 0.998. Also, the measure produces non-symmetric results for $sim(r_a, r_b)$ and $sim(r_b, r_a)$ (We will discuss this in detail in Sect. 3). Furthermore, it fails to calculate the similarities on general resource pairs (e.g., any resource used in DBpedia for representing a user's interests). For instance, both $sim(dbpedia:Doctor, dbpedia:Professor)$ and $sim(dbpedia:Doctor, dbpedia:Cucumber)$ will be 0, thus we cannot recommend items on some similar topics if *dbpedia:Doctor* is one of the interests of a user.

Leal et al. [13] presents an approach for computing the semantic relatedness of resources in DBpedia. In the paper, they proposed a similarity measure based on a notion of proximity, which measures how connected two resources are, rather than how distant they are. This means that the similarity measure considers both distance and the number of paths between two nodes. The similarity measure extends each step to find longer paths between two resources and penalizes proximity by steps, i.e., a longer path contributes less to the proximity and the extension is terminated by a defined value of maximum steps (max step). The similarity measure is implemented in a tool named "Shakti", which extracts an ontology for a given domain from DBpedia and uses it to compute the semantic relatedness of resources. We use *Shakti* to refer to this measure in the rest of the paper. However, they do not consider incoming nodes (resources) and properties of the resources as $LDSD_{sim}$ did. Furthermore, the proximity value for the same resources would be 0 since they will be removed before any extension. As a result, $sim(r_a, r_b) > sim(r_a, r_a)$ and thus violates the "minimality" axiom. In addition, the weights assigned to properties are defined manually and the authors pointed out the need for a sounder (automated) approach as future work.

Based on the *Shakti* measure, Strobin et al. [24] propose a method to find the weights automatically by using a genetic optimization algorithm based on a training dataset from Last.fm[5]. This method is quite efficient at learning the weights automatically. However, it needs a gold standard dataset (e.g., Last.fm dataset for music domain) to learn the weights of properties which is not always available in other domains.

For evaluation, every work proposes its own evaluation method for its measure and none of these studies have compared their proposed similarity measures to others. For example, some have evaluated the similarity measures in terms of specific domains of recommender systems [9,13,22,23] while others have evaluated them in terms of clustering problems [15].

In this work, we propose our similarity measure and also provide a comparative evaluation over $LDSD_{sim}$ and *Shakti* in terms of calculating the similarity of general resources (i.e., any resource in DBpedia without a domain restriction) and single-domain recommendations to examine the pros and cons of each measure.

---

[5] http://last.fm.

3   Resim Similarity Measure

In this section, we present a similarity measure named *Resim* (*Resource Similarity*) to calculate the similarity of resources in DBpedia. The method is built on top of the $LDSD_{sim}$ measure and resolve its aforementioned limitations. In this regard, we first discuss each component of $LDSD_{sim}$ in Sect. 3.1 and elaborate upon their limitations. Then we describe the components of *Resim* that resolve these limitations and also satisfy the axioms of "equal self-similarity", "minimality" (Sect. 3.2) and "symmetry" (Sect. 3.3). In addition, we present a method to calculate the property similarity of resources in Sect. 3.4 and present a final equation for *Resim* in Sect. 3.5 We use the definition of a dataset following the Linked Data principles outlined in [22].

Definition 1. A dataset following the Linked Data principles is a graph G such as $G = (R, L, I)$ in which $R = \{r_1, r_2, ...r_n\}$ is a set of resources identified by their URI, $L = \{l_1, l_2, ...l_n\}$ is a set of typed links identified by their URI and $I = \{i_1, i_2, ...i_n\}$ is a set of instances of these links between resources, such as $i_i = < l_j, r_a, r_b >$.

3.1   LDSD Similarity Measure

The $LDSD_{sim}$ measure (see Eqs. (1) and (2)) consists of two C_d functions with $C_{ii}(l_i, r_a, r_b)$ and $C_{io}(l_i, r_a, r_b)$. C_d is a function that computes the number of direct and distinct links between resources in a graph G. $C_d(l_i, r_a, r_b)$ equals 1 if there is an instance of l_i from resource r_a to resource r_b. Otherwise, if there is no instance of l_i from resource r_a to resource r_b, $C_d(l_i, r_a, r_b)$ equals to 0. By extension C_d can be the total number of distinct instances of the link l_i from r_a to any node ($C_d(li, r_a, n)$). For example, in the example graph (Fig. 1), we have:

$C_d(influences, Ariana\_Grande, Selena\_Gomez) = 1$
$C_d(influences, Ariana\_Grande, n) = 1$
$C_d(musicalguests, List\_of\_The\_Tonight\_Show\_with\_Jay\_Leno\_episodes\_$
$(2013 - 14), n) = 2$
      C_{ii} and C_{io} are functions that compute the number of indirect and distinct links, both incoming and outgoing, between resources in a graph G. $C_{ii}(l_i, r_a, r_b)$ equals 1 if there is a resource n that satisfy both $< l_i, r_a, n >$ and $< l_i, r_b, n >$, 0 if not. Similarly, $C_{io}(l_i, r_a, r_b)$ equals 1 if there is a resource n that is linked to both r_a and r_b via outgoing l_i, 0 if not. In the example (Fig. 1), we have $C_{ii}(musicalguests, Ariana\_Grande, Selena\_Gomez) = 1$ (via incoming property from List_of_The_Tonight_Show_with_Jay_Leno_episodes_(2013-14)) and $C_{io}(subject, Ariana\_Grande, Selena\_Gomez) = 1$ (via outgoing property to *Category:21st-century_American_singers*).

3.2   Equal Self-similarity and Minimality

"Equal self-similarity" denotes that the similarity of the same resources should be the same, while "minimality" denotes that the similarity of the same resources should be bigger than the similarity of two different resources. Formally, these axioms can be defined as below [7]:

- Equal self-similarity: $sim(r_a, r_a) = sim(r_b, r_b)$, for all resources r_a and r_b.
- Minimality: $sim(r_a, r_a) > sim(r_a, r_b)$, for all resources $r_a \neq r_b$.

In order to achieve "equal self-similarity", we can simply define conditions such as "the similarity of two resources r_a and r_b equals 1 if the two resources are exactly the same or r_a and r_b have the *owl:sameAs* relationship". Such an *owl:sameAs* statement indicates that two resources in DBpedia refer to the same thing. Hence, the first component of our method can be defined as below:

$$Resim(r_a, r_b) = 1, \text{ if } URI(r_a) = URI(r_b) \text{ or } r_a \text{ owl:sameAs } r_b \qquad (3)$$

The similarity measure thus scales from 0 to 1, and the similarity of r_a and r_b will be 1 if the two resources are exactly the same or if r_a and r_b have the *owl:sameAs* relationship. Otherwise, the similarity of two resources will be less than 1. As a result, the similarity measure satisfies both "equal self-similarity" and "minimality" axioms.

$$LDSD'(r_a, r_b) =$$

$$\frac{1}{\begin{array}{l} 1 + \sum_i \frac{C_d(l_i, r_a, r_b)}{1 + \log(C_d(l_i, r_a, n))} + \sum_i \frac{C_d(l_i, r_b, r_a)}{1 + \log(C_d(l_i, r_b, n))} \\ + \sum_i \frac{C_{ii}(l_i, r_a, r_b)}{1 + \log(\frac{C_{ii}(l_i, r_a, n) + C_{ii}(l_i, r_b, n)}{2})} + \sum_i \frac{C_{io}(l_i, r_a, r_b)}{1 + \log(\frac{C_{io}(l_i, r_a, n) + C_{io}(l_i, r_b, n)}{2})} \end{array}} \qquad (4)$$

$$LDSD'_{sim}(r_a, r_b) = 1 - LDSD'(r_a, r_b) \qquad (5)$$

3.3 Symmetry

The "symmetry" axiom denotes that the similarity of two resources r_a and r_b will be the same as the similarity for a reversed order of the two resources. Formally, it can be defined as:

- Symmetry: $sim(r_a, r_b) = sim(r_b, r_a)$, for all resources r_a and r_b.

As we can see from Eq. (1), the sum of two C_d functions produces the same results for the similarities of two resources and the reversed order of them. That is, $LDSD_{sim}(r_a, r_b) = LDSD_{sim}(r_b, r_a)$ while considering C_d functions only. The non-symmetric results occur due to the normalization parts of C_{ii} and C_{io} functions. The normalizations of the two functions are carried out using the logarithmic value of all incoming/outgoing nodes of r_a for $LDSD_{sim}(r_a, r_b)$. However, when calculating the similarity of resources with a reversed order - $LDSD_{sim}(r_b, r_a)$, the normalizations of C_{ii} and C_{io} are carried out using the logarithmic value of all incoming/outgoing nodes of r_b. This means that the different normalizations used when reversing the order of r_a and r_b causes non-symmetric results. Thus, we modify $LDSD$ as $LDSD'$ in Eq. (4) and $LDSD_{sim}$ as $LDSD'_{sim}$ in Eq. (5). The modified normalization considers incoming/outgoing nodes of both r_a and r_b. Hence, the similarities for two resources and their reversed order are the same, so this satisfies the "symmetry" axiom.

3.4 Property Similarity

Using the aforementioned definitions of property in Sect. 1, we add the property similarity for resources as an additional component to *Resim*. This property similarity is defined in Eq. (6). Our intuition behind this is that the property similarity of resources is important when the relationship of two resources is not available.

$$Property_{sim}(r_a, r_b) = \frac{\sum_i \frac{C_{sip}(l_i, r_a, r_b)}{C_d(l_i, n, n)}}{C_{ip}(r_a) + C_{ip}(r_b)} + \frac{\sum_i \frac{C_{sop}(l_i, r_a, r_b)}{C_d(l_i, n, n)}}{C_{op}(r_a) + C_{op}(r_b)} \tag{6}$$

Definition 2. C_{sip} and C_{sop} are functions that compute the number of distinct shared incoming and outgoing links (properties) between resources in a graph G. $C_{sip}(l_i, r_a, r_b)$ equals 1 if there is an incoming link l_i that exists for both r_a and r_b, and $C_{sop}(l_i, r_a, r_b)$ equals 1 if there is an outgoing link l_i that exists for both r_a and r_b. C_{ip} and C_{op} are functions that compute the number of incoming and outgoing links for a resource. $C_d(l_i, n, n)$ (see Sect. 3.1) denotes the total number of distinct instances of the link l_i between any two resources.

One thing to note is that we normalize the weight of C_{sip} (C_{sop}) by the total number of distinct instances of the link l_i between any two resources in G instead of the logarithm of the number. This will penalize frequently appearing properties more heavily, and we found that this approach as a part of *Resim* yields a better result for recommendations than the logarithm value of the number.

3.5 Resim Similarity Measure

Based on the components discussed above, *Resim* combines these components by calculating the weighted arithmetic mean of $Property_{sim}$ and $LDSD'_{sim}$. The final equation for *Resim* is defined as follows:

$$Resim(r_a, r_b) = \begin{cases} 1, & \text{if } URI(r_a) = URI(r_b) \text{ or } r_a \text{ } owl{:}sameAs \text{ } r_b \\ \frac{w_1 * Property_{sim}(r_a, r_b) + w_2 * LDSD'_{sim}(r_a, r_b)}{w_1 + w_2}, & \text{otherwise} \end{cases} \tag{7}$$

The weights may be adjusted according to the given dataset for which the measures should be applied (e.g. within our empirical evaluation we used a weight of 1 for w_1 and 2 for w_2 to give higher importance to the relationships between two resources).

4 Evaluation Setup

In this section, we describe an experiment to evaluate our similarity measure compared to $LDSD_{sim}$ and *Shakti*. For the *Shakti* similarity measure, we use the weights of properties manually assigned by the authors in [13]. In *Shakti*, seven properties related to the music domain have been considered such as *dbpedia-owl:genre*, *instrument*, *influences*, *associatedMusicalArtist*, *associatedBand*, *currentMember* and *pastMember*.

Firstly, we examine these similarity measures in terms of the three axioms: "equal self-similarity", "symmetry" and "minimality".

Secondly, we aim to evaluate the performance of calculating similarities on general resources without restricting to any domain, i.e., for any resources in DBpedia. For example, the similarity of the two resources $dbpedia{:}Cat$ and $dbpedia{:}Dog$ should be higher than that of $dbpedia{:}Cat$ and $dbpedia{:}Human$, and a test pair can be created as $sim(dbpedia{:}Cat, dbpedia{:}Dog) > sim(dbpedia{:}Cat, dbpedia{:}Human)$. In order to get the gold standard test pairs, we use the Word-Sim353 dataset [8]. WordSim353 is a dataset containing English word pairs along with human-assigned similarity judgements on a scale from 0 (totally unrelated words) to 10 (very much related or identical words), and is used to train and/or test algorithms implementing semantic similarity measures (i.e., algorithms that numerically estimate the similarity of natural language words). We retrieved word pairs from the dataset that satisfy $sim(W_a, W_b) > sim(W_a, W_c)$ where the difference is higher than 2. For instance, the word "car" appears several times with words such as "automobile" and "flight" among the word pairs, and $sim(car, automobile) = 8.49 > sim(car, flight) = 4.94$. We then retrieve the corresponding DBpedia resources (i.e., $dbpedia{:}Car$, $dbpedia{:}Automobile$, $dbpedia{:}Flight$) and construct a test pair as $sim(dbpedia{:}Car, dbpedia{:}Automobile) > sim(dbpedia{:}Car, dbpedia{:}Flight)$. In all, 28 test pairs of resources were retrieved (see Table 2). We evaluate the similarity measures on these test cases and see how many of them can be satisfied by each similarity measure.

Table 1. Similarity measures evaluated on axioms

Axiom	LDSDsim	Shakti	Resim
Equal self-similarity			√
Symmetry		√	√
Minimality	√		√

Finally, we evaluate the similarity measure by adopting it to item-based recommendations in the music domain. The recommender system recommends the top-N similar music artists for a music artist based on the similarities among all candidates and the music artist. Passant [22] evaluated the $LDSD$ measure in the music domain by comparing with a recommendations list from Last.fm. Last.fm offers a ranked list of similar artists/bands for each artist/band based on their similarities. They showed that in spite of a slight advantage for Last.fm, $LDSD$ based recommendations achieved a reasonable score, especially considering that it does not use any collaborative filtering approach, and relies only on links between resources. Similarly, we adopt the recommendations list from Last.fm to evaluate the performance of the recommendations. First of all, all the resources of type of $dbpedia\text{-}owl{:}MusicArtist$ or $dbpedia\text{-}owl{:}Band$ are extracted via the DBpedia SPARQL endpoint[6]. By doing so, 75,682 resources are obtained

---

[6] http://dbpedia.org/sparql.

Table 2. Evaluation on test pairs of resources based on extracted word pairs from WordSim353

Test pairs of resources			LDSDsim	Shakti	Resim
sim(dbpedia:Car, dbpedia:Automobile)	>	sim(dbpedia:Car, dbpedia:Flight)	✓	✓	✓
sim(dbpedia:Money, dbpedia:Currency)	>	sim(dbpedia:Money, dbpedia:Business_operations)		✓	✓
sim(dbpedia:Money, dbpedia:Cash)	>	sim(dbpedia:Money, dbpedia:Bank)			
sim(dbpedia:Money, dbpedia:Cash)	>	sim(dbpedia:Money, dbpedia:Demand_deposit)	✓		✓
sim(dbpedia:Professor, dbpedia:Doctor_of_Medicine)	>	sim(dbpedia:Professor, dbpedia:Cucumber)	✓	✓	✓
sim(dbpedia:Doctor_of_Medicine, dbpedia:Nursing)	>	sim(dbpedia:Doctor_of_Medicine, dbpedia:Bus_driver)		✓	✓
sim(dbpedia:Ocean, dbpedia:Sea)	>	sim(dbpedia:Ocean, dbpedia:Continent)	✓	✓	✓
sim(dbpedia:Computer, dbpedia:Keyboard)	>	sim(dbpedia:Computer, dbpedia:News)			
sim(dbpedia:Computer, dbpedia:Internet)	>	sim(dbpedia:Computer, dbpedia:News)	✓	✓	✓
sim(dbpedia:Computer, dbpedia:Software)	>	sim(dbpedia:Computer, dbpedia:Laboratory)	✓	✓	✓
sim(dbpedia:Cup, dbpedia:Drink)	>	sim(dbpedia:Cup, dbpedia:Article)		✓	✓
sim(dbpedia:Cup, dbpedia:Coffee)	>	sim(dbpedia:Cup, dbpedia:Substance)		✓	✓
sim(dbpedia:Drink, dbpedia:Mouth)	>	sim(dbpedia:Drink, dbpedia:Ear)			✓
sim(dbpedia:Drink, dbpedia:Eating)	>	sim(dbpedia:Drink, dbpedia:Mother)	✓		✓
sim(dbpedia:Football, dbpedia:Association_football)	>	sim(dbpedia:Football, dbpedia:Basketball)		✓	
sim(dbpedia:Monarch, dbpedia:Queen_consort)	>	sim(dbpedia:Monarch, dbpedia:Cabbage)	✓	✓	✓
sim(dbpedia:Tiger, dbpedia:Jaguar)	>	sim(dbpedia:Tiger, dbpedia:Organism)	✓	✓	✓
sim(dbpedia:Day, dbpedia:Night)	>	sim(dbpedia:Day, dbpedia:Summer)		✓	✓
sim(dbpedia:Coast, dbpedia:Shore)	>	sim(dbpedia:Coast, dbpedia:Forest)	✓		✓
sim(dbpedia:Coast, dbpedia:Shore)	>	sim(dbpedia:Coast, dbpedia:Hill)	✓		✓
sim(dbpedia:Governor, dbpedia:Office)	>	sim(dbpedia:Governor, dbpedia:Interview)			
sim(dbpedia:Food, dbpedia:Fruit)	>	sim(dbpedia:Food, dbpedia:Rooster)			✓
sim(dbpedia:Life, dbpedia:Death)	>	sim(dbpedia:Life, dbpedia:Term_(time))	✓	✓	✓
sim(dbpedia:Digital_media, dbpedia:Radio)	>	sim(dbpedia:Digital_media, dbpedia:Trade)	✓	✓	✓
sim(dbpedia:Planet, dbpedia:Moon)	>	sim(dbpedia:Planet, dbpedia:People)		✓	✓
sim(dbpedia:Opera, dbpedia:Performance)	>	sim(dbpedia:Opera, dbpedia:Industry)		✓	✓
sim(dbpedia:Nature, dbpedia:Environment)	>	sim(dbpedia:Nature, dbpedia:Man)		✓	
sim(dbpedia:Energy, dbpedia:Laboratory)	>	sim(dbpedia:Energy, dbpedia:Secretary)			✓
Total :			13	18	23

consisting of 45,104 resources of type *dbpedia-owl:MusicArtist* and 30,578 resources of type *dbpedia-owl:Band*. Then we randomly selected 10 resources from
these 75,682 resources. For each resource (a music artist or band in this case),
we manually get the top 10 recommendations list from Last.fm for each resource
which can be found in DBpedia. To construct a candidate list for recommendations, we create a candidate list with these top 10 recommendations from Last.fm
and 200 randomly selected resources among the 75,682 resources of type *dbpedia-
owl:MusicArtist* or *dbpedia-owl:Band*. For example, if a user is interested in the
music artist *dbpedia:Ariana_Grande*, the candidate list consists of the top 10
similar music artists recommended by Last.fm (that can be found in DBpedia)
and 200 randomly selected resources of type *dbpedia-owl:MusicArtist* or *dbpedia-
owl:Band*. Then we calculate the similarities between *dbpedia:Ariana_Grande*
and the candidate list with *Resim*, $LDSD_{sim}$ and *Shakti* to get the top-N recommendations. Our goal is to see the performance of the top-N recommendations
based on these similarity measures.

The performance of the recommendations was measured by means of R@N
and MRR (Mean Reciprocal Rank). For a resource (e.g., *dbpedia:Ariana_Grande*),
recall at N (R@N) is the fraction of resources that are relevant to the resource
that are successfully retrieved in the top-N recommendations and MRR indicates
at which rank the first item relevant to the resource occurs on average.

We use N = 5, 10 and 20 in the evaluation and report the results of averaged R@N over the 10 randomly selected resources of *dbpedia-owl:MusicArtist*
or *dbpedia-owl:Band* based on *Resim*, $LDSD_{sim}$, *Shakti*. Since the *Shakti* similarity measure uses the value of max step for the extension of the paths between
two resources, we use 3 and 5 for the value of max step and denote these variants
as *Shakti3* (*max step set to 3*) and *Shakti5* (*max step set to 5*).

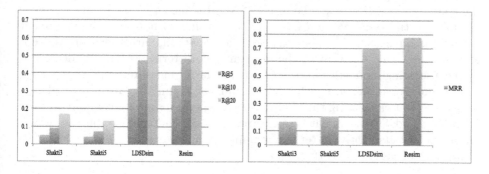

Fig. 2. Average recall at n and MRR for the recommendations of 10 random samples

5   Results

Table 1 shows the details of three similarity measures on the three axioms: "equal
self-similarity", "symmetry" and "minimality". As we can see from the table,

Resim satisfies all of the axioms while both $LDSD_{sim}$ and *Shakti* do not satisfy two of the axioms.

The details of results for test pairs of general resources are presented in Table 2. Among these test cases of general resources, *Resim* can correctly calculate the similarity of resources and satisfy 23 test pairs out of 28 on account of $Property_{sim}$, while $LDSD_{sim}$ and *Shakti* can satisfy 13 and 18 pairs respectively. In more detail, $LDSD_{sim}$ failed to calculate the similarities for many of these general resource pairs. That is, $sim(r_a, r_b) = 0$ since there was no relationship between them. In this case, a recommender system based on $LDSD_{sim}$ cannot recommend anything to a user. For instance, a user has an interest on the topic of *dbpedia:Money* and there are two news items on the topics of *dbpedia:Currency* and *dbpedia:Business_operations*. Based on $LDSD_{sim}$, the recommender system cannot recommend any news for these topics to the user since both $LDSD_{sim}(dbpedia : Money, dbpedia : Currency)$ and $LDSD_{sim}(dbpedia : Money, dbpedia : Business\_operations)$ are 0. From the results of the *Shakti* similarity measure, incorporating the number of paths between two resources improved the performance. However, it also generates some incorrect results by incorporating the number of paths in some cases such as $sim(dbpedia : Money, dbpedia : Cash)$ and $sim(dbpedia : Money, dbpedia : Demand\_deposit)$.

The results of R@N and MRR for recommendations based on the randomly selected music artists and bands are displayed in Fig. 2. The *paired t-test* is used for testing the significance where the significance level was set to 0.05 unless otherwise noted. Overall, *Resim* performed better than $LDSD_{sim}$, *Shakti3* and *Shakti5* and achieved 48 % recall for the top 10 recommendations. In more detail, both $LDSD_{sim}$ and *Resim* outperform *Shakti3* and *Shakti5* significantly in terms of R@N and MRR. In addition, the results of R@N are increased by 2 % and 1 % respectively when n is equal to 5 and 10 using *Resim* compared to the results using $LDSD_{sim}$. The result of MRR is increased by 8 % using *Resim* compare to the result using $LDSD_{sim}$. One thing to note in our experiment is that the higher value of max step for the *Shakti* similarity measure did not improve the recall. Conversely, the performance in terms of R@N is decreased by incorporating more steps in the *Shakti* similarity measure.

To summarise, *Resim* satisfies the axioms as a similarity measure and performs better at calculating the similarities of general resources, compared to the $LDSD_{sim}$ and *Shakti* similarity measures. For single-domain resources, *Resim* has a similar but slightly better performance compared to $LDSD_{sim}$ and significantly better performance than *Shakti*.

6   Study of Linked Data Sparsity Problem

During the experiment mentioned in the previous section, we found that some of the random samples with less incoming/outgoing links yielded poor recall. For instance, the recall at 10 of recommendations for *dbpedia:Jasmin_Thompson* is 0.1, which is one of the random samples that has 42 outgoing links and 3 incoming links. In contrast, the recall of recommendations for the *dbpedia:Dead_Kennedys* is 0.9, which has 117 outgoing links and 119 incoming links.

This observation motivates us to investigate if the performance of the item-based recommender system suffers from *"Linked Data sparsity"*. Here, the *Linked Data sparsity problem* means that the performance of the recommender system based on similarity measures of resources decreases when resources lack information (i.e., when they have a lesser number of incoming/outgoing relationships to other resources). In this regard, the null hypothesis to test can be defined as below:

H_0 : *The number(log) of incoming/outgoing links for resources has no relationship to the performance of a recommender system.*

We use the logarithm of the number (denote as number(log)) to decrease the variation in numbers. We reject the null hypothesis if the number(log) of incoming/outgoing links and the recall of recommendations have a strong relationship (*Pearson's correlation > 0.4*), otherwise we accept the null hypothesis.

To this end, we additionally selected 10 popular DBpedia resources of type *dbpedia-owl:MusicArtist* as samples, and then calculate the recall at 5, 10 and 20 in the same way as we did for the 10 randomly selected samples. The assumption here is that the popular samples tend to have more information (i.e., incoming/outgoing links) than random samples. This is because these resources in DBpedia are a reflection of the corresponding concepts/articles in Wikipedia, and usually popular music artists have more information thanks to a higher number of contributors.

First, we intend to see if the recommendation system performs better on popular samples than on random ones. On top of that, we aim to investigate the correlation by calculating the *Pearson's coefficient* between the number(log) of incoming/outgoing links for resources and the recall of the recommender system.

As we can see from Fig. 3, the recall results of the recommender system on popular samples are significantly better than the results on random samples. Following this finding, we calculate the correlation between the number(log) of incoming/outgoing links for resources and the performance (recall) of the recommender system. We report R@10 based on *Resim* here, and similar results can be observed by using other measures. The result shows the performance of the recommender system has a very strong positive relationship (Fig. 4, *Pearson's*

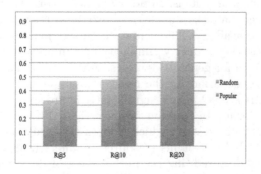

Fig. 3. Recall of recommendations on random samples and popular ones

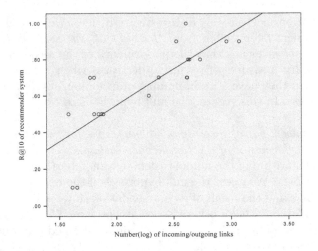

Fig. 4. Scatter plot of R@10 and number(log) of links, r=0.798

correlation of 0.798) with the total number(log) of incoming/outgoing links ($p<0.01$). Hence, the null hypothesis is rejected. In other words, the performance of the recommender system decreases for the resources with sparsity (i.e., less incoming/outgoing links). It also indicates that, on one hand, utilizing Linked Data to build a recommender system can mitigate the traditional sparsity problem [11] of collaborative recommender systems, but on the other hand, the system can also have a *Linked Data sparsity problem* for resources in the Linked Data set that the recommender system has adopted.

7   Conclusion and Future Work

In this paper, we introduced a similarity measure called *Resim* (*Resource Similarity*) to calculate semantic similarity for resources in DBpedia. Based on the work of $LDSD_{sim}$, we tackled some of the limitations of this similarity measure and constructed our similarity measure to resolve these limitations so as to satisfy some fundamental axioms. In addition, we incorporated property similarity in *Resim* to calculate the similarity of two resources when the relationship between them is not available. An evaluation on test pairs of general resources shows that incorporating property similarity can improve the performance of calculating similarities for general resources. Furthermore, an evaluation based on the top n recommendations in the music domain shows that our similarity measure outperforms $LDSD_{sim}$ and significantly improves performance over the *Shakti* similarity measure. In addition, we investigated if the performance of an item-based recommender system, which adopts similarity measures for calculating the similarity between items (resources), suffers from the "*Linked Data sparsity problem*", and proved that the performance of the recommender system has a very strong positive relationship with the number(log) of the total number of incoming/outgoing links ($p< 0.01$) for resources.

In future work, we propose to extend the current similarity measure by incorporating longer paths, while being mindful that a trade off between performance and accuracy might be a challenge. In addition, we plan to extend our similarity measure by incorporating paths to calculate the similarity of a user interest graph which can then be applied to social recommender systems.

Acknowledgments. This publication has emanated from research conducted with the financial support of Science Foundation Ireland (SFI) under Grant Number SFI/12/RC/2289 (Insight Centre for Data Analytics).

References

1. Abel, F., Gao, Q., Houben, G.J., Tao, K.: Analyzing temporal dynamics in twitter profiles for personalized recommendations in the social web. In: Proceedings of the 3rd International Web Science Conference, p. 2. ACM (2011)
2. Abel, F., Herder, E., Houben, G.J., Henze, N., Krause, D.: Cross-system user modeling and personalization on the social web. User Model. User Adap. Inter. **23**(2–3), 169–209 (2013)
3. Auer, S., Bizer, C., Kobilarov, G., Lehmann, J., Cyganiak, R., Ives, Z.G.: DBpedia: A nucleus for a web of open data. In: Aberer, K., et al. (eds.) ASWC 2007 and ISWC 2007. LNCS, vol. 4825, pp. 722–735. Springer, Heidelberg (2007)
4. Brickley, D., Guha, R.V.: RDF vocabulary description language 1.0: RDF schema (2004)
5. Di Noia, T., Mirizzi, R., Ostuni, V.C., Romito, D.: Exploiting the web of data in model-based recommender systems. In: Proceedings of the sixth ACM conference on Recommender systems, pp. 253–256. ACM (2012)
6. Di Noia, T., Mirizzi, R., Ostuni, V.C., Romito, D., Zanker, M.: Linked open data to support content-based recommender systems. In: Proceedings of the 8th International Conference on Semantic Systems, I-SEMANTICS 2012, pp. 1–8, NY, USA (2012). http://doi.acm.org/10.1145/2362499.2362501
7. Ennis, M.D., Ashby, F.G.: Similarity Measures (2007). http://www.scholarpedia. org/article/Similarity_measures
8. Finkelstein, L., Gabrilovich, E., Matias, Y., Rivlin, E., Solan, Z., Wolfman, G., Ruppin, E.: Placing search in context: the concept revisited. In: Proceedings of the 10th International Conference on World Wide Web, pp. 406–414. ACM (2001)
9. Groues, V., Naudet, Y., Kao, O.: Adaptation and evaluation of a semantic similarity measure for dbpedia: a first experiment. In: 2012 Seventh International Workshop on Semantic and Social Media Adaptation and Personalization (SMAP), pp. 87–91. IEEE (2012)
10. Heath, T., Bizer, C.: Linked data: Evolving the web into a global data space. Synth. Lect. Semant. Web: Theor. Technol. **1**(1), 1–136 (2011)
11. Heitmann, B., Hayes, C.: Using linked data to build open, collaborative recommender systems. In: AAAI Spring Symposium: Linked Data Meets Artificial Intelligence, pp. 76–81 (2010)
12. Jeh, G., Widom, J.: SimRank: A measure of structural-context similarity. In: Proceedings of the Eighth ACM SIGKDD International Conference on Knowledge Discovery and Data Mining, KDD 2002, pp. 538–543. ACM, NY, USA, New York (2002)

13. Leal, J.P., Rodrigues, V., Queirós, R.: Computing semantic relatedness using dbpedia(2012)
14. Lee, S., Yang, J., Park, S.-Y.: Discovery of hidden similarity on collaborative filtering to overcome sparsity problem. In: Suzuki, E., Arikawa, S. (eds.) DS 2004. LNCS (LNAI), vol. 3245, pp. 396–402. Springer, Heidelberg (2004)
15. Maedche, A., Zacharias, V.: Clustering ontology-based metadata in the semantic web. In: Elomaa, T., Mannila, H., Toivonen, H. (eds.) PKDD 2002. LNCS (LNAI), vol. 2431, pp. 348–360. Springer, Heidelberg (2002)
16. Meng, L., Huang, R., Gu, J.: A review of semantic similarity measures in wordnet. Inter. J. Hybrid Inf. Technol. **6**(1), 1–12 (2013)
17. Miller, G.A.: WordNet: a lexical database for english. Commun. ACM **38**(11), 39–41 (1995)
18. Musto, C., Basile, P., Lops, P., de Gemmis, M., Semeraro, G.: Linked open data-enabled strategies for Top-N recommendations. In: CBRecSys 2014, p. 49 (2014)
19. Noy, N.F., McGuinness, D.L.: Ontology development 101: A guide to creating your first ontology (2001)
20. Orlandi, F., Breslin, J., Passant, A.: Aggregated, interoperable and multi-domain user profiles for the social web. In: Proceedings of the 8th International Conference on Semantic Systems (2012)
21. Ostuni, V.C., Di Noia, T., Di Sciascio, E., Mirizzi, R.: Top-n recommendations from implicit feedback leveraging linked open data. In: Proceedings of the 7th ACM Conference on Recommender Systems, pp. 85–92. ACM (2013)
22. Passant, A.: dbrec — music recommendations using DBpedia. In: Patel-Schneider, P.F., Pan, Y., Hitzler, P., Mika, P., Zhang, L., Pan, J.Z., Horrocks, I., Glimm, B. (eds.) ISWC 2010, Part II. LNCS, vol. 6497, pp. 209–224. Springer, Heidelberg (2010)
23. Passant, A.: Measuring Semantic Distance on Linking Data and Using it for Resources Recommendations. In: AAAI Spring Symposium: Linked Data Meets Artificial Intelligence, vol. 77, p. 123 (2010)
24. Strobin, L., Niewiadomski, A.: Evaluating semantic similarity with a new method of path analysis in RDF using genetic algorithms. Comput. Sci. **21**(2), 137–152 (2013)

Identifying an Agent's Preferences Toward Similarity Measures in Description Logics

Teeradaj Racharak[1,2](✉), Boontawee Suntisrivaraporn[1], and Satoshi Tojo[2]

[1] School of Information, Computer and Communication Technology,
Sirindhorn International Institute of Technology,
Thammasat University, Pathum Thani, Thailand
r.teeradaj@gmail.com, sun@siit.tu.ac.th
[2] School of Information Science,
Japan Advanced Institute of Science and Technology, Ishikawa, Japan
{racharak,tojo}@jaist.ac.jp

Abstract. In Description Logics (DLs), concept similarity measures (CSMs) aim at identifying a degree of commonality between two given concepts and are often regarded as a generalization of the classical reasoning problem of equivalence. That is, any two concepts are equivalent if their similarity degree is one, and vice versa. When two concepts are not quite equivalent but similar, nevertheless, a problem may arise as to which aspects of commonality should play more important role than others. This work presents the so-called *preference profile*, which is design guidelines for an agent's preferences and points out to our preliminary developing stage of sim^π [1], in which an agent's preferences can influence the calculation of CSM in DL $\mathcal{ELH}$.

Keywords: Preference profile · Concept similarity measures · Non-standard reasoning services · Description logics

1 Introduction and Motivation

Preferences are used in a variety of related, but not identical, ways in human beings' daily life: to express what they like and dislike, to express their desired goals when choosing routes for travelling [2], etc. In psychology, preferences may be conceived of as an individual's attitude towards a set of objects when making decisions [3]. Alternatively, it can be interpreted as a judgment in a sense of liking or disliking an object [4].

In Description Logics (DLs), concept similarity measures (CSMs) aim at identifying a degree of commonality between two given concept names and are often regarded as a generalization of the classical reasoning problem of equivalence. That is, any two concepts are equivalent if their similarity degree is one, and vice versa. To date, many elegant CSMs have been developed (cf. Subsect. 2.2). These developments can induce efficient similarity-oriented DL reasoning services, i.e., to measure if two concepts are similar, to check if an individual is a relaxed instance of a concept, and to retrieve those individuals similar to a given individual.

© Springer International Publishing Switzerland 2016
G. Qi et al. (Eds.): JIST 2015, LNCS 9544, pp. 201–208, 2016.
DOI: 10.1007/978-3-319-31676-5_14

Unfortunately, those similarity measures may be counter-intuitive when it comes to human perception. A counterexample could be the similarity between animals in accordance with human beings' preferences. A person may perceive that Frog is similar to TreeLizard because they belong to the same class and their skin colors are similar. On the other hand, another person may rather perceive that Frog is similar to TreeLizard because they belong to the same class and their natural habitation and current living environment are near. These scenarios reveal that human beings always have *bias* or *preferences* when making judgments about concepts in question.

Example 1. Aforementioned concepts could be modeled in DL as follows:

$$\text{Frog} \sqsubseteq \text{Reptile} \sqcap \exists\text{hasColor.Green} \sqcap \exists\text{hasHabitat.Forest}$$
$$\text{TreeLizard} \sqsubseteq \text{Reptile} \sqcap \exists\text{hasColor.Yellow} \sqcap \exists\text{liveIn.Forest}$$
$$\text{hasHabitat} \sqsubseteq \text{hasLocation}$$
$$\text{liveIn} \sqsubseteq \text{hasLocation}$$

Reasonable perception when considering on the DLs is that Frog and TreeLizard are not much similar. However, in reality, the similarity value between each of them could be varied, by an agent's preferences as stated above.   □

In this paper, we study and formulate essential aspects of preferences which can be expressed by an agent when measuring the similarity between two concept descriptions. A numerical degree yielded from a computation of concept similarity measures under an agent's preferences will thereby comply more with the agent's intuition than that from the base measures do.

The structure of this paper is organized as follows. Section 2 reviews Description Logics, particularly $\mathcal{ELH}$, and the definition of concept similarity measures (CSMs). Section 3 presents potential aspects of preference expressions formalized as *guidelines for developing concept similarity measures under preferences* in DLs and discusses that, to the best of our knowledge, none of existing measures for DLs satisfy all elements of the so-called *preference profile*. Note here that all the proposed preference expressions are each represented as functions and are considered collectively as the preference profile. Finally, Sect. 4 presents the conclusion and our future work.

2   Preliminaries

2.1   Description Logics

In Description Logics (DLs), *concept descriptions* are inductively defined by the help of a set of *constructors*, a set of concept names CN, and a set of role names RN. $\mathcal{ELH}$ concept descriptions are formed using the conjunction ($\sqcap$), existential restrictions (e.g., ($\exists r.C$) where $r \in$ RN and $C \in$ CN), and the top concept ($\top$). The set of concept descriptions, or simply concepts, for a specific DL $\mathcal{L}$ is denoted by $\mathsf{Con}(\mathcal{L})$. For instance, $\mathsf{Con}(\mathcal{ELH})$ is the set of all $\mathcal{ELH}$ concept descriptions.

Conventionally, concept names are denoted by A and B, concept descriptions are denoted by C and D, and role names are denoted by r and s.

A terminology or TBox $\mathcal{T}$ is a finite set of concept definitions (e.g., $A \sqsubseteq D$ or $A \equiv D$ where $A, D \in \mathsf{CN}$) and role hierarchy axioms (e.g., $r \sqsubseteq s$ where $r, s \in \mathsf{RN}$). A TBox is called *unfoldable* if it contains at most one concept definition for each concept name in CN and does not contain cyclic dependencies. Concept names occurring on the left-hand side of a concept definition are called defined concept names (denoted by $\mathsf{CN}^{\mathsf{def}}$), the other concept names are primitive concept names (denoted by $\mathsf{CN}^{\mathsf{pri}}$). Primitive concept definitions are commonly found in realistic terminologies in which necessary conditions of concepts are merely known. Such a primitive definition $A \sqsubseteq D$ can easily be transformed into a semantically equivalent full definitions $A \equiv X \sqcap D$ where X is a fresh concept name. When a TBox $\mathcal{T}$ is unfoldable, concept names can be expanded by exhaustively replacing all defined concept names by their definitions until only primitive concept names remain. Such concept names are called *fully expanded concept names*. In this work, we assume that concepts are fully expanded since TBox can be completely disregarded from decision procedures. Furthermore, a set of statements about the characteristics of roles can be axiomatized by a role hierarchy. Like primitive definitions, a role hierarchy axiom $r \sqsubseteq s$ can be transformed in to a semantically equivalent role definition $r \equiv t \sqcap s$ where t is a fresh role name. Role names occurring on the left-hand side of a role definition are called defined role names (denoted by $\mathsf{RN}^{\mathsf{def}}$).

In order to defined a formal semantics for a specific DL $\mathcal{L}$, we consider an *interpretation* $\mathcal{I} = \langle \Delta^{\mathcal{I}}, \cdot^{\mathcal{I}} \rangle$, which consists of a nonempty set $\Delta^{\mathcal{I}}$ as the domain of the interpretation and an interpretation function $\cdot^{\mathcal{I}}$ which assigns to every concept name A a set $A^{\mathcal{I}} \subseteq \Delta^{\mathcal{I}}$ and to every role name r a binary relation $r^{\mathcal{I}} \subseteq \Delta^{\mathcal{I}} \times \Delta^{\mathcal{I}}$ (cf. [5,6] for more details). An interpretation $\mathcal{I}$ is a *model* of a TBox $\mathcal{T}$ if, for each axiom in $\mathcal{T}$, the conditions corresponding to their semantics are satisfied. One of the main classical reasoning problems is the *subsumption problem*. That is, given two concept descriptions C and D and a TBox $\mathcal{T}$, C is subsumed by D w.r.t. a TBox $\mathcal{T}$ (written as $C \sqsubseteq_{\mathcal{T}} D$) if $C^{\mathcal{I}} \subseteq D^{\mathcal{I}}$ in every model $\mathcal{I}$ of $\mathcal{T}$. Furthermore, C and D are equivalent w.r.t. $\mathcal{T}$ (written as $C \equiv_{\mathcal{T}} D$) if $C \sqsubseteq_{\mathcal{T}} D$ and $D \sqsubseteq_{\mathcal{T}} C$. When a TBox $\mathcal{T}$ is empty or is clear from the context, we omit to denote $\mathcal{T}$, i.e. $C \sqsubseteq D$ and $C \equiv D$.

2.2   Concept Similarity Measure

Concept similarity measure (CSM) is one of non-standard DL reasoning services. It determines how similar two concepts are. Formally, given $C, D \in \mathsf{Con}(\mathcal{L})$ be two concept descriptions for a specific DL $\mathcal{L}$. Then, a *concept similarity measure* w.r.t. a TBox $\mathcal{T}$ is a function $\sim_{\mathcal{T}} : \mathsf{Con}(\mathcal{L}) \times \mathsf{Con}(\mathcal{L}) \to [0, 1]$ such that $C \sim_{\mathcal{T}} D = 1$ iff $C \equiv_{\mathcal{T}} D$ (total similarity) and $C \sim_{\mathcal{T}} D = 0$ indicates total dissimilarity between C and D. When a TBox $\mathcal{T}$ is empty or is clear from the context, we simply write $C \sim D$.

There exist many state-of-the-art measures and those can be seen as actual instances of CSMs. For instance, two elegant measures, viz. $\sim^s$ and $\sim^c$, based on an automata-theoretic characterization of subsumption in $\mathcal{FL}_0$ are defined in [7] to calculate the similarity degree between two $\mathcal{FL}_0$ concept descriptions w.r.t. different levels of strongness. The measure sim from [5] for $\mathcal{EL}$ concept descriptions is defined based on a characterization of subsumption by tree homomorphism. The work from [5] is continued to define the similarity-based instance checking [8] and to define for measuring the similarity between $\mathcal{ELH}$ concept descriptions [6,9] together with two concrete algorithms for implementing the proposed measure. Another measure for DL $\mathcal{ELH}$ is the parameterizable measure called *simi* [10] which allows calibrating via various parameters of the measure to fit the expectation. A set of well-defined properties for CSMs is also collected and introduced in [10]. Those formally defined properties are believed to desirable properties for CSMs, i.e., actual instances of CSMs complying with those properties can produce predictable outcomes for CSM users. Fortunately, sim [6,9] and *simi* [10] are theoretically proven to fulfill most of those properties. To illustrate an application of CSMs, applying sim [6,9] on concepts given in Example 1 yields sim(Frog, TreeLizard) = 0.475 (See more details in [1]).

3   Preference Profile

A numerical degree value obtained by CSMs indicates the similarity degree value between two concept descriptions. For instance, sim(Frog, TreeLizard) = 0.475 indicates that the similarity between Frog and TreeLizard is 47.5 %. Unfortunately, the finding reported by CSMs might not be intuitive and reasonable concerning different perceiving agents. Consider two aforementioned agents from Sect. 1:

> **Agent 1:** Frog is similar to TreeLizard due to classes and skin color;
> **Agent 2:** Frog is similar to TreeLizard due to classes and surrounding.

Most modern CSMs including sim reveal Frog and TreeLizard are not quite similar. Hence, they per se are not appropriate to be used in our scenario. In this section, we explore various aspects of preference expressions which can be seamlessly captured in CSMs and then formalize them as guidelines for *CSMs under preferences*. These aspects of preferences are compiled together as the *preference profile* π. Any CSMs which expose a syntax and satisfy semantics of these aspects are appropriate to be used under an agent's preferences.

In the following, we present five aspects of preference expressions, which can be adopted in and thereby influencing the calculation of CSMs. The syntax and semantics for each aspect of a preference profile are given in term of partial functions since the exact domain of each aspect is varied from agents to agents.

1. Primitive concept importance;
2. Role importance;
3. Primitive concepts similarity;
4. Primitive roles similarity; and
5. Role discount factor.

Definition 1 (Primitive Concept Importance). *Let* $\mathsf{CN}^{\mathsf{pri}}(\mathcal{T})$ *be a set of primitive concept names occurring in* $\mathcal{T}$. *Then,* a primitive concept importance *is a partial function* $\mathfrak{i}^{\mathsf{c}} : \mathsf{CN} \to \mathbb{R}_{\geq 0}$, *where* $\mathsf{CN} \subseteq \mathsf{CN}^{\mathsf{pri}}(\mathcal{T})$.

For any $A \in \mathsf{CN}^{\mathsf{pri}}(\mathcal{T})$, $\mathfrak{i}^{\mathsf{c}}(A) = 1$ captures an expression of normal importance for A, $\mathfrak{i}^{\mathsf{c}}(A) > 1$ (and $\mathfrak{i}^{\mathsf{c}}(A) < 1$) indicates that A has higher (and lower, respectively) importance, and $\mathfrak{i}^{\mathsf{c}}(A) = 0$ indicates that A is entirely ignored by an agent. For example, suppose both agents consider heavily whether the two concepts are in the same class, i.e., Reptile. Therefore, they might express as $\mathfrak{i}^{\mathsf{c}}(\mathsf{Reptile}) = 2$ for their own preference profiles.

Practically, many primitive concept names might be not assigned the important values by agents, i.e. those concept names are not mapped to the corresponding values. Hence, we assign the default importance value of 1 for $A \in \mathsf{CN}^{\mathsf{pri}}(\mathcal{T})$ in case $\mathfrak{i}^{\mathsf{c}}(A)$ is not defined. Furthermore, the total function $\mathfrak{i}_0^{\mathsf{c}}(A) = 1$ for all $A \in \mathsf{CN}^{\mathsf{pri}}(\mathcal{T})$ is called the *default primitive concept importance*.

Definition 2 (Role Importance). *Let* $\mathsf{RN}(\mathcal{T})$ *be a set of role names occurring in* $\mathcal{T}$. *Then,* a role importance *is a partial function* $\mathfrak{i}^{\mathsf{r}}: \mathsf{RN} \to \mathbb{R}_{\geq 0}$, *where* $\mathsf{RN} \subseteq \mathsf{RN}(\mathcal{T})$.

For any $r \in \mathsf{RN}(\mathcal{T})$, $\mathfrak{i}^{\mathsf{r}}(r) = 1$ captures an expression of normal importance for r, $\mathfrak{i}^{\mathsf{r}}(r) > 1$ (and $\mathfrak{i}^{\mathsf{r}}(r) < 1$) indicates that r has higher (and lower, respectively) importance, and $\mathfrak{i}^{\mathsf{r}}(r) = 0$ indicates that r is entirely ignored by an agent. For example, **Agent 1** may consider heavily on their skin colors, i.e., $\mathfrak{i}^{\mathsf{r}}(\mathsf{hasColor}) = 2$. In addition, **Agent 2** may consider heavily on their surrounding, i.e., $\mathfrak{i}^{\mathsf{r}}(\mathsf{hasHabitat}) = 2$ and $\mathfrak{i}^{\mathsf{r}}(\mathsf{liveIn}) = 2$.

Practically, many role names might be not assigned the important values by agents, i.e. those role names are not mapped to the corresponding values. Hence, we use the default importance value of 1 for $r \in \mathsf{RN}(\mathcal{T})$ in case $\mathfrak{i}^{\mathsf{r}}(r)$ is not defined. Furthermore, the total function $\mathfrak{i}_0^{\mathsf{r}}(r) = 1$ for all $r \in \mathsf{RN}(\mathcal{T})$ is called the *default role importance*.

Definition 3 (Primitive Concepts Similarity). *Let* $\mathsf{CN}^{\mathsf{pri}}(\mathcal{T})$ *be a set of primitive concept names occurring in* $\mathcal{T}$. *For* $A, B \in \mathsf{CN}^{\mathsf{pri}}(\mathcal{T})$, *a primitive concepts similarity is a partial function* $\mathfrak{s}^{\mathsf{c}} : \mathsf{CN} \times \mathsf{CN} \to [0, 1]$, *where* $\mathsf{CN} \subseteq \mathsf{CN}^{\mathsf{pri}}(\mathcal{T})$, *such that* $\mathfrak{s}^{\mathsf{c}}(A, B) = \mathfrak{s}^{\mathsf{c}}(B, A)$ *and* $\mathfrak{s}^{\mathsf{c}}(A, A) = 1$.

For $A, B \in \mathsf{CN}^{\mathsf{pri}}(\mathcal{T})$, $\mathfrak{s}^{\mathsf{c}}(A, B) = 1$ captures an expression of total similarity between A and B and $\mathfrak{s}^{\mathsf{c}}(A, B) = 0$ captures an expression of total dissimilarity between A and B. For example, **Agent 1** may feel that there is similarity between Green and Yellow. Hence, $\mathfrak{s}^{\mathsf{c}}(\mathsf{Green}, \mathsf{Yellow}) = 0.5$.

Practically, many pairs of primitive concept names might be not assigned the similarity values by agents, i.e. those pairs of concept names are not mapped to the corresponding values. Hence, we assign the default similarity value of 0 for $(A, B) \in \mathsf{CN}^{\mathsf{pri}}(\mathcal{T}) \times \mathsf{CN}^{\mathsf{pri}}(\mathcal{T})$ in case $\mathfrak{s}^{\mathsf{c}}(A, B)$ is not defined. Furthermore, the total function $\mathfrak{s}_0^{\mathsf{c}}(A, B) = 0$ for all $(A, B) \in \mathsf{CN}^{\mathsf{pri}}(\mathcal{T}) \times \mathsf{CN}^{\mathsf{pri}}(\mathcal{T})$ is called the *default primitive concept similarity*.

Definition 4 (Primitive Roles Similarity). *Let* $\mathsf{RN}^{\mathsf{pri}}(\mathcal{T})$ *be a set of primitive role names occurring in* $\mathcal{T}$*. For* $r, s \in \mathsf{RN}^{\mathsf{pri}}(\mathcal{T})$*, a primitive roles similarity is a partial function* $\mathfrak{s}^{\mathfrak{r}} : \mathsf{RN} \times \mathsf{RN} \to [0,1]$*, where* $\mathsf{RN} \subseteq \mathsf{RN}^{\mathsf{pri}}(\mathcal{T})$*, such that* $\mathfrak{s}^{\mathfrak{r}}(r, s) = \mathfrak{s}^{\mathfrak{r}}(s, r)$ *and* $\mathfrak{s}^{\mathfrak{r}}(r, r) = 1$*.*

For $r, s \in \mathsf{RN}(\mathcal{T})$, $\mathfrak{s}^{\mathfrak{r}}(r, s) = 1$ captures an expression of total similarity between r and s and $\mathfrak{s}^{\mathfrak{r}}(r, s) = 0$ captures an expression of total dissimilarity between r and s. For example, **Agent 2** may feel that there is similarity between knowing the natural habitat of the animals, i.e., hasHabitat, and knowing the living environment of the animals, i.e., liveIn. Hence, expressing the similarity between their corresponding new primitive role names can be exploited, i.e., $\mathfrak{s}^{\mathfrak{r}}(t, u) = 0.1$.

Practically, many pairs of primitive role names might be not assigned the similarity values by agents, i.e. those pairs of role names are not mapped to the corresponding values. Hence, we assign the default similarity value of 0 for $(r, s) \in \mathsf{RN}^{\mathsf{pri}}(\mathcal{T}) \times \mathsf{CN}^{\mathsf{pri}}(\mathcal{T})$ in case $\mathfrak{s}^{\mathfrak{r}}(r, s)$ is not defined. Furthermore, the total function $\mathfrak{s}^{\mathfrak{r}}_0(r, s) = 0$ for all $(r, s) \in \mathsf{RN}^{\mathsf{pri}}(\mathcal{T}) \times \mathsf{RN}^{\mathsf{pri}}(\mathcal{T})$ is called the *default primitive role similarity*.

Definition 5 (Role Discount Factor). *Let* $\mathsf{RN}(\mathcal{T})$ *be a set of role names occurring in* $\mathcal{T}$*. Then, a* role discount factor *is a partial function* $\mathfrak{d} : \mathsf{RN} \to [0,1]$*, where* $\mathsf{RN} \subseteq \mathsf{RN}(\mathcal{T})$*.*

For any $r \in \mathsf{RN}(\mathcal{T})$, $\mathfrak{d}(r) = 1$ captures an expression of total importance on a role (over a corresponding nested concept) and $\mathfrak{d}(r) = 0$ captures an expression of total importance on a nested concept (over a corresponding role). This notion is inspired by [6] in which sim is used with different values of the discount factors in the similarity application on SNOMED CT (cf. Section 5 of [1]). For example, **Agent 2** may believe that knowing actual surrounding information is more important. Thus, $\mathfrak{d}(\mathsf{hasHabitat}) = 0.3$ and $\mathfrak{d}(\mathsf{liveIn}) = 0.3$ might be expressed this situation.

Like others, many role names might be not assigned the discount values by agents, i.e. those role names are not mapped to the corresponding values. Here, we use the default discount value of 0.4 for $r \in \mathsf{RN}(\mathcal{T})$ in case $\mathfrak{d}(r)$ is not defined. This amount of fixed value is influenced by sim [5] where the value of 0.4 is used for the discount factor when the similarity between two existential restrictions is considered. The total function $\mathfrak{d}_0(r) = 0.4$ for all $r \in \mathsf{RN}(\mathcal{T})$ is called the *default role discount factor*.

Hence, we now conclude that a preference profile π is a quintuple of preference functions, viz. $\mathfrak{i}^{\mathfrak{c}}, \mathfrak{i}^{\mathfrak{r}}, \mathfrak{s}^{\mathfrak{c}}, \mathfrak{s}^{\mathfrak{r}}$, and $\mathfrak{d}$. When a preference profile π is given, π can thereby influence the calculation of CSMs (cf. Section 5 of [1]).

Definition 6 (Preference Profile). *A* preference profile*, in symbol* π*, is a quintuple* $\langle \mathfrak{i}^{\mathfrak{c}}, \mathfrak{i}^{\mathfrak{r}}, \mathfrak{s}^{\mathfrak{c}}, \mathfrak{s}^{\mathfrak{r}}, \mathfrak{d} \rangle$ *where* $\mathfrak{i}^{\mathfrak{c}}, \mathfrak{i}^{\mathfrak{r}}, \mathfrak{s}^{\mathfrak{c}}, \mathfrak{s}^{\mathfrak{r}}$*, and* $\mathfrak{d}$ *are as defined above and the* default preference profile*, in symbol* π_0*, is the quintuple* $\langle \mathfrak{i}^{\mathfrak{c}}_0, \mathfrak{i}^{\mathfrak{r}}_0, \mathfrak{s}^{\mathfrak{c}}_0, \mathfrak{s}^{\mathfrak{r}}_0, \mathfrak{d}_0 \rangle$ *where* $\mathfrak{i}^{\mathfrak{c}}_0, \mathfrak{i}^{\mathfrak{r}}_0, \mathfrak{s}^{\mathfrak{c}}_0, \mathfrak{s}^{\mathfrak{r}}_0$*, and* $\mathfrak{d}_0$ *are as defined above.*

For example, let denote a preference profile of the **Agent 1** and the **Agent 2** by π_1 and π_2, respectively. We conclude that $\pi_1 = \langle \mathfrak{i}^{\mathfrak{c}}, \mathfrak{i}^{\mathfrak{r}}, \mathfrak{s}^{\mathfrak{c}}, \mathfrak{s}^{\mathfrak{r}}, \mathfrak{d} \rangle$, where $\mathfrak{i}^{\mathfrak{c}}(\mathsf{Reptile}) = 2$,

i^r(hasColor) $= 2$, and s^c(Green, Yellow) $= 0.5$ indicating that (i) *Reptile* is important, (ii) *Having skin color* is important, and (iii) *Green* and *Yellow* are similar, respectively. In addition, we conclude that $\pi_2 = \langle i^c, i^r, s^c, s^r, \partial \rangle$, where i^c(Reptile) $= 2$, i^r(hasHabitat) $= 2$, i^r(liveIn) $= 2$, s^r(t, u) $= 0.1$, ∂(hasHabitat) $= 0.3$, and ∂(liveIn) $= 0.3$ indicating that (i) *Reptile* is important, (ii) *Having habitat* is important; (iii) *Having living environment* is important, respectively.

Concept Similarity Measures under Preferences

Preference profile π intends to be a generic guideline for a development of concept similarity measures used under an agent's preferences. It suggests concept similarity measures for any DLs (e.g., $\mathcal{FL}_0$, $\mathcal{ELH}$, $\mathcal{ALC}$, and so on) which permit an agent's preferences to influence the calculation should expose all elements of π. Given an arbitrary CSM $\sim$, a *concept similarity measure under preference profile* π is a function $\overset{\pi}{\sim} : \mathsf{Con}(\mathcal{L}) \times \mathsf{Con}(\mathcal{L}) \to [0,1]$. A CSM $\sim$ is called *preference invariant w.r.t. equivalence* if $C \overset{\pi}{\sim} D = 1$ iff $C \equiv D$ for any π (cf. Sect. 4 of [1] for its formal definition and properties).

Developing such functions can be done from scratch or by generalizing existing CSMs. For example, we have generalized sim as a function called sim^{π} and published our theoretical development in [1]. Our sim^{π} is also *preference invariant w.r.t. equivalence*, meaning that similarity between two equivalent $\mathcal{ELH}$ concepts is always one regardless of agents' preferences. When π_0 (cf. Definition 6) has been used as the value of a preference profile, $\mathsf{sim}^{\pi_0}(C, D) = \mathsf{sim}(C, D)$ for $C, D \in \mathsf{Con}(\mathcal{ELH})$. Table 1 shows our investigation on existing CSMs and found that none of them, to the best of our knowledge, comply with our preference profile.

Table 1. State-of-the-art CSMs intrinsically use a preference profile

CSM	i^c	i^r	s^c	s^r	∂
sim^{π}	✔	✔	✔	✔	✔
sim [6]					✔
simi [10]	✓		✔		

4 Conclusion and Future Work

We present the preference profile π as design guidelines for a development of concept similarity measures under preferences in DLs. CSMs which exposes all elements of preference profile will be appropriate to use when human perceptions are involved (See more details in [1]). This work is still in preliminary stage. We have intended to explore the possibility of implementations on realistic ontologies formulated in DLs, especially DL $\mathcal{ELH}$. We are also interested in investigating deeply desirable properties concept similarity measures under preference profiles must have. It would also be interesting to explore preference profile π when used beyond the realm of similarity measures, i.e., relaxed instance checking and relaxed instance retrieval.

Acknowledgments. This research is partially supported by Thammasart University Research Fund under the TU Research Scholar, Contract No. TOR POR 1/13/2558; the Center of Excellence in Intelligent Informatics, Speech and Language Technology, and Service Innovation (CILS), Thammasat University; and the JAIST-NECTEC-SIIT dual doctoral degree program.

References

1. Racharak, T., Suntisrivaraporn, B., Tojo, S.: Are Frog and TreeLizard similar? guidelines for concept similarity measures under an agent's preferences in dls. Technical-Report BS2015-01, School of Information, Computer and Communication Technology, Sirinthorn International Institute of Technology, Thammasart University, Sirinthorn International Institute of Technology, Thailand (2015). http://ict.siit.tu.ac.th/~sun/dw/doku.php?id=reports
2. Son, T.C., Pontelli, E.: Planning with preferences using logic programming. In: Lifschitz, V., Niemelä, I. (eds.) LPNMR 2004. LNCS (LNAI), vol. 2923, pp. 247–260. Springer, Heidelberg (2003)
3. Lichtenstein, S., Slovic, P. (eds.): The Construction of Preference. Cambridge University Press, New York (2006)
4. Scherer, K.: What are emotions? and how can they be measured? Soc. Sci. Inf. **44**(4), 693–727 (2005)
5. Suntisrivaraporn, B.: A similarity measure for the description logic el with unfoldable terminologies. In: INCoS, pp. 408–413 (2013)
6. Tongphu, S., Suntisrivaraporn, B.: Algorithms for measuring similarity between elh concept descriptions: a case study on snomed ct. J. Comput. Inf. (2015). Accessed 7 May 2015 (to appear)
7. Racharak, T., Suntisrivaraporn, B.: Similarity measures for FL0 concept descriptions from an automata-theoretic point of view. In: Proceedings of The 6th Annual International Conference on Information and Communication Technology for Embedded Systems (ICICTES 2015) (2015)
8. Tongphu, S., Suntisrivaraporn, B.: A non-standard instance checking for the description logic ELH. In: Proceedings of the International Conference on Knowledge Engineering and Ontology Development (KEOD 2014) (2014)
9. Tongphu, S., Suntisrivaraporn, B.: On desirable properties of the structural subsumption-based similarity measure. In: Supnithi, T., Yamaguchi, T., Pan, J.Z., Wuwongse, V., Buranarach, M. (eds.) Semantic Technology. Lecture Notes in Computer Science (LNCS), pp. 19–32. Springer, Heidelberg (2015)
10. Lehmann, K., Turhan, A.-Y.: A Framework for Semantic-Based Similarity Measures for elh-concepts. In: Cerro, L.F., Herzig, A., Mengin, J. (eds.) JELIA 2012. LNCS, vol. 7519, pp. 307–319. Springer, Heidelberg (2012)

A Contrastive Study on Semantic Prosodies of Minimal Degree Adverbs in Chinese and English

Zhong Wu and Lihua Li[✉]

Wuhan University of Engineering Science, Wuhan 430200, Hubei, China
zhongwu2000@163.com, liliua0806@sina.com

Abstract. From the perspective of cross-language, this paper, by using Chinese and English corpus, examines the semantic prosody similarities and differences between *shaowei* (稍微 'a little') and "a little", the minimal degree adverbs in these two languages. The research findings indicate that the semantic prosody of *shaowei* is neutral while "a little" is negative.

Keywords: Semantic prosody · Minimal degree adverb · "shaowei" (稍微 'a little') · A little

1 Introduction

Semantic prosody describes the way in which nodes can be perceived with positive or negative associations through frequent occurrences with particular collocations. From the perspective of function, semantic prosody is usually divided into three categories, namely, positive prosody, neutral prosody and negative prosody [1].

Negative prosody associates with words referring to the unpleasant things or qualities and positive prosody connects words with positive or pleasant connotations while neutral prosody tends to co-occur with words have no obvious tendency. Naturally, neutral prosody is also called mixed prosody by Wei Nai-xing and with which the collocated words are neither positive nor negative, i.e. it shows a mixed semantic features because some of the collocation words are positive while others are negative [2].

Semantic prosody, with the property of probability, can not eliminate exception, but the probability is relatively low [3]. For example, by searching the corpus, it was found that the collocation possibility of "rather" with words of negative connotations is much higher than the other two categories. Therefore, "rather" was regarded as a word with negative semantic prosody. However, this does not exclude the possibility that it occasionally co-occurred with positive or neutral words, for instance, with "fine".

The previous theoretical studies and empirical researches on semantic prosody are mostly from monolingual perspective, especially in English. And several other studies, focused on bilingual semantic prosody comparison, mainly concern the contrast between English and other European languages and few attempts have been made to compare the semantic prosody between Chinese and English, because they belong to different language family. This paper, from the perspective of cross-language, will

© Springer International Publishing Switzerland 2016
G. Qi et al. (Eds.): JIST 2015, LNCS 9544, pp. 209–215, 2016.
DOI: 10.1007/978-3-319-31676-5_15

analyze the prosody feature of minimal degree adverbs in Chinese and English, revealing the speaker's implicit attitude and intention in communication.

2   The Magnitude Characteristic of the Minimal Degree Adverbs–*Shaowei* and "a Little"

As a subcategory of adverb in both Chinese and English, degree adverbs can be further divided according to their semantic and syntactic features. The meaning degree adverbs play a key role and they can be re-categorized by order of magnitude.

This study, based on typological framework, will divide adverbs in three categories, namely, minimal degree adverbs, high degree adverbs and excessive degree adverbs [4]. After that, the most typical minimal degree adverbs in Chinese, *shaowei*, and its English equivalent, "a little", were selected as examples to find the similarity and difference of their semantic prosody. This paper also tries to find out the corresponding rules of Chinese-English minimal degree adverbs and give a reasonable explanation by analyzing data derived from corpus-based software.

"*Shaowei*" is a common degree adverb representing minimal degree in modern Chinese. According to Modern Chinese Dictionary (2010), "*shaowei*" is an adverb and denotes "a small number or minimal degree." Similarly, in Modern Chinese Dictionary of Function Words (2005), "*shaowei*" indicates minimal degree and it tends to collocate with positive adjectives and exclude negative words. However, from the perspective of typology, Liu holds that minimal degree adverbs, like *shaowei* or "*shao*" (稍 'a little), often modify negative adjectives for the expression of euphemism [4]. But there is no conclusion that "*shaowei*" co-occurs with negative words or positive ones.

According to Modern Chinese-English Dictionary (1993), "*shaowei*" is an adverb and its English equivalents are "a little", "a bit", "a bit of" and "a trifle". Longman English Grammar (2011) also indicates that "a little" is a degree adverb and collocates with adjectives, adverbs, comparatives and verbs. Obviously, "a little", as a minimal degree adverb in English, equates with "*shaowei*" to a certain extent.

3   The Design of the Research

This research adopts keyword searching method KWIC (keywords in context) to search words co-occurred with keywords. The retrieval software is Antconc3.2.1. The study is based on the exploration of CCL (developed by the Center for Chinese Linguistics of PKU), a Chinese on line corpus with over 477 million Chinese characters. The Brown Corpus (hereinafter referred to as BROWN), a classical corpus with 1 million words, is also used in the course of the investigation. In this study, the node words are *shaowei* and "a little" respectively, and the span is - 4 ~ + 4. After extracting concordance with keywords in Chinese and English corpus, the colligation is established after observation and analysis. Then, collocation words within the colligation will be classified into positive, neutral and negative categories and the characteristics of their semantic prosodies will be summarized.

4 Analysis and Findings

4.1 A Study on the Semantic Prosody of "Shaowei"

By searching the CCL, 2218 effective concordances containing *shaowei*, the key word, are hit. After downloading the first 200 and deleting those occurred in the oral text, 195 concordances are left. The colligation of *"shaowei"* falls into three types as follow:

(1) adv + adj
Eg. 1：大白菜　　价钱　〈稍微〉　好　一点。
　　　dabaicai　　jiaqian　<shaowe>i　hao　yi dian
　　　'The price of cabbage is <slightly> higher.'

(2) adv + v
Eg. 2：只要　　〈稍微〉　　推　一下　就　行了，　　无需　费 多少　　力气。
　　　zhiyao　<shaowei>　tui　yixia　jiu　xingliao , wuxu　fei duoshao　liqi
　　　'Yao Ming looks like a stone on the edge of cliff, just <a little> push, without much effort....'

(3) adv + adv
Eg 3：由于　　　〈稍微〉　　有些　早产，　所以　母亲 "甜甜"　的　奶水　不足。
　　　youyu　　<shaowei >　youxie　zaochan ,suoyi　muqin "tian tian " de　naishui　buzu
　　　'Because of <a little> premature, milk of mother, *"tiantian"*, is insufficient'

It should be noted that *"shaowei"* mainly collocates with minimal-degree adverbs, such as "a bit" or "a bit of."

Because the software does not automatically recognize the negative, neutral and positive connotation, classification has to be done manually by the researcher. Some collocations, like "slightly stable", "slightly comfortable" can be classified as positive while others, such as "slightly lag behind", "slightly lower" can be classified as negative. Some other neutral words, which often describe the property of things and actions, can not be judged as positive or negative within their colligation, such as "large, small, high, low, hot, cool" and so on. In that case, clauses containing such colligation have to be taken into account necessarily. For example:

Eg. 4：后者　的　头球　〈稍微〉　　高　出　横梁。
　　　houzhe　de　tou qiu　<shaowei>　gao　chu　hengliang
　　　'The latter's heading　<slightly>　over the crossbar.'

Eg. 5：：想　　为　　层次　〈稍微〉　高　一点　的　观众　　拍　　些　有
　　　xiang wei　cen ci　<shaowei>　gao　yidian　de　guanzhong　pai　xie　you
　　　品味　的　　片子。
　　　pinwei　de　pian zi
　　　'Someone wants to make some better films for <slightly> higher civilized viewers.'

The associative meaning of *"shaowei* + high" can be judged neither positive nor negative by the structure itself. However, in example 4, "the heading over the crossbar" is negative because the heading misses the goal. In example 5, "slightly higher civilized viewers" are better from the speaker's point of view. Thus, the structure is classified as positive.

According to the above description, the associative meaning of *shaowei* collocation can be classified in Chart 1:

As shown in Chart 1, the statistical results support neither the hypothesis that *shaowei* has negative semantic prosody nor that it has positive one, because positive and negative collocations are between 42 %–44 %. That is to say, they are all close to 50 % of the total and there is no significant difference. According to Wei, *shaowei* has neutral semantic prosody or mixed prosody [2].

4.2   A Study on the Semantic Prosody of "a Little"

In the BROWN online corpus, 301 concordances containing the keyword "a little" are hit. After getting rid of 109 concordances, such as the adjective "little" and fixed phrase like "a little while", 192 are left. They are all minimal degree adverbs and fall into three colligations as bellow:

(1) ADV+ADJ
     Eg. 6 to be <a little> reckless
(2) V + ADV
     Eg. 7: had neglected them <a little>
(3) ADV+ADV
     Eg. 8: he arrived <a little> late

It's worthy of noting that "a little" generally follows verbs, so its structure is "V + ADV".

Just like the Chinese pattern, if the semantic prosody of "a little" can not be judged within its colligation, the researchers should determine the associative meaning within a larger range. As the example shown below:

Eg. 9: "fortunately, the house is <a little> far from here"

Eg. 10: "…the house is <a little> far from here, we have to walk a long path…"

If judged within the colligation, it is not clear that the collocation unit is positive or negative, because the expression is just an objective description of traits, for instance, "the house a little far away from here". However, if judged within a larger unit, it is well known that "a little far" is positive in example 9 because of the parenthesis, "fortunately", in the sentence. In contrast, "a little far" is negative in example 10 because someone "have to walk a long path".

According to the classification principles above, the associative meaning of "a little" collocation can be classified in Chart 2:

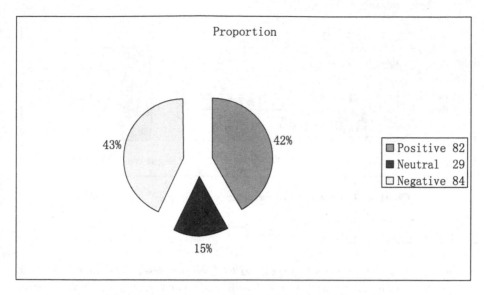

Chart 1. The pie chart of the collocation of '*shaowei*'

As shown in Chart 2, "a little" modifies verbs, adjectives and adverbs and their corresponding comparatives, indicating low degree. The positive collocation of "a little" accounts for over 61 % and the neutral one is around 18 %, while the negative collocation of "a little" is 20 % only. According to Sinclair [5], the conclusion can be reached that "a little" has negative prosody, which supports the previous hypothesis.

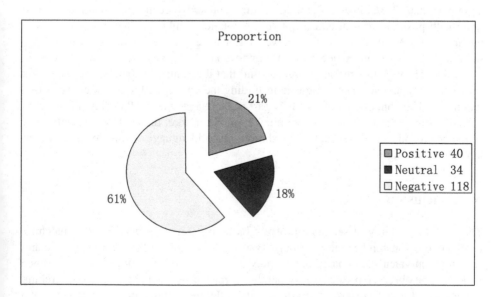

Chart 2. The pie chart of the collocation of 'a little'

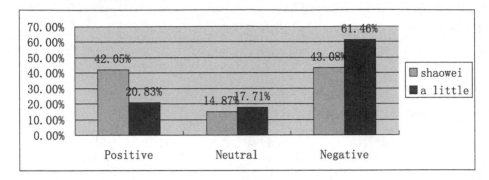

Chart 3. The histogram of the prosody of '*shaowei*' and 'a little'

5   Contrastive Analysis

Through observing the BROWN corpus and CCL corpus, researchers found that both *shaowei* and "a little" have three types of colligation, namely, they can both co-occur with verbs, nouns and adverbs. However, "a little" frequently collocates with comparatives of English adjectives and adverbs, because English belongs to Indo-European language family and has morphological changes, while Chinese, a Sino-Tibetan language, can only express comparatives by means of lexical devices.

In order to facilitate the comparison of *shaowei* and "a little", analyzing the distribution and the proportion of their semantic prosody, Chart 3 is drawn as below:

The contrastive analysis, by means of corpus software, shows that the semantic prosodies of *"shaowei"* and "a little" do not fully correspond with each other. As shown in Chart 3, *shaowei* has neutral or mixed prosody because it can collocate with units with positive, neutral and negative connotation while "a little" tends to have a negative one. Although the authoritative dictionary equates *shaowei* to "a little", we hold that the equation is debatable after the contrastive analysis of their semantic prosodies. Through this research, we can find that the contrastive analysis of semantic prosodies play an ever increasing role in finding the corresponding rules between two languages. The connotation of words is part of language and insufficiency may lead to difficulties in understanding and communication. Therefore, the contrastive analysis of semantic prosodies is significant in translation, second language acquisition and foreign language teaching [6].

6   Conclusion

The contrastive study of semantic prosody based on corpus reveals that the matching level and relationship between language systems are no longer limited to syntactic and semantic equivalence, but based on a variety of real and specific mutual correspondence which can be observed on the basis of parallel corpus. This study, by means of searching Chinese and English corpus, has examined the differences and similarities between two minimal degree adverbs in Chinese and English, namely, *shaowei* and "a little". It has been found that *shaowei* possesses neutral semantic prosody while "a little" tends to have negative one.

However, the collocation words within the colligation were examined manually in present study, which is not suit for a lager scale investigation. In future study, a new dictionary should be compiled in which every word was evaluated and graded according to its positive or negative tendency. For instance, "love" was evaluated as "+1" because it is positive and "hate" was evaluated as "−1" because of its negative tendency while neutral word was evaluated as "0". After extracting data from corpus, all collocated words of the key word, based on the dictionary, will be graded and calculated by computer. Therefore, according to the result and the preset standard, the prosody of the key word can be determined automatically.

References

1. Stubbs, M.: Collocations and semantic profiles: On the cause of the trouble with quantitative methods. Funct. Lang. **2**(1), 23–55 (1995)
2. Naixing, W.: A Definition and Research System of Words Collocation. Shanghai Jiaotong University Press, Shanghai (2002)
3. Partington, A.: Utterly Content in Each other's company: Semantic prosody and semantic preference. Int. J. Corpus Linguist. **9**, 131–156 (2004)
4. Danqing, L.: A Handbook of Grammar Investigation, p. 534. Shanghai Education Press, Shanghai (2008)
5. Sinclair, J.: Corpus, concordance and collocation. Oxford University Press, Oxford (1991)
6. Xiulian, Z., Shujuan, L.: A corpus driven study on cross-language semantic prosody-taking "too" and "tai" as an example. Mod. Chin. **12**, 78–80 (2010)

Question Answering

A Graph Traversal Based Approach to Answer Non-Aggregation Questions over DBpedia

Chenhao Zhu[1]($\boxtimes$), Kan Ren[1], Xuan Liu[1], Haofen Wang[2],
Yiding Tian[1], and Yong Yu[1]

[1] Shanghai Jiao Tong University, Shanghai, China
{chzhu,kren,liuxuan0526,killa,yyu}@apex.sjtu.edu.cn
[2] East China University of Science and Technology, Shanghai, China
whfcarter@ecust.edu.cn

Abstract. We present a question answering system over DBpedia, filling the gap between user information needs expressed in natural language and a structured query interface expressed in SPARQL over the underlying knowledge base (KB). Given the KB, our goal is to comprehend a natural language query and provide corresponding accurate answers. Focusing on solving the non-aggregation questions, in this paper, we construct a subgraph of the knowledge base from the detected entities and propose a graph traversal method to solve both the semantic item mapping problem and the disambiguation problem in a joint way. Compared with existing work, we simplify the process of query intention understanding and pay more attention to the answer path ranking. We evaluate our method on a non-aggregation question dataset and further on a complete dataset. Experimental results show that our method achieves best performance compared with several state-of-the-art systems.

Keywords: Question Answering · Non-aggregation questions · Linked data · Graph traversal · Path ranking

1 Introduction

Nowadays great volume of linked data has been produced efficiently in both research and industrial areas, such as DBpedia [12], YAGO [19], Freebase [3], Google's Knowledge Graph and Microsoft's Satori. Each of them contains a wealth of valuable knowledge stored in the form of predicate-argument structures, e.g., (*subject, predicate, object*) triples. Meanwhile, the quality of linked data (coverage and accuracy) is also increasing effectively with the help of well-designed research work and community efforts. Consequently, a lot of work for various purposes have been developed by taking linked data as the underlying knowledge base.

However, surfing linked data requires ontological knowledge beforehand. Even SPARQL is the most common query language of RDF data, reading and surfing linked data web requires professional skills and extra learning cost, which makes

© Springer International Publishing Switzerland 2016
G. Qi et al. (Eds.): JIST 2015, LNCS 9544, pp. 219–234, 2016.
DOI: 10.1007/978-3-319-31676-5_16

common people unwilling to deeply browse. It is crucial to propose techniques to fill the gap between users information needs and implicit data models including schema and instances.

Natural language Question Answering over Linked Data (QALD) may commendably achieve this goal while maintaining advantages of knowledge base. Lopez et al. [14] surveyed the trend of question answering in semantic web and revealed some challenges as well as opportunities in natural language question answering.

Generally speaking, the main challenge of understanding a query intention in a structural form is to solve two problems, which are semantic item mapping and semantic item disambiguation. Semantic item mapping is recognizing the semantic relation topological structures in the natural language questions and then semantic item disambiguation is instantiating these structures regarding a given knowledge base. Unger et al. [20] relies on parsing a question to formulate a SPARQL template to capture the intention of a user query. This template is then instantiated using statistical entity identification and predicate detection. He et al. [9] combines Markov networks with first-order logic in a probabilistic framework to achieve the goal of semantic mapping. Zou et al. [24] proposes a method to jointly solve semantic item mapping and disambiguation problems by reducing question answering to a subgraph matching problem. However, they all focus on the semantic item mapping problem by adopting well-designed templates or complex model. In contrast, we simplify the semantic item mapping problem and pay more attention to solving the disambiguation problem.

In this paper, we propose a graph traversal-based method to solve semantic item mapping and disambiguation problems. We solve the semantic item mapping problem in a simple way by parsing the question text to generate matched topological structures. Based on these structures, we start from the detected entities in a question text. Then we traverse from these entities to find connected predicates and resources in the knowledge base. Next our approach uses a jointly ranking algorithm to solve the disambiguation problem. Meanwhile we implement a constraint matching assessment of the answer type to find the best answer.

Since entities contain the most important semantic information in a natural language query, the intuition is that we may get the right answer by finding the most suitable path in the knowledge base around the detected entities.

Fig. 1. Graph traversal example

Take the question *"Who is the mayor of Berlin?"* as an example. As is shown in Fig. 1, firstly we find the mention *"Berlin"* and link it to the resource **res:Berlin** in the knowledge base. Then our system traverses the subgraph around the resource and calculates ranking scores of the connected predicates. In this case, we find that **dbo:leader** matches better since its label *"leader"* has higher relatedness with *"mayor"*. Next our system makes a judgement that the traversing process stops here and **res:Klaus_Wowereit** matches the answer type constraint. So we return **res:Klaus_Wowereit** as the final answer.

The contributions in this paper are summarized as follows:

- We present an approach for answering non-aggregation questions over DBpedia, which fills the gap between user information needs expressed in natural language and a structured query interface expressed in SPARQL.
- We present an approach which is simple in structure and employs relatively lightweight machinery compared with existing work concentrating on complex models and training.
- We compare our approach with several state-of-the-art systems on public dataset and achieve the best performance on non-aggregation questions. Moreover, we extend our approach to the complete dataset and also achieve the best performance.

The rest of this paper is organized as follows. Section 2 briefly describes related work. Section 3 introduces the proposed graph traversal approach in detail. Section 4 shows our experiments, including dataset collection, evaluation metric, comparison between our approach and several state-of-the-art work and error analysis. Finally, Sect. 5 concludes the paper and points out the future work.

2   Related Work

Question answering (QA), which is to return the exact answers to a given natural language question, is a challenging task and has been advocated as a key problem for advancing web search. Previous work is mainly dominated by keyword based approaches, while recent blossom of large-scale knowledge bases have enable numerous KB-based systems. A KB-based QA system answers a question by directly querying structured knowledge, which can be retrieved using a structured query engine.

Recently many works have been published in this field. Apart from [9,20,24] discussed in Sect. 1, PowerAqua [13] proposes a natural language user interface making people query and explore semantic web content more convenient. Two research work [22,23] present an ILP(Integer Linear Programming) method to translate a natural language question into a structured SPARQL query. The Paralex system [6] studies question answering as a machine learning problem and induces a function that maps open-domain questions to queries over a database of web extractions. Meanwhile, Shekarpour et al. [18] presents an approach for question answering over a set of interlinked data sources. This approach firstly employs a Hidden Markov Model to determine the most suitable resources for a use-supplied query and secondly constructs a federated formal query using the disambiguated resources and linking structure of underlying datasets. And two research work [1,2] develop semantic parsing techniques that map natural language utterances into logical form queries, which can be executed on a knowledge base. Xu et al. [21] develops a transition-based parsing model to do semantic parsing for aggregation questions.

However, most of these methods focus on translating a question to a SPARQL query. Meanwhile many methods need many well designed manual rules. Also lots of them focus on recognizing the inherent structure of user's query intention using different semantic parsing techniques implemented by complex models. In contrast, we aim to finding the most appropriate path rather than generate SPARQL query templates directly. And we simplify the process of query intention understanding and pay more attention to the answer path ranking. Our approach is simple in structure but effective in terms of performance.

3   Framework

The overall framework is shown in Fig. 2. Our approach aims to find the most appropriate path in the knowledge base rather than generate SPARQL query templates directly. The whole process contains three phases:

- **Question Understanding**: In this phase, the system detects the query's topological pattern, trying to capture its intention. To achieve this goal, we use an entity linking method to detect the mention-entity pairs, of which the mention is used for the phrase boundary identification. And next we build a list of topological patterns to discover the structure by taking advantage of the parsing result of the query.

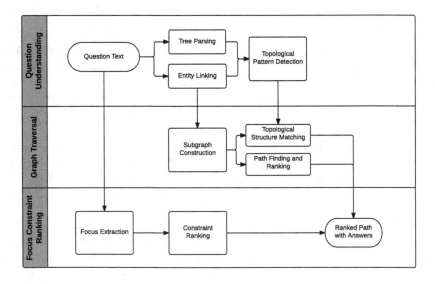

Fig. 2. Overall framework

- **Graph Traversal**: In this phase, we firstly build a subgraph of the underlying knowledge base rooted from entities we've found in last step. Then we use a jointly ranking method to find the most appropriate traversal path in the subgraph. The topological structure is used for semantic item mapping and judging traversal stop condition.
- **Focus Constraint**: We extract a phrase describing the answer directly from the query, which is called a *focus*. Then we use this information to help modify final path ranking scores.

We solve the semantic item mapping problem during the question understanding phrase and the disambiguation problem in the next two phrases. After above three phases, the overall path candidates ranking list is obtained. The answers found along the path with highest score will be returned.

3.1   Question Understanding

In this phase, we focus on question parsing and topological structure extraction. Our system digests a natural language query and outputs its corresponding topological structure. Figure 3 shows the whole process and results along with an example.

Given a question text, we firstly use an entity linking method to detect mentions in the query and link them to the resources in the knowledge base. In this step we consider class entities and category entities such as **res:Actor** are of little importance. So in the entity linking phrase, we discard those corresponding mentions. Meanwhile, each mention-entity pair will be attached with a confidence score. We set a global threshold to discard those linking results with relatively low

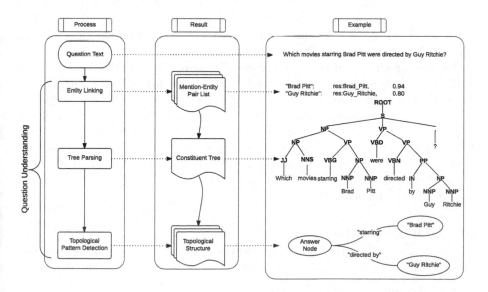

Fig. 3. Process of question understanding with question example

confidences. As in the example, our system detects two mentions *"Brad Pitt"* and *"Guy Ritchie"* with corresponding entity results and confidence scores. In our experiment, we use the Wikipedia Miner tool[1] [16] to detect the mentions from the question text and get the corresponding entity linking results. Empirically the parameter of min-Probability is set as 0.15.

In the next step, we extract topological structure of user intention with regard to our topological patterns. We start from the constituent tree of a question text. And Table 1 lists our topological patterns.

Each pattern captures one form of relationship between two arguments. For example, the pattern $VB \rightarrow VB+NP$ means the VB and NP on the right side are the children of the VB on the left side with regard to the constituent tree. We may derive the binary relation with arguments from it. The third column in the table demonstrates the extraction result of the given example. While ANSNODE is a wildcard representing the current answer we are looking for.

Table 1. Topological pattern list

ID	Pattern	Example	Extraction Result
1	VB → VB+NP	Who produces Orangina?	ANSNODE - "produces" - "Orangina"
2	VP → VB+PP	Which television shows were created by John Cleese?	ANSNODE - "created by" - "John Cleese"
3	NP → NP+PP	Who is the mayor of Berlin?	ANSNODE - "mayor of" - "Berlin"
4	SQ → VB+NP+VP	When was Alberta admitted as province?	ANSNODE - "admitted as province" - "Alberta"

In our running example, our system extracts two relationships using Pattern 1 and 2. The extraction results are presented in Fig. 3.

We use a recursive method to discover all the relations in the question text. Note that there is one case we should handle carefully. If the entity linking phase produces a mention that is fit for one topological pattern, our extraction stops the recursive process immediately. Take question *"Who wrote the book The Pillars of the Earth?"* as an example. Pattern 3 is matched in the phrase *"The Pillars of the Earth"* while it is a piece of a mention detected by the entity linking method. So our recursive algorithm will skip processing it. In our experiment, we use the Stanford Parser[2] [15] to parse the question text and generate the corresponding constituent tree.

3.2 Graph Traversal

Our approach does not generate any SPARQL templates for a given question text. We just traverse the knowledge base from entities found in the entity

---

[1] http://wikipedia-miner.cms.waikato.ac.nz/.

[2] http://nlp.stanford.edu/software/index.shtml.

linking phase. The main three steps are subgraph construction, path finding and topological structure matching. The whole process has been presented in Fig. 4.

In the first step, we begin from linked entities in the knowledge base and construct a subgraph surrounding with them. Given an entity e in the mention-entity list, we root the graph from e and expand one layer. By this means we may obtain all the entities including resources and classes around e within 1 step distance. Then we retrieve more entities around entity e with 2 step distance. In the end, we have K layers subgraphs in the knowledge base around entity e. Here K is the longest distance between two nodes in topological structure.

In our running example, the entity list provides two resources **res:Brad_Pitt** and **res:Guy_Ritchie**. We begin from each of them and construct a subgraph on the right part in Fig. 4. Here K is 1. And the two subgraphs of the linked entities here are combined into a single one.

After the subgraph construction, we implement a path finding and ranking algorithm to get the answer. The method also begins from detected entities. The goal is to find the path which is the most appropriate to match the information in the question text.

We start from a known entity e and make one step ahead. After that we obtain many one step pathes rooted at e. Our system calculates the semantic similarity between the predicates around the entity e and the phrase text from the edge in the topological structure. Then we will make one more step outward and get many two-length paths. For these new predicates, we repeat the same calculation of semantic similarity between predicate labels and the corresponding phrase text. To the end we may obtain many paths starting from entity e with ranking scores of each edge.

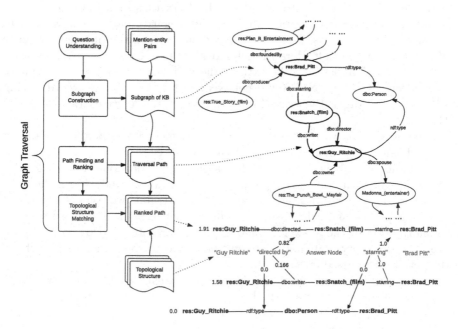

Fig. 4. Process of graph traversal with question example

One question is how we judge that we will stop expanding outward. Here we make two stop conditions. The first one is simple that we will stop finding when it gets to the outmost layer of the subgraph. The second condition is based on the topological structure. Our system will decide that if the path obtained above matches the topological structure, and discard those not matching ones.

As is shown in Fig. 3, the topological structure is triangular. Two known entities link to the answer node with two pieces of phrase text. The answer node is a wildcard representing what we search for in the knowledge base. We use the phrase text of the edge in topological structure to match a predicate.

In Fig. 4, our approach finds three different pathes. Each path contains two known entities. Our ranking module calculates scores for each predicate with phrase texts. At last our system will jointly rank each path with regard to ranking scores for predicates. The ranking list shows that the most appropriate path is the first one and **res:Snatch_(film)** is our answer (till now).

3.3   Focus Constraint

As is defined in Sect. 3, a *focus* is a phrase in the question text describing the answer directly. For example, *"television shows"* is the focus of *"Which television shows were created by John Cleese?"*. It implies that the answer is a type of television show. In this part, we extract the focus from a question text for the calculation of additional ranking scores to pathes obtained above.

Intuitively, we may derive focus information from interrogatives. More specifically, we extract *person* and *organization* from *"who"*, *place* from *"where"* and *date* from *"when"*.

Besides, we extract a focus based on the POS tags of a question. In our approach, the longest noun phrase after the interrogative part of a question will be considered as the focus. The interrogative part means the phrase in a query used to start the question. Here the interrogative part could be *"Give me all"*, *"What"*, *"Which"* and so on. We extract the first word after the interrogative part with the POS tag "NN" (or other NN-like tags) to the last word having continuous "NN" tag as the focus. In our running example, we extract *"movies"* as the focus.

In the next step, we use the focus phrase to modify the predicate ranking result and calculate a matching score of the answer type.

For predicate ranking, we use the focus phrase as additional information to rank the last predicate. The last predicate is the nearest predicate in the path to get to the answer entities. For answer type constraint, we use the headword of the focus phrase and calculate the similarity between the headword and the type information of the answers. Answer entities are obtained by ranked path above. The similarity matching score will be added to the ranking score of the path. Detail is discussed in Sect. 3.4.

Now that we've calculated the ranking score for each path with answer type similarity score, we may obtain the best answer of the question text.

3.4   Path Ranking

A path score is composed of two parts, including a predicate score and a type score. Moreover, the predicate score is related to all predicates matched in the topological structure. We can formulate a path score as Eq. 1.

$$PathScore = \frac{1}{m} \sum_{i=1}^{m} PredicateScore_i + TypeScore \qquad (1)$$

Next in this section, we will discuss the details about the predicate ranking and jointly path score calculation.

For predicate ranking, we consider the semantic similarity between a predicate and the phrase text extracted from the topological structure. More specially, to the predicates leading to the final answers we add some information from focus constraint since it also contributes to the predicate identification. The detail is showed in Algorithm 1. We use the UMBC Semantic Similarity Service[3] [8] to calculate the semantic similarity between two words.

Algorithm 1. Predicate Ranking

Input: A predicate p; Phrase text t;
Output: The Predicate p's ranking score s;
 1: Find all labels $Labels$ of p in RDF Repository;
 2: **for** each label $l \in Labels$ **do**
 3:    **for** each word w of l **do**
 4:      **for** each word tw of t **do**
 5:        Calculate the word semantic similarity wss between w and tw;
 6:      **end for**
 7:      Set the maximal wss as w's score ws;
 8:    **end for**
 9:    Set l's score ls as the arithmetic mean of each word's wss;
10: **end for**
11: Set s as the maximal ls;

For path ranking, we combine the information from the possible predicates, the topological pattern and the answer type. If the question satisfies the one step path, the path ranking score is simply the sum of the predicate ranking score and the answer type score generating from the focus constraint. While, for the two step path ranking, it needs more work. The detail is show in Algorithm 2. Since the knowledge base is in a large scale and an entity has too many predicates, we need to restrict the candidate size for efficiency. In our experiment the maximal size of candidate predicates is set to 5 in each step during the phrase of path ranking.

---

[3] http://swoogle.umbc.edu/SimService/.

Algorithm 2. Two Step Path Ranking

Input: Predicate pairs *pps* matching the corresponding topological pattern; Phrase
   Texts *pts* of the corresponding predicates; Minimal semantic similarity threshold τ
   for each predicate;
Output: Path ranking scores;
 1: **for** each predicate pair $pp \in pps$ **do**
 2:    Calculate the two predicates' ranking score $s1$, $s2$ using *pts*;
 3:    **if** $s1 < \tau || s2 < \tau$ **then**
 4:       Remove this predicate pair;
 5:       **Continue**;
 6:    **end if**
 7:    Set *pp*'s predicate pair Score *ps* as the arithmetic mean of $s1$ and $s2$;
 8:
 9:    Find all answers according to the path constructed by *pp* and calculate the type
       constraint score *ts* according to their matching degree;
10:
11:    Set the path's ranking score as the sum of *ps* and *ts*;
12: **end for**

---

4 Experiment

4.1 Dataset

QALD task [4] is the only benchmark for the KB-based QA problem. It includes
both the underling knowledge base and the natural language questions. QALD
is based on DBpedia knowledge base.

We use the QALD-3 test dataset[4] in our experiments. This dataset has
99 questions in total. And it includes various types of questions. Table 2 lists the
details. To build a non-aggregation questions dataset (QALD-3-NA), we filter the
questions which need operations such as count, filter or order by, meanwhile, we
filter the questions which have no answers in DBpedia or can not be solved using
DBpedia singly. Then our new dataset QALD-3-NA has 61 questions in total.

Table 2. Question type in QALD-3 test dataset

Type	Non-aggregation	Count	Filter	Order By	Boolean	Out Of Scope
Num	61	4	7	4	7	16

Considering the public dataset QALD-3 used by other state-of-the-art systems also contains aggregation questions while our approach focuses on answering non-aggregation questions, meanwhile few of state-of-the-art systems are
publicly avaiable, we design two aspects of experiments to verify the validity of
our method.

---

[4] http://greententacle.techfak.uni-bielefeld.de/cunger/qald/3/data/dbpedia-test.xml.

Firstly, we collect a dataset by selecting all non-aggregation questions from the public dataset. Then we compare our method with one state-of-the-art system gAnswer [24] on the new dataset, since gAnswer [24] also uses DBpedia as the underlying KB and offers an publicly online demo[5], which makes it possible to serve as a contrast. This experiment is used to indicate our performance on non-aggregation questions directly.

Secondly, we make an experiment on the complete public dataset and compare our method with two state-of-the-art systems DEANNA [22], gAnswer [24] and all participating systems in the QALD-3 competition. If we achieve better performance, it can verify the validity of our method on non-aggregation questions from another point of view.

4.2   Evaluation Metrics

To enable the comparison with other state-of-the-art systems and the systems in QALD-3 competition, we adopt the same evaluation metrics used in QALD-3. That is to say, firstly, for each of the questions, we evaluate its precision, recall and F1-measure. Next we compute the overall precision and recall taking the average mean of all single precision and recall values, as well as the overall F1-measure [4].

4.3   Evaluation Results

On the QALD-3-NA dataset, we compare our method with one state-of-the-art system gAnswer [24]. Table 3 shows the evaluation result of average precision, average recall and average F-1 score. Meanwhile, it shows the number of question our system can answer, the number of right and partially right answers among them. We report the 30 questions which we can answer correctly in Table 6 and the 13 questions which we can answer partially in Table 7.

Table 3. Evaluation result on QALD-3-NA test dataset

	Total	Processed	Right	Partial	Avg.Recall	Avg.Precision	Avg.F-1
gAnswer demo	61	38	21	7	0.41	0.45	0.42
Ours	61	53	30	13	**0.67**	**0.61**	**0.61**

To further indicate our performance, we apply our approach on the QALD-3 dataset compared with two state-of-the-art systems, namely, gAnswer [24] and DEANNA [22]. Also, we compare our approach with all the participating systems in QALD-3 competition, whose results are reported in the QALD-3 overview paper [4]. Table 4 shows the results.

From Table 3, we can see that our approach achieves much better performance on the non-aggregation questions. Meanwhile, the result in Table 4 also verifies

---

[5] http://59.108.48.18:8080/gAnswer/ganswer.jsp.

Table 4. Evaluation result on QALD-3 test dataset

	Total	Processed	Right	Partial	Avg.Recall	Avg.Precision	Avg.F-1
Ours (NA)	99	53	30	13	0.42	0.38	0.38
Ours (Total)	99	60	31	17	**0.46**	**0.40**	**0.40**
gAnswer demo	99	50	23	11	0.30	0.30	0.28
gAnswer [24]	99	76	32	11	0.40	**0.40**	**0.40**
DEANNA [22]	99	27	21	0	0.21	0.21	0.21
CASIA [10]	99	52	29	8	0.36	0.35	0.36
Scalewelis [11]	99	70	32	1	0.33	0.33	0.33
RTV [7]	99	55	30	4	0.34	0.32	0.33
Intui2 [5]	99	99	28	4	0.32	0.32	0.32
SWIP [17]	99	21	15	2	0.16	0.17	0.17

the validity of our approach from another point of view. On the QALD-3 dataset, we show our two results. **Ours (NA)** is evaluated by adapting our performance on QALD-3-NA to the complete QALD-3 dataset, which simply multiplies the ratio of the question numbers of two dataset according to the evaluation metrics. That is to say, we set the precision, recall and F1-score as 0 on the other types questions and then get a global result. Although it is a little unfair to us, **Ours (NA)** outperforms most of state-of-the-art systems and only has a narrow gap to the best one of state-of-the-art systems. However, **Ours (Total)** achieves best performance, especially in the evaluation of recall.

The reason why our F-1 scores on two datasets are not equal to the ratio of the question numbers of two dataset has two explanations. Firstly, for those questions which need count-operations, using our approach, we can get the results and what we need to do further is to simply count the number. Secondly, for those questions which need filter-operations, our approach can get right answers together with wrong answers which should be filtered. The details of the contribution from different types of questions are showed in Table 5.

Table 5. Contribution of different types of questions on QALD-3 dataset

Type	Total	Processed	Right	Partial	Avg.Recall	Avg.Precision	Avg.F-1
Non-aggregation	61	53	30	13	0.42	0.38	0.38
Count	4	3	1	0	0.01	0.01	0.01
Filter	7	4	0	4	0.03	0.01	0.02
Order By	4	0	0	0 ·	0	0	0
Boolean	7	0	0	0	0	0	0
Out Of Scope	16	0	0	0	0	0	0
Sum	99	60	31	17	0.46	0.40	0.41[a]

[a]Compared to 0.40 in Table 4, the inconsistency here is the result of a round-off error.

Table 6. The QALD-3-NA Questions that can be answered correctly in our system

ID	Questions
Q3	Who is the mayor of Berlin?
Q4	How many students does the Free University in Amsterdam have?
Q7	When was Alberta admitted as province?
Q19	Give me all people that were born in Vienna and died in Berlin
Q20	How tall is Michael Jordan?
Q22	Who is the governor of Wyoming?
Q24	Who was the father of Queen Elizabeth II?
Q30	What is the birth name of Angela Merkel?
Q35	Who developed Minecraft?
Q38	How many inhabitants does Maribor have?
Q42	Who is the husband of Amanda Palmer?
Q43	Give me all breeds of the German Shepherd dog
Q44	Which cities does the Weser flow through?
Q45	Which countries are connected by the Rhine?
Q53	What is the ruling party in Lisbon?
Q54	What are the nicknames of San Francisco?
Q56	When were the Hells Angels founded?
Q58	What is the time zone of Salt Lake City?
Q65	Which instruments did John Lennon play?
Q66	Which ships were called after Benjamin Franklin?
Q68	How many employees does Google have?
Q71	When was the Statue of Liberty built?
Q74	When did Michael Jackson die?
Q76	List the children of Margaret Thatcher
Q81	Which books by Kerouac were published by Viking Press?
Q83	How high is the Mount Everest?
Q85	How many people live in the capital of Australia?
Q86	What is the largest city in Australia?
Q98	Which country does the creator of Miffy come from?
Q100	Who produces Orangina?

4.4 Error Analysis

Here we provide the error analysis of our approach. There are four key reasons for the error of some questions in our approach. The first one is the entity linking error. In some cases, we fail to find the correct entities in a question text. The second one is the semantic item mapping error. It contains two aspects of reasons. In some cases we fail to extract the structure correctly, and in other cases,

Table 7. The QALD-3-NA Questions that can be answered partially in our system

ID	Questions
Q2	Who was the successor of John F. Kennedy?
Q8	To which countries does the Himalayan mountain system extend?
Q17	Give me all cars that are produced in Germany
Q21	What is the capital of Canada?
Q28	Give me all movies directed by Francis Ford Coppola
Q29	Give me all actors starring in movies directed by and starring William Shatner
Q41	Who founded Intel?
Q48	In which UK city are the headquarters of the MI6?
Q64	Give me all launch pads operated by NASA
Q67	Who are the parents of the wife of Juan Carlos I?
Q72	In which U.S. state is Fort Knox located?
Q84	Who created the comic Captain America?
Q89	In which city was the former Dutch queen Juliana buried?

since our current algorithm of finding predicates around an entity does not consider the subject or object role of entities detected at the beginning or generated as the intermediate result, we make some mistakes. For example, in the case of *"Who are the parents of the wife of Juan Carlos I?"*, we not only correctly get the parents of Juan Carlos I's wife, but also make a mistake by getting her children at the same time. This is the key reason that we have a relatively high recall compared with the precision. In this case, the recall equals 1 while the precision has a loss and only gives the value 0.40. The third one is the path ranking error. We fail to rank the right path in the first place in some cases. The fourth one we call it the restriction error. In our method we use the focus in a query to grade predicates and answer type constricts. However, a focus should be transferred to part of the final SPARQL query which restricts the answer type. The percentage of each reason is showed in Table 8.

Table 8. Error analysis

Type	Percentage
Entity linking error	26 %
Structure extraction error	16 %
Semantic role error	13 %
Path ranking error	13 %
Restriction error	29 %
Others	3 %

5   Conclusion and Future Work

In this paper, we propose a graph traversal-based approach to answer non-aggregation natural language questions over linked data. Our system starts from the detected entities and puts more attention on ranking the predicate paths. By translating the natural language question to a topological structure and mapping the structure to the linked data utilizing both the semantic features of the phrase similarity and type constraints. Compared with existing work, our method employs relatively lightweight machinery but has good performance. In the future, we will adapt our method to answer aggregation questions, meanwhile we will try to answer a question by combining multiple knowledge bases to make our system more adaptable and more powerful.

Acknowledgments. This work was partially supported by the National Science Foundation of China (project No: 61402173) and the Fundamental Research Funds for the Central Universities (Grant No: 22A201514045).

References

1. Berant, J., Chou, A., Frostig, R., Liang, P.: Semantic parsing on freebase from question-answer pairs. In: EMNLP, pp. 1533–1544 (2013)
2. Berant, J., Liang, P.: Semantic parsing via paraphrasing. In: Proceedings of ACL. vol. 7, p. 92 (2014)
3. Bollacker, K., Evans, C., Paritosh, P., Sturge, T., Taylor, J.: Freebase: A collaboratively created graph database for structuring human knowledge. In: Proceedings of the 2008 ACM SIGMOD International Conference on Management of Data, pp. 1247–1250. ACM (2008)
4. Cimiano, P., Lopez, V., Unger, C., Cabrio, E., Ngonga Ngomo, A.-C., Walter, S.: Multilingual question answering over linked data (QALD-3): Lab overview. In: Forner, P., Müller, H., Paredes, R., Rosso, P., Stein, B. (eds.) CLEF 2013. LNCS, vol. 8138, pp. 321–332. Springer, Heidelberg (2013)
5. Dima, C.: Intui2: A prototype system for question answering over linked data. In: Proceedings of the Question Answering over Linked Data lab (QALD-3) at CLEF (2013)
6. Fader, A., Zettlemoyer, L.S., Etzioni, O.: Paraphrase-driven learning for open question answering. In: ACL (1), pp. 1608–1618. Citeseer (2013)
7. Giannone, C., Bellomaria, V., Basili, R.: A hmm-based approach to question answering against linked data. In: Proceedings of the Question Answering over Linked Data lab (QALD-3) at CLEF (2013)
8. Han, L., Kashyap, A., Finin, T., Mayfield, J., Weese, J.: Umbc ebiquity-core: Semantic textual similarity systems. Proc. Sec. Joint Conf. Lexical Comput. Seman. 1, 44–52 (2013)
9. He, S., Liu, K., Zhang, Y., Xu, L., Zhao, J.: Question answering over linked data using first-order logic. In: Proceedings of Empirical Methods in Natural Language Processing (2014)
10. He, S., Liu, S., Chen, Y., Zhou, G., Liu, K., Zhao, J.: Casia@ qald-3: A question answering system over linked data. In: Proceedings of the Question Answering over Linked Data lab (QALD-3) at CLEF (2013)

11. Joris, G., Ferré, S.: Scalewelis: a scalable query-based faceted search system on top of sparql endpoints. In: Work Multilingual Question Answering over Linked Data (QALD-3) (2013)
12. Lehmann, J., Isele, R., Jakob, M., Jentzsch, A., Kontokostas, D., Mendes, P.N., Hellmann, S., Morsey, M., van Kleef, P., Auer, S., et al.: Dbpedia-a large-scale, multilingual knowledge base extracted from wikipedia. Semantic Web (2014)
13. Lopez, V., Fernández, M., Motta, E., Stieler, N.: Poweraqua: Supporting users in querying and exploring the semantic web. Seman. Web 3(3), 249–265 (2012)
14. Lopez, V., Uren, V., Sabou, M., Motta, E.: Is question answering fit for the semantic web?: A survey. Seman. Web 2(2), 125–155 (2011)
15. Manning, C.D., Surdeanu, M., Bauer, J., Finkel, J., Bethard, S.J., McClosky, D.: The stanford corenlp natural language processing toolkit. In: Proceedings of 52nd Annual Meeting of the Association for Computational Linguistics: System Demonstrations, pp. 55–60 (2014)
16. Milne, D., Witten, I.H.: An open-source toolkit for mining wikipedia. Artif. Intell. **194**, 222–239 (2013)
17. Pradel, C., Peyet, G., Haemmerlé, O., Hernandez, N.: Swip at qald-3: Results, criticisms and lesson learned. Valencia, Spain (2013)
18. Shekarpour, S., Ngonga Ngomo, A.C., Auer, S.: Question answering on interlinked data. In: Proceedings of the 22nd International Conference on World Wide Web, pp. 1145–1156. International World Wide Web Conferences Steering Committee (2013)
19. Suchanek, F.M., Kasneci, G., Weikum, G.: Yago: A core of semantic knowledge. In: Proceedings of the 16th International Conference on World Wide Web, pp. 697–706. ACM (2007)
20. Unger, C., Bühmann, L., Lehmann, J., Ngonga Ngomo, A.C., Gerber, D., Cimiano, P.: Template-based question answering over rdf data. In: Proceedings of the 21st International Conference on World Wide Web, pp. 639–648. ACM (2012)
21. Xu, K., Zhang, S., Feng, Y., Huang, S., Zhao, D.: What is the longest river in the usa? semantic parsing for aggregation questions. In: Twenty-Ninth AAAI Conference on Artificial Intelligence (2015)
22. Yahya, M., Berberich, K., Elbassuoni, S., Ramanath, M., Tresp, V., Weikum, G.: Natural language questions for the web of data. In: Proceedings of the 2012 Joint Conference on Empirical Methods in Natural Language Processing and Computational Natural Language Learning, pp. 379–390. Association for Computational Linguistics (2012)
23. Yahya, M., Berberich, K., Elbassuoni, S., Weikum, G.: Robust question answering over the web of linked data. In: Proceedings of the 22nd ACM International Conference on Conference on Information and Knowledge Management, pp. 1107–1116. ACM (2013)
24. Zou, L., Huang, R., Wang, H., Yu, J.X., He, W., Zhao, D.: Natural language question answering over rdf: A graph data driven approach. In: Proceedings of the 2014 ACM SIGMOD International Conference on Management of Data, pp. 313–324. ACM (2014)

Answer Type Identification
for Question Answering
Supervised Learning of Dependency Graph Patterns from Natural Language Questions

Andrew D. Walker[1(✉)], Panos Alexopoulos[2], Andrew Starkey[1], Jeff Z. Pan[1],
José Manuel Gómez-Pérez[2], and Advaith Siddharthan[1]

[1] University of Aberdeen, Aberdeen, UK
andrew.walker.05@aberdeen.ac.uk,
{a.starkey,jeff.z.pan,advaith}@abdn.ac.uk
[2] Expert System, Amsterdam, Netherlands
{palexopoulos,jmgomez}@expertsystem.com

Abstract. Question Answering research has long recognised that the identification of the type of answer being requested is a fundamental step in the interpretation of a question as a whole. Previous strategies have ranged from trivial keyword matches, to statistical analyses, to well-defined algorithms based on shallow syntactic parses with user-interaction for ambiguity resolution. A novel strategy combining deep NLP on both syntactic and dependency parses with supervised learning is introduced and results that improve on extant alternatives reported. The impact of the strategy on QALD is also evaluated with a proprietary Question Answering system and its positive results analysed.

1 Introduction

Question Answering (QA) technologies were envisioned early on in the artificial intelligence community. Indeed, at least 15 experimental English language QA systems were described by Simmons (1965). Notable early attempts include BASEBALL (Green Jr., et al., 1961) and LUNAR (Woods 1973; 1977) with new technologies and resources often prompting a new wave of QA solutions using them. For example: relational databases (Codd 1970) with PLANES (Waltz 1978); Google (Brin and Page 1998) with AskMSR and AskMSR2 (Brill et al. 2001); the semantic web (Berners-Lee et al. 2001) by Bernstein et al. (2005); and Wikipedia (Wales and Sanger 2001) by Buscaldi and Rosso (2006).

Depending on the structure of the data being used to derive the answers, there are two high-level strategies typically employed: that of information retrieval with large corpora of text where tokens and patterns in the query are matched to documents and either the document itself or a snippet thereof is returned as the answer; and that of natural language processing (NLP) where the grammatical and/or syntactic structure of the questions are analysed against structured data like relational databases or ontologies. As both approaches have shown potential

© Springer International Publishing Switzerland 2016
G. Qi et al. (Eds.): JIST 2015, LNCS 9544, pp. 235–251, 2016.
DOI: 10.1007/978-3-319-31676-5_17

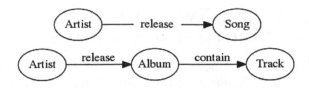

Fig. 1. Simple (top) and complex (bottom) song/artist entity relationship

there is no clear superior approach upon which to build. This study concerns the NLP approach.

Consider a question such as "What songs has Elvis Costello released?". The relevant data might preferably be encoded as in Fig. 1 (top). But perhaps instead songs will be listed as tracks, belong to albums, and the albums will have the relation to the artist, as in Fig. 1 (bottom).

In order to answer the question then, several tasks need to be undertaken. "Elvis Costello" must be identified with a member of the **artist** entity type, "songs" must be mapped to the entity type **track**, and the path connecting the two via the **album** entity type must be recognised. It may also be necessary to realise that the question is asking for every entity that matches this pattern, and not simply whether or not such a pattern exists, as would be the case for "Has Elvis Costello released any songs?", but this is outwith the scope of this study.

The task being examined in this study is that of establishing the answer-type identifier (ATI): a single word in the question that best identifies the semantic class of entity being sought by the question for its answers. This is a critical and non-trivial step that has so far largely been treated as a means to other ends rather than a subject of research in its own right. Note we take a single word as the ATI rather than an n-gram because of the approach used. See Sect. 3 for why.

Simple strategies, such as using the head of the first noun-phrase following a *wh*-word, have been shown to perform inadequately (Krishnan et al. 2005). Many QA systems have taken note of the type of question, as indicated by the lead question word such as *What, Why, Who, Where*, etc. This works fairly well for *Where* (locations), *When* (date-times) and *Who* (people or organisations) but in general seems to be insufficient (Damljanovic et al. 2010; Krishnan et al. 2005).

More advanced strategies include: the application of part-of-speech patterns (Sect. 2.1); and support vector machines (SVMs) and conditional random fields (CRFs) (Sect. 2.2). However, in neither case is an ATI strictly the goal. In the former, it is one possible output with another notion constituting the preference and in the latter they seek the related, but distinct, notion of an informer span.

Our contributions are in defining a general purpose task in answer-type identification that requires only natural language processing; i.e. no knowledge base. We design a supervised learning method that performs better than baselines found implemented in previous work and which is fully automated – requiring no interaction with the user from initiation to completion.

2 Related Work

2.1 Question Focus

The QA system called "FREyA" – **F**eedback, **R**efinement and **E**xtended Vocabulary **A**ggregation (Damljanovic et al. 2012) – made use of a user-interactive algorithm drawing from syntactic heuristics and ontology-based lookup to find, preferably, the question "focus" and secondarily an ATI (Damljanovic et al. 2010), working in the context of Question Answering over Linked Data (QALD).

A question focus was defined by Moldovan et al. (2000) as "a word or sequence of words which define the question and disambiguate the question by indicating what the question is looking for". Often, the question focus corresponds with the type of answer being sought (as in "What is the capital of Uruguay?", with focus "capital"), and in others where the type is not explicit as a noun-phrase it corresponds to an entity or concept of which the desired data should be found (as in "When did the Jurassic Period end?", with focus "Jurassic Period" even though the answer-type is a date-time).

FREyA examined the syntactic parse of the question to find pre-pre-terminals[1]. It examined each pre-pre-terminal in sequence and examined its part-of-speech (POS) to decide if it should be accepted as an ATI, question focus, or be skipped. The algorithm is best understood as put by Damljanovic et al. (2010).

The output of the algorithm could be: an ATI, a question focus, or `null`. In evaluation, an ATI was returned for 18 % of a 250-question gold standard, a correct question focus for 69.6 %, an incorrect question focus for 11.6 %, and `null` for 0.8 %.

Further processing involving examination of the target ontology on which the questions were being posed and possibly soliciting user feedback culminated in the identification of the answer type. After this consolidation step, the answer types for 53.2 % had been automatically identified correctly – that is, with no user-interaction. A single interaction with the user augmented this with a further 45.6 % to 98.8 %.

2.2 Informer Span

Krishnan et al. (2005) introduced the notion of "informer spans", which are short contiguous sequences of 1–3 tokens that constitute adequate clues for question classification (e.g. with informer spans in *italics*: *Who* wrote Hamlet?, How much does a rhino *weigh*? or What *country*'s president was shot at Ford's Theater). Classifications were to some predefined taxonomy of labels – such as `LENGTH`, `POPULATION`, `DISEASE`, `CONSTELL` (-ation), and `ROLE`. Human-annotated informer spans performed much better than common simple heuristics from QA systems, so SVMs were trained using various features from the questions and

---

[1] A pre-pre-terminal is a node for which every child is a pre-terminal. A pre-terminal is a node with a single child which is itself a leaf.

their informer spans. They found best results from SVMs trained on question bigrams and WordNet (Miller 1995) hypernyms of n-grams from the informer span, for all possible n. A strategy to automatically identify informer spans using CRFs achieved an accuracy of 86.2 %.

2.3 Statistical Question Classification

A statistical approach (Prager et al. 2002) to question classification worked on the premise that in natural text an ATI will often be found in conjunction with other entities of the same or similar (by hypernymy or meronymy) semantic class. Entities were annotated with one or more labels from a fixed taxonomy of semantic classes – just like before (Krishnan et al. 2005). Four measures for any given term were derived from a corpus, describing co-occurrence with other labelled entities within a 3-line scope (the line with the term plus the previous and subsequent lines), within the same sentence, within a phrase separated by at most one other word and the average of the other three.

It was found that the correct classification was normally listed as the first or second result, with errors typically being due to polysemy (such as "plant" possibly referring to living organisms from the kingdom Plantae or a factory, among others) or stylistic or domain-dependent factors. Performance ranged from 33 % to 89 %, depending on the kind of question, the corpus and measure used.

2.4 Summary

These three strategies use terminology describing distinct concepts from this study so here we draw attention to the differences. An answer-type identifier ATI is a word (or its lemma) in the question that refers to the semantic class of the answer being sought. A question focus (Damljanovic et al. 2012; Moldovan et al. 2000) is a word, or sequence of words, describing what is being sought or the entity of which a property is being sought; the ATI would be the property itself. Informer spans (Krishnan et al. 2005) pertain to the distinct but similar problem

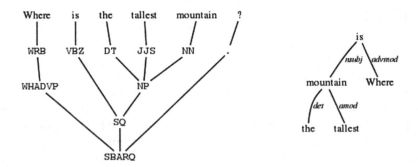

Fig. 2. Syntactic (left) and dependency (right) parses of "Where is the tallest mountain?"

of question classification but are in practice very similar to ATIs. Prager et al.'s 2002 study assumes an ATI is pre-determined in order to classify questions.

We make use of dependency parses, provided by the Stanford Parser (De Marneffe et al. 2006). These attempt to portray the dependency relationships between tokens in the input sentence, as opposed to syntactic parses which show their grammatical structure. The difference is portrayed in Fig. 2.

Rather than manually explore the various parses for common patterns to single out the ATIs, we designed and implemented a strategy to learn them automatically; described in Sect. 3. An evaluation of the results of this strategy is in Sect. 4, including discussion on its limitations. These lead naturally to a small set of manual augmentations to the overall strategy, detailed in Sect. 5 with a re-evaluation. The impact of using our ATI strategy is evaluated against a smaller dataset in Sect. 6. We conclude and suggest future work in Sect. 7.

3 Pattern Learning

The strategy seeks to learn grammatical structures for the identification of the ATI automatically, rather than by manual exploration. Patterns have already been described for syntactic constituent parses (Damljanovic et al. 2010) but the smaller structure of dependency graphs might yield better results. These dependency parses were provided by the Stanford Parser (De Marneffe et al. 2006). Although the dependency graphs are derived from the syntactic parses, there may be multiple syntactic structures that yield the same dependency graph, or similar ones. Dependency graphs constitute a higher level of abstraction.

3.1 Patterns

A pattern is a graph G consisting of a finite non-empty set V of pairs $\langle lemma, pos \rangle$ that are called *vertices*, together with a set E of distinct triples $\langle gov \in V, dep \in V, label \rangle$ called *edges* (where *gov* is the "governor" of a relation and *dep* is its "dependent"), a root node $r \in V$ and a target $t \in V$. Lemmas[2], parts-of-speech (POS)[3] and edge labels can all be wild-cards (denoted as ? in figures).

An extension of Ullmann's 1976 algorithm for sub-graph isomorphism was implemented allowing for directed graphs, and the extra criteria of lemmas, parts of speech and edge labels that must also match, possibly as wild-cards. A pattern G_α is *isomorphic* to a sub-graph of a semantic graph G_β if and only if there is an injective function f which maps the vertices and edges of G_α to those of this sub-graph such that directed adjacency is preserved.

---

[2] A word's root form without any morphological indications of tense, number, mood etc. E.g., the lemma of 'children' is 'child', of 'quickest' is 'quick', of 'processing' is 'process'.

[3] A category to which a word is assigned in accordance with its syntactic function, such as verb, noun and others depending on language. In this study we use POS abbreviations from the Penn Treebank tag set (Marcus et al. 1993).

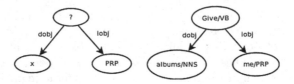

Fig. 3. Graph pattern and example match of "Give me all albums by Elvis."(Note that "by Elvis" is omitted from the image as not being relevant to the match. It would be attached to the `albums/NNS` node.)

Figure 3 exemplifies a situation in which the ATI is the dependent of a direct-object (`dobj`) relation governed by the root, which itself must also govern an indirect-object (`iobj`) relation to a personal pronoun (`PRP`). Note that the qualification that the albums be "by Elvis" is not considered pertinent.

A pattern can be measured against a corpus of questions (or more specifically, their dependency graphs). A pattern may or may not match a dependency graph at all. A matched pattern may or may not identify the correct target (but will always match *something* as the target). There is no sense in dividing those unmatched further. Thus, there are three values: true positive (TP) – a correct match from the application of a pattern to a dependency graph ($f(t_\alpha) = t_\beta$); false positive (FP) – an incorrect match from the application of a pattern to a dependency graph ($f(t_\alpha) \neq t_\beta$); and unanswered (U) – a non-match from the application of a pattern to a dependency graph (no f was found).

A number of operations are possible on patterns that might change the TP, FP and U counts. Making a pattern more specific cannot cause it to find matches on more dependency graphs, but might eliminate previous matches, both TPs and FPs (which would increase Us). Such operations could include specifying the value of a wild-card POS, lemma or edge-label; or adding a new edge. Generalising a pattern cannot eliminate previous matches but might introduce new matches, again both TPs and FPs (which would decrease Us). Generalising operations could include wild-carding a POS, lemma or edge-label; or removing an edge (and any consequently isolated non-root vertices).

3.2   Method

We learn patterns by iteratively testing every possible pattern against a corpus and selecting the best, removing the matched questions for the next iteration. Each question in the corpus is parsed with the Stanford Parser to generate the POS and dependency relations. Dijkstra's 1959 algorithm is used to find the shortest path from the parser-given root to the labelled target, as both of these are required for the patterns, and lemmas and edge-labels are defined as wild-cards. These are then extended by adding edges as much as possible to minimise FPs. The best scoring of these is logged, the matched questions removed (both TPs and FPs) and the process restarted, until there are no questions unanswered.

On a given corpus, a pattern with one or more FPs is always considered worse than another with none. Two patterns with identical TP and FP counts are equivalent regardless of the actual numbers. We want to minimise FPs so having fewer FPs is more important than more TPs. For two patterns with the same FP count, the one with more TPs is considered better.

A pattern and its resultant match counts imply a hypothetical *best possible* extension, whereby its TPs remain the same but its FPs are eliminated to zero. While navigating the search space of pattern extensions, a pattern's best possible extension is compared with the current best pattern found. If the best possible extension is still worse than the current best found by the aforementioned criteria, it is not explored further.

4   Evaluation

As we have found no other study exactly comparable with this (FREyA finds ATIs but prefers question foci when possible), we show comparisons with four baselines. Our sources of data are described in Sect. 4.1 along with how and why they were filtered. The annotation strategy is documented in Sect. 4.2, describing how ATIs were chosen for training. The baselines are described and their performances detailed in Sect. 4.3 before our learning method (Sect. 3.2) is evaluated in Sect. 4.4. Limitations are discussed in Sect. 4.5 and an attempt at ascertaining the resultant coverage in Sect. 5.2.

4.1   Data Sources

A series of open challenges on Question Answering over Linked Data (QALD) have run[4] providing sets of manually written questions for training and evaluation, the training sets also offering answers. A total of 300 questions from the 3 annual challenges were available at the time of writing. However, a number of these are duplicates, the elimination of which results in 232. Another 2 are duplicates with just a change in punctuation (exclamation mark and period). These were also removed along with one that appears to have been entered with a typographic error, leaving 229 unique questions. Another two questions differed only in capitalisation, but both of these were left included as capitalisation is taken into account by most parsers and may yield different results. Some other questions were also basically the same but as they were formulated in different grammatical structures they could highlight limitations of the process described here, and thus were left unchanged.

Li and Roth (2002) produced a training set of 5500 questions labelled with query types, as per their own classification system, for the University of Illinois at Urbana- Champaign (UIUC) shared corpora[5]. We took a subset of 1000 of these and removed manner questions (e.g. "How do I ...") and reason questions

---

[4] http://greententacle.techfak.uni-bielefeld.de/cunger/qald/.

[5] http://cogcomp.cs.illinois.edu/Data/QA/QC/.

(e.g. "Why is ..."), as these questions typically require extensive explanations. Some more entries had not been parsed and/or presented correctly in the corpus and were also removed, yielding a total of 870 labelled questions. Examples of these latter include "What country did the Nazis occupy for 1, CD NNS IN NNP NNP NNP." where the last words are represented as their POS and "What Boris Pasternak book sold 5,0 copies to become the fiction best-seller of 1958 ?" where "5,000" has been incorrectly tokenized to "5,0".

The QALD corpus will be referred to as C_Q and the UIUC as C_U hereafter. The union of the two will be denoted C_{all}. (Note that $C_U \cap C_Q = \varnothing$.)

It will be important to bear in mind that the entries in C_Q were hand-crafted. This means that they are not questions that have been posed by genuine users of a system seeking a response, but designed to test candidate systems on recognised grammatical and syntactic variations that English questions can take. Walker et al. (2014) showed these question sets can be skewed towards favouring certain grammatical structures more than real end-users would. The entries in C_U on the other hand have nothing to do with computer systems – those questions are more *natural* that might be put to another person. They may contain errors and may be out of context (e.g. "What was her real name?").

4.2 Annotation

Each of the questions in C_Q and C_U were annotated manually with the words that represent the best ATI – that is, the word that refers to the semantic class of the desired answer most exclusively. Only single words were selected, not phrases and no alternatives.

Typically, the ATI was the head of a noun-phrase, like "Give me all live *albums* by Michael Jackson.". For some questions, though, no such term exists. For example, in "For how *long* was Tina Turner married to Ike Turner?" it is the adjective "long" that describes the type of answer sought by the querent (i.e. a duration of time[6]). In "*Who* created Goofy?" only the word "Who" refers to the individual requested. Conversely, for a paraphrase such as "*Who* was the *creator* of Goofy?" there is also the noun "creator" which is considered a *better* description by virtue of the fact it refers to a narrower category of individuals.

Of the 229 questions in C_Q, 69 could not be annotated with a noun. Of these, 45 were *wh-* questions. That is, questions beginning with *wh-* words. 21 were *who* questions – though as noted before, some *who* questions do have better ATIs elsewhere. Just one was a *where* question – normally these questions contain noun ATIs. The remaining 23 began either "*when*", "*since when*" or "*until when*". A further 19 of these 69 were verbs "*did*", "*is*" and "*was*"; all corresponding to boolean questions. The remaining 5 were of the form "*how [ADJ]*".

Some special cases are trivial. For "*When* did Finland join the EU?", with the ultimate answer type of a date, we have only "when" as a possible ATI. This is almost universal for "when" questions. Similarly, for boolean questions

---

[6] http://wordnetweb.princeton.edu/perl/webwn?o0=1&o8=1&o1=1&s=long&i=10#c.

like "*Is* Natalie Portman an actress?" only "is" reflects this boolean quality (the same for "did" and "was").

Examples follow, with the ATIs in *italics*:

1. How *far* can a man travel in outer space?
2. How *many* colleges are in Wyoming?
3. In what Olympic *Games* did Nadia Comaneci become popular?
4. Italy is the largest producer of *what*?
5. Name a *tiger* that is extinct.
6. What *age* followed the Bronze Age?
7. *Where* do chihuahuas come from?
8. Who is Snoopy's *arch-enemy*?

Example 2 demonstrates a situation where the question is arguably unclear. Strictly speaking, the question is requesting the **number** of colleges that meet the defined criterion, but perhaps a user would phrase their question in this way actually seeking a **list** of all such colleges. We chose to annotate it strictly, as a QA system could still present the list as justification for the numeric answer. Example 3 highlights a situation where the single-word selection for ATI training seems inappropriate: the question seeks an instance of an Olympic Games, not any games event. However, assuming the parser works as intended, compound nouns like this one are identifiable by the "noun compound modifier" (nn) dependency.

4.3   Baselines

Four simple baselines are: WH – first *wh*-word; FI – first instance of a *wh*-word, "do" or "be" (when lemmatised); NPH – head of the first noun-phrase; and NPHFI – the first occurrence of either NPH or FI. The results of applying these to both corpora are presented in Table 1.

Table 1. Results of applying 4 baselines to both corpora: C_Q and C_U.

Baseline	C_Q			C_U		
	TP %	FP %	U %	TP %	FP %	U %
WH	19.65	38.43	41.92	33.33	53.45	13.22
FI	29.69	49.78	20.52	33.79	63.10	3.10
NPH	55.02	44.98	0.00	53.91	45.98	0.11
NPHFI	48.47	51.53	0.00	35.40	64.60	0.00

4.4   Cross-Validation

The strategy was tested with cross-validation on each corpus C. We selected p questions from C at random for training and tested on the remainder. This was done 10 times and the mean average results are shown in Fig. 4.

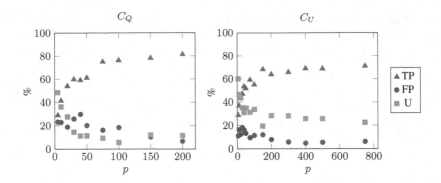

Fig. 4. Cross-validation results; x-axis denotes sizes of training sets

The results for C_Q seem to tend towards 85 % TPs, 6 % FPs and 9 % Us, whereas for C_U the trend seems to be towards 73 % TPs, 5 % FPs and 22 % Us. The difference in performance shown between these corpora is interesting. As mentioned before (Sect. 4.1), each entry in C_Q was created for the QALD problem – that is, addressing a computer system – whereas C_U contains any form of question a person might pose, to a computer system or otherwise (also having grammatical issues in some cases). As an aside, it is interesting that the questions designed to be put to computer systems are being interpreted better by our system but it may simply be due to the greater diversity of question forms in C_U, or both.

4.5   Error Analysis

The system relies on accurate outputs from the parser. Cer et al. (2010) report precision and recall at 87.3 % and 87.1 % respectively for unlabelled attachment, and 84.2 % and 84.1 % respectively for labelled attachment. Parser errors in training can result in inappropriate pattern sequences being learned. Parser errors in testing, however, might correspond to the errors in training and hence yield an accurate response nevertheless. Alternative, more probable, consequences include the novel dependency graphs being unmatched at all or matching to inappropriate patterns.

Questions of the form "Give me. . . " matched correctly with high consistency, suggesting the parser handles this structure (or at least the relevant parts of it) consistently and that it constitutes a useful strategy for end-users. Questions of the form "How many. . . " occasionally failed; in one case due to a bad parse and the other from matching a more general pattern first. The boolean questions beginning "Is. . . ", "Was. . . " and "Did. . . " were matched inconsistently.

5   Manual Augmentations

Given the analyses presented in Sect. 4.5 a number of deliberate amendments were made to the process to improve accuracy. The question indicators "Is. . . ",

"Was. . ." and "Did. . ." were taken as keywords for the boolean answer-type – C_{all} contains no counter-examples. These questions were detected and returned with the keywords' lemmas ("be" and "do") as the ATIs. The "How many. . ." questions similarly were captured returning the ATI "many". These were applied before the original approach – in both training and testing. A system making use of this process would catch these as special cases, along with *wh*-words.

The system was also augmented with last-resort post-processing where the original approach and pre-processing augmentation return a U for a given question. The typical *wh*- question words, "who", "where", "which", "when" and "what" are often themselves the desired ATI, so for U questions beginning with these terms, the system returns them instead.

Finally, as a simple measure to reduce diversity in question structure, questions leading with prepositions were paraphrased to begin with their internal *wh*-term and end with the preposition. Due to time constraints this was done manually, but studies on similar paraphrasing have already been conducted (McKeown 1983; Siddharthan 2006) suggesting it is perfectly possible to be automated.

In which century was Galileo born? → Which century was Galileo born in?

5.1 Re-Evaluation

These augmentations yield significant improvements. Figure 5 portrays these data for comparison. Training on C_Q then seems to tend towards 93 % TPs, 4 % FPs and 3 % Us; and for C_U towards 85 % TPs, 13 % FPs and 2 % Us.

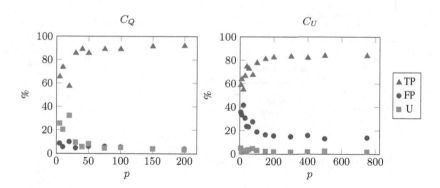

Fig. 5. Cross-validation results with manual augmentations; *x*-axis denotes sizes of training sets

5.2 Coverage

We used a sequence of 113 rules obtained from training on C_{all} on 329,510 English entries from a Yahoo! Answers corpus[7] to derive an estimate of the

---

[7] http://webscope.sandbox.yahoo.com/catalog.php?datatype=l – L6.

augmented strategy's coverage. 55.52 % of the questions returned a match – either TPs or FPs. 0.42 % failed to parse leaving 44.06 % which parsed but did not match any of the patterns. This figure can partly be accounted for by the ungrammatical content of the corpus, the presence of manner and why questions which were deliberately filtered from C_U and were not found in C_Q: 18,344 questions (5.6 %) began with "why"; 39,430 (12.0 %) with "how"), and a great many entries which do not constitute questions at all – such as "Cascading Style Sheets", "Flapjacks problem", and "Hotel in San Diego".

6 Impact on QALD

6.1 System Architecture

We built a QALD system to act as a baseline, with components for parsing, entity mapping, answer type identification, routing and translation (formalisation), depicted in Fig. 6. This pipeline is that extended by Alexopoulos et al. (2014) for handling vagueness in QALD.

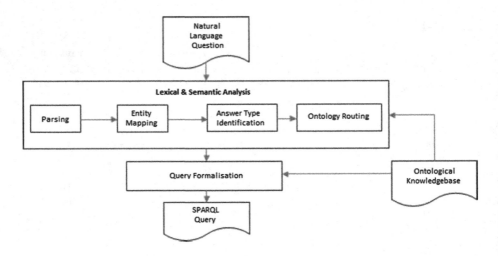

Fig. 6. Baseline QALD system architecture and information flow

The pipeline follows the annotator pattern, incrementally adding more data to interpretations as they flow through the system.

The parsed question is annotated with lemmas and synonyms (via Word-Net) forming the search space for entity mapping. Thus, if a question spoke of the "US", a synonym would be found as "United States" which is one of the `rdfs:labels` of the corresponding DBpedia resource. Similarly, if it spoke of "United States" then the syntactic node (which should be a NP noun-phrase) containing those words would be annotated with the DBpedia resource rather than trying to map the individual terms.

This mapping stage is exploratory and exhaustive. That is, we do not assume prior knowledge of the contents of the ontology beyond certain ubiquitous vocabularies (`rdfs:label`, `foaf:name` and, indeed, `dbp:name`) and so must query the ontology to find candidate matches. Matches may be classes, instances, relations or literals. We seek classes and relations with `rdfs:labels` containing the term (always a lemma); and instances with either a `rdfs:label`, `foaf:name` or `dbp:name` containing the term. Mappings also include fundamental question concepts, such as locations ("Where...?"), times ("When...?") and lists ("List...!", "Give...!", "Which...?" etc.). The cross-product of all these mappings is generated and passed on through to the disambiguation process (our query type identification module).

The query type identification process used for the baseline takes the head of the first noun-phrase (denoted `NPH` in Table 1), which performed best of the candidate baselines on C_Q and C_U. Thus, if the head of the first noun-phrase in a question were "country" and the mapper had given `dbo:Country` as its ontological concept, then the baseline would give `dbo:Country` as the query type.

The router tries to find out how these concepts, entities and relations connect together in the target ontology, and does this in an exploratory manner. Each concept is, in turn, linked to the root concept (the ATI, when available; the first class mapped otherwise, and the first entity mapped if there are no classes) by the shortest route possible.

6.2   Data

Though we would ideally evaluate the impact of the ATI strategy with the C_Q questions used in the core ATI analysis in Sects. 4 and 5, this would require physical resources unavailable for this study to load all of the data (DBpedia) and process it efficiently, and a more robust QA system able to handle the huge number of interpretations that would result from the noisy data.

Instead, we measure the impact of the use of ATIs on the Mooney Geoquery dataset[8]. These questions use a much more limited vocabulary and range of grammatical structures, but the data is much more tractable for semantic search. Putting the 250 questions through the ATI pattern extraction process derived a sequence of just 12 patterns, indicating these questions are more formulaic than those in C_{all}, which yielded 113.

The focus of this study is not to build a competitive QALD system, however, so some common features of natural language questions that can't be handled with mapping and routing are not supported by our pipeline. The 250 questions were therefore filtered manually in the following ways:

- 102 questions for including **superlatives** like "highest" point or "most" states.
- 17 questions were removed for requiring **aggregations** to answer, such as those asking for the combined populations of sets of locations, or "how many"

---

[8] http://www.cs.utexas.edu/users/ml/nldata/geoquery.html.

entities meet some criteria (but we kept "how many people live in..." questions as the solution would be found by mapping to the population property of subjects).

– 1 question had a **negation** that we have not provisioned for.

Finally, it is interesting and should be borne in mind that the questions were published all in lower case and occasionally with grammatical errors (most commonly, missing apostrophes to indicate genitive relationships), both of which can cause the parser to produce inappropriate trees and graphs consumed by the rest of the pipeline.

6.3   Results

We ran the same 130 questions through our pipeline three times - once with no analysis component, once with the baseline, and again with a component using our ATI strategy.

Table 2. Juxtaposition of results running the pipeline against 130 questions from the Mooney Geoquery set on the corresponding dataset with no ATI component, a baseline and with the ATI strategy detailed in this study. The first section details the ATI component results and the second its impact on QALD. 'Multiple ATs found' indicates that the correct word in the question was identified as representing the answer type, but other interpretations were also found; 'correct' indicates one or more appropriate SPARQL queries were generated and ranked the highest, with no inappropriate queries ranked equally; 'joint' indicates an appropriate query was generated and ranked best but an inappropriate query was also scored the same; 'rank' indicates the number of questions for which an appropriate query was found but inappropriate ones were ranked higher; 'incidental' indicates a query that would retrieve the right answers incidentally but the query itself was not a good representation of the original question; 'incorrect' means only inappropriate queries were generated; and 'none' means no queries were generated at all.

	None	Baseline	ATI
AT ID'd correctly	0	66	69
AT ID'd incorrectly	0	12	7
No AT found	130	44	54
Multiple ATs found	0	8	0
Correct	28	34	34
Joint	4	1	3
Rank	21	20	22
Incidental	0	0	2
Incorrect	56	56	51
None	21	19	18

The results (in Table 2) show both a benefit to the concept – the baseline and ATI results are both better than those without an analysis component – and further with the ATI strategy we have described. The benefit is tempered however by the simplicity and lack of grammatical diversity in the questions (which caused excessive favouring of certain ATI patterns). In many cases, the router's strategy of connecting the nodes and only anonymising the nodes and edges that were not already mapped worked well enough not to need an ATI-tagged term to anchor its search. Similarly the baseline performs well on these questions as the required type often was indeed referred to by the head of the first noun-phrase.

Some of the failures to find an appropriate formalisation are explicable by the rudimentary nature of the pipeline (a fully functional system was not the goal) and the nature of the questions themselves. For example, some of the questions do not have answers (i.e. the set of correct answers to the question is the empty set; such as for "which states border hawaii?"), but as the routing component seeks real examples of relation chains between entities will fail and fall back on weaker interpretations. The mappings with WordNet were occasionally insufficient, such as for questions that speak of "citizens living" in states and cities which would ideally be mapped to the "population" ontological property. The main benefit of using ATI patterns on simple data sets such as this appears to be in error reduction (56 to 51) rather than increased coverage.

7   Conclusion and Future Work

Identifying the type of answer sought by a question is a crucial preliminary step in modern QA, regardless of domain or data source. Other research has been conducted incorporating some form of answer-type identification or question classification as part of the documented process but little focus has been given to this important task itself. We introduced a strategy to iteratively explore a search space of dependency sub-graphs in order to extract answer-type identifiers ATIs from questions and detail subsequent augmentations that collectively yielded results improving on fully-automated strategies for those similar notions. As this method relies strictly on the grammatical structure of the questions it is trained on, it should be clear that performance should improve for controlled languages, and more so for the stricter of these.

A modular question answering pipeline was introduced and run against a well-known dataset showing improvements over baselines in producing formal queries for Question Answering over Linked Data.

This technique may also apply to question classification and its results can be used as a basis for semantic question answering. Indeed, this work is part of a larger work on QALD, forming a foundation upon which more research will be conducted. We intend to use this work as a basis for faithfully capturing the semantic content of user questions for mapping to linked data resources.

Acknowledgement. This research has been partly funded by the European Commission within the 7th Framework Programme/Marie Curie Industry-Academia Partnerships and Pathways schema/PEOPLE Work Programme 2011 project K-Drive number 286348 (cf. http://www.kdrive-project.eu).

References

Alexopoulos, P., Walker, A., Gomez-Perez, J.M., Wallace, M.: Towards ontology-based question answering in vague domains. In: 2014 9th International Workshop on Semantic and Social Media Adaptation and Personalization (SMAP), pp. 26–31. IEEE (2014)

Berners-Lee, T., Hendler, J., Lassila, O., et al.: The semantic web. Sci. Am. **284**(5), 28–37 (2001)

Bernstein, A., Kaufmann, E., Göhring, A., Kiefer, C.: Querying ontologies: a controlled english interface for end-users. In: Gil, Y., Motta, E., Benjamins, V.R., Musen, M.A. (eds.) ISWC 2005. LNCS, vol. 3729, pp. 112–126. Springer, Heidelberg (2005)

Brill, E., Lin, J., Banko, M., Dumais, S., Ng, A., et al.: Data-intensive question answering. In: Proceedings of the Tenth Text REtrieval Conference (TREC 2001) (2001)

Brin, S., Page, L.: The anatomy of a large-scale hypertextual web search engine. Comput. Netwo. ISDN Syst. **30**(1), 107–117 (1998)

Buscaldi, D., Rosso, P.: Mining knowledge from wikipedia for the question answering task. In: Proceedings of the International Conference on Language Resources and Evaluation (2006)

Cer, D.M., De Marneffe, M.-C., Jurafsky, D., Manning, C.D.: Parsing to stanford dependencies: trade-offs between speed and accuracy. In: LREC (2010)

Codd, E.F.: A relational model of data for large shared data banks. Commun. ACM **13**(6), 377–387 (1970)

Damljanovic, D., Agatonovic, M., Cunningham, H.: Identification of the question focus: combining syntactic analysis and ontology-based lookup through the user interaction. In: 7th Language Resources and Evaluation Conference (LREC), ELRA, La Valletta, Malta. Citeseer (2010)

Damljanovic, D., Agatonovic, M., Cunningham, H.: FREyA: an interactive way of querying linked data using natural language. In: García-Castro, R., Fensel, D., Antoniou, G. (eds.) ESWC 2011. LNCS, vol. 7117, pp. 125–138. Springer, Heidelberg (2012)

De Marneffe, M.-C., MacCartney, B., Manning, C.D.: Generating typed dependency parses from phrase structure parses. In: Proceedings of LREC, vol. 6, pp. 449–454 (2006)

Dijkstra, E.W.: A note on two problems in connexion with graphs. Nume. Math. **1**(1), 269–271 (1959)

Green Jr., B.F., Wolf, A.K., Chomsky, C., Laughery, K.: Baseball: an automatic question-answerer. In: 1961 Western Joint IRE-AIEE-ACM Computer Conference Papers Presented at the May 9–11, pp. 219–224. ACM (1961)

Krishnan, V., Das, S., Chakrabarti, S.: Enhanced answer type inference from questions using sequential models. In: Proceedings of the Conference on Human Language Technology and Empirical Methods in Natural Language Processing, pp. 315–322. Association for Computational Linguistics (2005)

Li, X., Roth, D.: Learning question classifiers. In: Proceedings of the 19th International Conference on Computational Linguistics, vol. 1, pp. 1–7. Association for Computational Linguistics (2002)

Marcus, M.P., Marcinkiewicz, M.A., Santorini, B.: Building a large annotated corpus of english: the penn treebank. Comput. Linguist. **19**(2), 313–330 (1993). ISSN 0891–2017, URL http://dl.acm.org/citation.cfm?id=972470.972475

McKeown, K.R.: Paraphrasing questions using given and new information. Comput. Linguist. **9**(1), 1–10 (1983)

Miller, G.A.: Wordnet: a lexical database for english. Commun. ACM **38**(11), 39–41 (1995)

Moldovan, D., Harabagiu, S., Pasca, M., Mihalcea, R., Girju, R., Goodrum, R., Rus, V.: The structure and performance of an open-domain question answering system. In: Proceedings of the 38th Annual Meeting on Association for Computational Linguistics, pp. 563–570. Association for Computational Linguistics (2000)

Prager, J., Chu-Carroll, J., Czuba, K.: Statistical answer-type identification in open-domain question answering. In: Proceedings of the Second International Conference on Human Language Technology Research, pp. 150–156. Morgan Kaufmann Publishers Inc. (2002)

Siddharthan, A.: Syntactic simplification and text cohesion. Res. Lang. Comput. **4**(1), 77–109 (2006)

Simmons, R.F.: Answering english questions by computer: a survey. Commun. ACM **8**(1), 53–70 (1965). ISSN 0001-0782, doi: 10.1145/363707.363732, URL http://doi.acm.org/10.1145/363707.363732

Ullmann, J.R.: An algorithm for subgraph isomorphism. J. ACM (JACM) **23**(1), 31–42 (1976)

Wales, J., Sanger, L.: Wikipedia, the free encyclopedia (2001). Accessed April 22, 2013, URL http://en.wikipedia.org/w/index.php?title=Wikipedia&oldid=551616049

Walker, A., Starkey, A., Pan, J.Z., Siddharthan, A.: Making test corpora for question answering more representative. In: Kanoulas, E., Lupu, M., Clough, P., Sanderson, M., Hall, M., Hanbury, A., Toms, E. (eds.) CLEF 2014. LNCS, vol. 8685, pp. 1–6. Springer, Heidelberg (2014)

Waltz, D.L.: An english language question answering system for a large relational database. Commun. ACM **21**(7), 526–539 (1978)

Woods, W.A.: Progress in natural language understanding: an application to lunar geology. In: Proceedings of the June 4–8, National Computer Conference and Exposition, AFIPS 1973, pp. 441–450. ACM, New York (1973). doi:10.1145/1499586.1499695, URL http://doi.acm.org/10.1145/1499586.1499695

Woods, W.A.: Lunar rocks in natural english. Linguist. Struct. Process. **5**, 521–569 (1977)

Ontologies, Semantics, and Reasoning

PROSE: A Plugin-Based Paraconsistent OWL Reasoner

Wenrui Wu, Zhiyong Feng, Xiaowang Zhang$^{(\boxtimes)}$, Xin Wang, and Guozheng Rao

School of Computer Science and Technology, Tianjin University, Tianjin, China
{wenruiwu,zyfeng,xiaowangzhang,wangx,rgz}@tju.edu.cn

Abstract. The study of paraconsistent reasoning with ontologies is especially important for the Semantic Web since knowledge is not always perfect within it. Quasi-classical semantics is proven to rationally draw more meaningful conclusions even from an inconsistent ontology with the stronger inference power of paraconsistent reasoning. In our previous work, we have conceived a quasi-classical framework called PROSE to provide rich paraconsistent reasoning services for OWL ontologies, whose architecture contains three parts: a classical OWL reasoner, a quasi-classical transformer, and OWL API connecting with them. This paper finally implements PROSE where quasi-classical transformer is bulit as a plugin for paraconsistent reasoning on classical reasoners. Additionally, we select three popular classical OWL reasoners (i.e., Pellet, HermiT, and FaCT++) and two typical kinds of reasoning services (i.e., QC-consistency checking and QC-classification) for users. As we excepted, PROSE does exactly enable current classical OWL reasoners to tolerate inconsistency in a simple and convenient way. Furthermore, we evaluate the three reasoners in three dimensions (class, property, individual) and, as a result, those results can amend the analysis of the three reasoners on inconsistent ontologies.

1 Introduction

As an extension of the World Wide Web (WWW), the Semantic Web [3] becomes more constantly changing and highly collaborative. Ontologies considered one of the pillars of the Semantic Web will rarely be perfect due to many reasons, such as modeling errors, migration from other formalisms, merging ontologies, and ontology evolution [7,10,28,33]. As a fragment of predicate logic, description logic (DL), which is the logical foundation of the Web Ontology Language [21] (e.g., sublanguages OWL Lite and OWL DL correspond to $\mathcal{SHIF}(\mathbf{D})$ and $\mathcal{SHOIN}(\mathbf{D})$ respectively), is unable to tolerate inconsistencies occurring in ontologies. Thus, the topic of inconsistency handling in OWL and DL has received extensive interests in the community in recent years [19,28,33].

There are several approaches to handling inconsistencies in DLs. All of them can be functionally roughly classified into two different types. One type is based on the assumption that inconsistencies indicate erroneous data which are to be removed in order to obtain a consistent ontology [6,10,12,13,17,31,33]. In these

© Springer International Publishing Switzerland 2016
G. Qi et al. (Eds.): JIST 2015, LNCS 9544, pp. 255–270, 2016.
DOI: 10.1007/978-3-319-31676-5_18

approaches, researchers hold a common view that ontologies should be completely free of inconsistencies, and thus try to eliminate inconsistencies from them to recovery consistency immediately by any means possible. However, there are some different opinions about the first type of treating inconsistency. And [4] argues that inconsistencies in knowledge are the norm in the real world, and so should be formalized and used, rather than always rejected. The other, called inconsistency-tolerant (or paraconsistent) approaches, is not to simply avoid inconsistencies but to apply non-standard reasoning methods (e.g., non-standard inference or non-classical semantics) to obtain meaningful answers [9,15,18,19,23,25,36,39,41,43,44]. In the second type of approaches, inconsistency treated as a natural phenomenon in realistic data, should be tolerated in reasoning. So far, the main idea of existing paraconsistent methods for handling inconsistency is introducing either non-standard inference or non-classical semantics to draw meaningful conclusions from inconsistent KBs [36]. Those paraconsistent approaches with non-standard inference presented by [9,39] are employing argument principles where consistent subsets are selected from an inconsistent KB as substitutes in reasoning. Those paraconsistent approaches are based on multi-valued semantics (a popular kind of non-classical semantics) such as four-valued DL studied by [18,19,25,36] based on Belnap's four-valued semantics [2], paradoxical DL presented by [41] based on Priest's paradoxical semantics, three-valued DL discussed by [23] based on Kleene's three-valued semantics, and [15] based on a dual interpretation semantics.

Multi-valued logics, as a family of non-classical logics, are successful in handling inconsistency and uncertainty in DL such as four-valued DL [18,19]. Because four-valued logic is a basic member of the family of multi-valued logics, four-valued semantics of DL has got a lot of attention [18,19,36]. However, the inference power of four-valued DL is rather weak as noted/argued by [15,18,19,41] although three kinds of implications (namely, *material implication*, *internal implication*, and *strong implication*) are introduced in four-valued DL to improve inference power. Some important properties about inference such as *disjunctive syllogism*, *resolution*, and *intuitive equivalence* are invalid in four-valued DL.

Indeed, the weak inference power is one of common characteristics of the family of paraconsistent logics where some important inference rules are prohibited in order to avoid the explosion of inference [4]. As a result, this topic of making more properties about inference valid under preventing the explosion of inference becomes interesting and important since more useful information can be inferred from inconsistent ontologies [24,25]. To avoid the shortcomings of four-valued DL, in our pervious work [40,42], we presented a *quasi-classical description logic* (QCDL), based on quasi-classical semantics proposed in [14]. It is proven that QCDL can be applied to tolerate inconsistency in reasoning on OWL ontologies with the stronger inference power of paraconsistent reasoning comparing with four-valued DL. In other words, QCDL can bring more conclusions than four-valued DL.

In our pervious work [42], we have developed a framwork of paraconsistent reasoning based on quasi-classical transformation whose aims is to employ off-the-shelf DL reasoners to implement a prototype reasoning system named PROSE. In this paper, we have successfully implemented such a framework in which three classical OWL reasoners, namely, Pellet [35] (which has been implemented in [42]), HermiT [34], and FaCT++[38] can be enriched in inconsistency-tolerant reasoning on OWL ontologies. In this sense, our implementation can be seen as a significant extension of current classical DL reasoners. Futhermore, we evaluate PROSE by employing the three classical reasoners for inconsistent ontologies in two reasoning problems and three dimensions (class, property, individual). Those results provide a way to observe performances of the three classical reasoners in the case of inconsistent ontologies to amend existing results. Additionally, we offer several alternatives for classical reasoners to meet mutliple requirements in a practical world.

Compared to the previous version of PROSE which is implemented only in Pellet, the current version of PROSE has several advantages:

- PROSE is a self-decision system. For classical consistent ontologies, PROSE can give answers in considerable time. However, PROSE has the ability to judge whether the input ontology is consistent or not and then decide to whether to shutdown the process of transformation or not. PROSE is more powerful in terms of reasoning abilities.
- PROSE supports three widely used classical OWL reasoners. We can select the best candidate for our reasoning tasks among Pellet, HermiT, and FaCT++. So each reasoner can do its adept jobs as possible.
- The implementation of PROSE is commonly used. Since PROSE strongly relies on the OWL-API, it can be easily integrated with other OWL reasoners which are compatible with OWL-API.

The rest of this paper is organized as follows: Sect. 2 introduces briefly DLs and quasi-classical semantics. Section 3 introduces quasi-classical transformer and Sect. 4 implements PROSE. Section 5 evaluates experiment results. Finally, Sect. 6 summarizes the paper.

2   Preliminaries

In this section, we briefly introduce descriptipn logics, as a logical foundation of OWL, and quasi-classical semantics. For more comprehensive background knowledge of DLs and quasi-classical semantics, we refer the reader to some basic references [1, 42].

2.1   Syntax of Description Logics

In description logics (DLs), elementary descriptions are *concept names* (unary predicates) and *role names* (binary predicates). Complex descriptions are built from them inductively using concept and role constructors provided by the particular DLs under consideration. In this section, we review the syntax and semantics of DLs.

Let N_C, N_R, and N_I be countably infinite sets of concept names, role names, and individual names. $N_R = \mathbf{R}_A \cup \mathbf{R}_D$ where $\mathbf{R}_A$ is a set of abstract role names and $\mathbf{R}_D$ is a set of concrete role names. The set of *roles* is then $N_R \cup \{R^- \mid R \in N_R\}$ where R^- is the inverse role of R. The function $Inv(\cdot)$ is defined on the sets of roles as follows, where R is a role name: $Inv(R) = R^-$ and $Inv(R^-) = R$. For roles R_1 and R_2, a *role axiom* is either a role inclusion, which is of the form $R_1 \sqsubseteq R_2$ for $R_1, R_2 \in \mathbf{R}_A$ or $R_1, R_2 \in \mathbf{R}_D$, or a transitivity axiom, which is of the form $\mathrm{Trans}(R)$ for $R \in \mathbf{R}_A$. A *role hierarchy* $\mathcal{R}$ (or an *RBox*) is a finite set of role axioms. Let $\underset{\mathcal{R}}{\overset{*}{\sqsubseteq}}$ be the reflexive-transitive closure of $\sqsubseteq$ on $\mathcal{R}$ as follows: $\{(R_1, R_2) \mid R_1 \sqsubseteq R_2 \in \mathcal{R}$ or $Inv(R_1) \sqsubseteq Inv(R_2) \in \mathcal{R}\}$. A role R is *transitive* in $\mathcal{R}$, if a role R' exists such that $R' \underset{\mathcal{R}}{\overset{*}{\sqsubseteq}} R$, $R \underset{\mathcal{R}}{\overset{*}{\sqsubseteq}} R'$, and either $\mathrm{Trans}(R') \in \mathcal{R}$ or $\mathrm{Trans}(Inv(R')) \in \mathcal{R}$. A role S is *simple* if no transitive role R exists such that $R \underset{\mathcal{R}}{\overset{*}{\sqsubseteq}} S$. R^{tc} denotes the transitive closure of R.

Concrete datatypes are used to represent literal values such as numbers and strings. A type system typically defines as a set of "primitive" datatypes, such as *string* or *integer*, and provides a mechanism for deriving new datatypes from existing ones. To represent concepts such as "a person whose age is at least 21", a set of concrete datatypes $\mathbf{D}$ is given, and, with each $d \in \mathbf{D}$, a set $d^{\mathbf{D}} \subseteq \Delta_{\mathbf{D}}$ is associated, where $\Delta_{\mathbf{D}}$ is the domain of all datatypes.

A set of datatypes is *conforming* if it satisfies the above criteria. The set of concepts is the smallest set such that each concept name $A \in N_C$ is a concept, and complex concept in OWL are formed according to the following syntax rules by using the operators shown in Table 1.

Note that the disjunction of nominals $\{o_1\} \sqcup \cdots \sqcup \{o_m\}$, where o_i ($1 \le i \le m$) and m is a positive integer, is still taken as a nominal, denoted by $\{o_1, \ldots, o_m\}$. Indeed, nominals can be technically treated as complex concepts.

A *terminology* or a *TBox* $\mathcal{T}$ is a finite set of *general concept inclusion axioms* (GCIs) $C \sqsubseteq D$ (possibly contains nominals and datatypes in the language of $\mathcal{O}$). In an ABox, one describes a specific state of affairs of an application domain in terms of concept and roles. It is the statement about how concepts are related to each other. An ABox $\mathcal{A}$ is a finite set of assertions of the forms $C(a)$ (*concept assertion*), $R(a,b)$ (role assertion), $a \doteq b$ (*equality assertion*), and $a \not\doteq b$ (*inequality assertion*). A knowledge base (KB) $\mathcal{K}$ (or ontology) is a triple $(\mathcal{R}, \mathcal{T}, \mathcal{A})$.

The semantics is given by means of interpretations. An interpretation $\mathcal{I}_c = (\Delta^{\mathcal{I}_c}, \cdot^{\mathcal{I}_c})$ consists of a non-empty domain $\Delta^{\mathcal{I}_c}$, disjoint from the concrete domain $\Delta_{\mathbf{D}}$, and a mapping $\cdot^{\mathcal{I}_c}$ which maps concepts, roles, and nominals according to Table 1 ($\sharp$ denotes set cardinality).

Let $\mathcal{K}$ be a KB. We use $Mod(\mathcal{K})$ to denote the collection of all models of $\mathcal{K}$. $\mathcal{K}$ is *consistent* if $Mod(\mathcal{K}) \ne \emptyset$ and *inconsistent* otherwise. Consistency is an important property of KBs. For instance, consider a KB $\mathcal{K} = (\{A \sqsubseteq B\}, \{A(a), \neg B(a)\})$. Clearly, $\mathcal{K}$ is inconsistent. Note that most reasoning services can be reduced to the consistency problem.

Table 1. Syntax and Semantics of DLs

Elements	Syntax	Semantics
individual (N_I)	a	$a^{\mathcal{I}_c} \in \Delta^{\mathcal{I}_c}$
atomic concept (N_C)	A	$A^{\mathcal{I}_c} \subseteq \Delta^{\mathcal{I}_c}$
abstract role ($\mathbf{R}_A$)	R	$R^{\mathcal{I}_c} \subseteq \Delta^{\mathcal{I}_c} \times \Delta^{\mathcal{I}_c}$
concrete role ($\mathbf{R}_D$)	T	$T^{\mathcal{I}_c} \subseteq \Delta^{\mathcal{I}_c} \times \Delta_{\mathbf{D}}^{\mathcal{I}_c}$
datatype ($\mathbf{D}$)	d	$d^{\mathbf{D}} \subseteq \Delta_{\mathbf{D}}$
inverse abstract role	$Inv(R)$	$\{(x,y) \mid (y,x) \in R^{\mathcal{I}_c}\}$
transitive abstract role	$\text{Trans}(R)$	$R^{\mathcal{I}_c} = (R^{\mathcal{I}_c})^+$
Complex Concepts		
top concept	$\top$	$\top^{\mathcal{I}_c} = \Delta^{\mathcal{I}_c}$
bottom concept	$\bot$	$\bot^{\mathcal{I}_c} = \emptyset^{\mathcal{I}_c}$
negation	$\neg C$	$(\neg C)^{\mathcal{I}_c} = \Delta^{\mathcal{I}_c} \setminus C^{\mathcal{I}_c}$
conjunction	$C \sqcap D$	$(C \sqcap D)^{\mathcal{I}_c} = C^{\mathcal{I}_c} \cap D^{\mathcal{I}_c}$
disjunction	$C \sqcup D$	$(C \sqcup D)^{\mathcal{I}_c} = C^{\mathcal{I}_c} \cup D^{\mathcal{I}_c}$
exist restriction	$\exists R.C$	$\{x \mid \exists y.(x,y) \in R^{\mathcal{I}_c} \text{ and } y \in C^{\mathcal{I}_c}\}$
value restriction	$\forall R.C$	$\{x \mid \forall y.(x,y) \in R^{\mathcal{I}_c} \text{ implies } y \in C^{\mathcal{I}_c}\}$
existential restriction	$\exists R.C$	$(\exists R.C)^{\mathcal{I}_c} = \{x \mid \exists y, (x,y) \in R^{\mathcal{I}_c} \text{ and } y \in C^{\mathcal{I}_c}\}$
function ($\mathcal{F}$)	$\leq 1R$	$\{x \mid \forall y.\forall z.((x,y) \in R^{\mathcal{I}_c} \text{ and } (x,z) \in R^{\mathcal{I}_c}) \text{ implies } y = z\}$
nominal ($\mathcal{O}$)	$\{o\}$	$\{o\}^{\mathcal{I}_c} \subseteq \Delta^{\mathcal{I}_c}, \sharp(\{o\}^{\mathcal{I}_c}) = 1$
number restriction ($\mathcal{N}$)	$\geq nR$	$\{x \mid \sharp(\{y.(x,y) \in R^{\mathcal{I}_c}\}) \geq n\}$
	$\leq nR$	$\{x \mid \sharp(\{y.(x,y) \in R^{\mathcal{I}_c}\}) \leq n\}$
atleast restriction	$\geq nS.C$	$\{x \mid \sharp(\{y.(x,y) \in S^{\mathcal{I}_c}\} \cap C^{\mathcal{I}_c}) \geq n\}$
atmost restriction	$\leq nS.C$	$\{x \mid \sharp(\{y.(x,y) \in S^{\mathcal{I}_c}\} \cap C^{\mathcal{I}_c}) \leq n\}$
datatype exists	$\exists T.d$	$\{x \in \Delta^{\mathcal{I}_c} \mid \exists y.(x,y) \in T^{\mathcal{I}_c} \text{ and } y \in d^{\mathbf{D}}\}$
datatype value ($\mathbf{D}$)	$\forall T.d$	$\{x \in \Delta^{\mathcal{I}_c} \mid \forall y.(x,y) \in T^{\mathcal{I}_c} \text{ implies } y \in d^{\mathbf{D}}\}$
equality(UNA)	$a \doteq b$	$a^{\mathcal{I}_c} = b^{\mathcal{I}_c}$
inequality(UNA)	$a \neq b$	$a^{\mathcal{I}_c} \neq b^{\mathcal{I}_c}$
Axioms		
concept inclusion ($\mathcal{AL}$)	$C \sqsubseteq D$	$C^{\mathcal{I}_c} \subseteq D^{\mathcal{I}_c}$
concept definition ($\mathcal{AL}$)	$C \equiv D$	$C^{\mathcal{I}_c} = D^{\mathcal{I}_c}$
nominals inclusion ($\mathcal{O}$)	$\{o_1,\ldots,o_s\} \sqsubseteq \{o_1',\ldots,o_t'\}$	$\{o_1^{\mathcal{I}_c},\ldots,o_s^{\mathcal{I}_c}\} \subseteq \{o_1'^{\mathcal{I}_c},\ldots,o_t'^{\mathcal{I}_c}\}$
role inclusion ($\mathcal{H}$)	$R \sqsubseteq S$	$R^{\mathcal{I}_c} \subseteq S^{\mathcal{I}_c}$

2.2 Quasi-classical Semantics

In syntax, quasi-classical description logic (QCDL) slightly extends the syntax of classical DLs by introducing the *quasi-classical negation* (QC negation) of a concept. The QC negation of a concept C is denoted by $\overline{C}$.

The quasi-classical semantics is characterized by two interpretations, namely, *weak interpretation* and *strong interpretation*, which are built on *base interpretation* defined as follows.

A *base interpretation* $\mathcal{I}$ is a pair $(\Delta^{\mathcal{I}}, \cdot^{\mathcal{I}})$ where the domain $\Delta^{\mathcal{I}}$ is a set of individuals, $\Delta_{\mathbf{D}}$ a concrete domain of datatypes and the assignment function $\cdot^{\mathcal{I}}$ assigns each individual to an element of $\Delta^{\mathcal{I}}$ and assigns: $\top^{\mathcal{I}} = \langle \Delta^{\mathcal{I}}, \emptyset \rangle$; $\bot^{\mathcal{I}} = \langle \emptyset$; $A^{\mathcal{I}} = \langle +A, A \rangle$; $R^{\mathcal{I}} = \langle +R, -R \rangle$; $(R^-)^{\mathcal{I}} = \langle +R^-, -R^- \rangle$; $(R^{tc})^{\mathcal{I}} = \langle +R^{tc}, -R^{tc} \rangle$; $d^{\mathcal{I}} = \langle +d, -d \rangle$; and $T^{\mathcal{I}} = \langle +T, -T \rangle$. Here $\pm A, N \subseteq \Delta^{\mathcal{I}}$, $\pm R, \pm R^-, \pm R^{tc} \subseteq \Delta^{\mathcal{I}} \times \Delta^{\mathcal{I}}$, $\pm T \subseteq \Delta^{\mathcal{I}} \times \Delta_{\mathbf{D}}$, and $\pm d^{\mathbf{D}} \subseteq \Delta_{\mathbf{D}}$.

Note that $+X$ and $-X$ are not necessarily disjoint when $X \in \{C, R, T, d\}$. Intuitively, $+X$ is the set of elements known to be in the extension of X while $-X$ is the set of elements known to be not in the extension of X.

Weak interpretations and strong interpretations are base interpretations, to be deleted here due to the limitation of space, and can be found in [42].

Let $\mathcal{K}$ be a QC KB and ϕ be a QC axiom. We call $\mathcal{K}$ quasi-classically entails (QC entails) ϕ, denoted $\mathcal{K} \models_Q \phi$, if for every base interpretation $\mathcal{I}$, $\mathcal{I} \models_s \mathcal{K}$ implies $\mathcal{I} \models_w \phi$. In this case, $\models_Q$ is called *QC entailment*.

Lemma 1 *[42]. Let $\mathcal{T}$ be a terminology, $\mathcal{R}$ a role hierarchy, $\mathcal{A}$ an ABox, and C, D concepts. For any base interpretation $\mathcal{I}$, we interpret $U^{\mathcal{I}} = \langle \Delta^{\mathcal{I}} \times \Delta^{\mathcal{I}}, \emptyset \rangle$. Then*

1. $(\mathcal{T}, \mathcal{R}, \mathcal{A}) \models_Q C(a)$ *if and only if* $(\mathcal{T}, \mathcal{R}, \mathcal{A} \cup \{\overline{C}(a)\})$ *is QC inconsistent w.r.t.* $\mathcal{R}_U$;
2. $(\mathcal{T}, \mathcal{R}, \emptyset) \models_Q C \sqsubseteq D$ *if and only if* $(\mathcal{T}, \mathcal{R}, \{(C \sqcap \overline{D})(\iota)\})$ *is QC inconsistent w.r.t.* $\mathcal{R}_U$ *for some new individual* $\iota \in \Delta$.

By Lemma 1, we can find that some reasoning problems (instance checking and subsumption (or classification)) of QCDL can be reduced to the QC-consistency problem.

3 Quasi-classical Transformer

In this section, we introduce quasi-classical transformer serving for transformation as a plugin in the architecture of PROSE [42].

Note that the quasi-classical transformer contains two transformers: *weak transformer* and *strong transformer*. The weak transformation is stated in Table 2 where NA is a new concept name.

Table 2. Weak Transformation Rules for OWL

Syntax	Weak transformation $\mathcal{W}(\cdot)$
$\top$	$NA \sqcup \neg NA$
$\bot$	$NA \sqcap \neg NA$
$\neg\top$	$NA \sqcap \neg NA$
$\neg\bot$	$NA \sqcup \neg NA$
X $(X \in \{A, R, S, T, d\})$	X^+
$\neg X$ $(X \in \{A, R, S, T, d\})$	X^-
$\{o\}$	$\{o\}$
$\neg\{o\}$	A_o: a new concept name
$\overline{C}$	$\neg\mathcal{W}(C)$
$\overline{\overline{C}}$	$\mathcal{W}(C)$
$\neg\neg C$	$\mathcal{W}(C)$
$C \sqcap D$	$\mathcal{W}(C) \sqcap \mathcal{W}(D)$
$\neg(C \sqcap D)$	$\mathcal{W}(\neg C) \sqcup \mathcal{W}(\neg D)$
$C \sqcup D$	$\mathcal{W}(C) \sqcup \mathcal{W}(D)$
$\neg(C \sqcup D)$	$\mathcal{W}(\neg C) \sqcap \mathcal{W}(\neg D)$
$\exists R.C$	$\exists \mathcal{W}(R).\mathcal{W}(C)$
$\forall R.C$	$\forall \mathcal{W}(R).\mathcal{W}(C)$
$\neg(\exists R.C)$	$\forall \mathcal{W}(R).\mathcal{W}(\neg C)$
$\neg(\forall R.C)$	$\exists \mathcal{W}(R).\mathcal{W}(\neg C)$
$\exists T.d$	$\exists \mathcal{W}(T).\mathcal{W}(d)$
$\forall T.d$	$\forall \mathcal{W}(T).\mathcal{W}(d)$
$\neg(\exists T.d)$	$\forall \mathcal{W}(T).\mathcal{W}(\neg d)$
$\neg(\forall T.d)$	$\exists \mathcal{W}(T).\mathcal{W}(\neg d)$
$\geq n\, S$	$\geq n\, \mathcal{W}(S)$
$\leq n\, S$	$\leq n\, \mathcal{W}(S)$
$\neg(\geq n\, S)$	$\leq (n-1)\, \mathcal{W}(S)$
$\neg(\leq n\, S)$	$\geq (n+1)\, \mathcal{W}(S)$
$\geq n\, R.C$	$\geq n\, \mathcal{W}(R).\mathcal{W}(C)$
$\leq n\, R.C$	$\leq n\, \mathcal{W}(R).\mathcal{W}(C)$
$\neg(\leq n\, R.C)$	$\geq (n+1)\, \mathcal{W}(R).\mathcal{W}(\neg C)$
$\neg(\geq n\, R.C)$	$\leq (n-1)\, \mathcal{W}(R).\mathcal{W}(\neg C)$
$C(a)$	$\mathcal{W}(C)(a)$
$C \sqsubseteq D$	$\mathcal{W}(C) \sqsubseteq \mathcal{W}(D)$
$R(a, b)$	$\mathcal{W}(R)(a, b)$
$R_1 \sqsubseteq R_2$	$\mathcal{W}(R_1) \sqsubseteq \mathcal{W}(R_2)$
$Trans(R)$	$Trans(\mathcal{W}(R))$
$a \neq b$	$a \neq b$
$a \doteq b$	$a \doteq b$

The strong transformer $\mathcal{S}(\cdot)$ is identical with the weak transformation $\mathcal{W}(\cdot)$ except for disjunctions, the negation of conjunctions, and GCIs which are defined as follows: where $X_i \in \{C_i, R_i\}$ $(i = 1, 2)$,

$$\mathcal{S}(\neg(C \sqcap D)) = \mathcal{S}(\neg C \sqcup \neg D);$$
$$\mathcal{S}(C \sqcup D) = (\mathcal{S}(C) \sqcup \mathcal{S}(D)) \sqcap (\neg\mathcal{S}(\neg C) \sqcup \mathcal{S}(D)) \sqcap (C \sqcup \neg\mathcal{S}(\neg D)); \qquad (1)$$
$$\mathcal{S}(X_1 \sqsubseteq X_2) = \{\mathcal{S}(X_1) \sqsubseteq \mathcal{S}(X_2), \mathcal{S}(\neg X_2) \sqsubseteq \mathcal{S}(\neg X_1), \neg\mathcal{S}(\neg X_1) \sqsubseteq X_2\}.$$

We have the following lemma [42].

Lemma 2 *[42]. Let $\mathcal{K}$ be a QC KB and φ a QC axiom. Then*

1. $\mathcal{K}$ *is QC consistent if and only if $\mathcal{S}(\mathcal{K})$ is consistent;*
2. $\mathcal{K} \models_Q \varphi$ *if and only if $\mathcal{S}(\mathcal{K}) \models \mathcal{W}(\varphi)$.*

By Lemma 2, we can reduce the reasoning problems of QCDL to the reasoning problems of DL.

4   PROSE: A Plugin-Based Paraconsistent Reasoner

Based on the quasi-classical transformer introduced in the previous section, we have implemented a prototype system for OWL, called PROSE (*paraconsistent reasoning on semantic web*). The architecture of PROSE is shown in Fig. 1 (see [42]). PROSE is designed in the Decorator Pattern [8] extending the inner classical OWL reasoner so that paraconsistent reasoning on inconsistent KBs can be performed. The module *Strong Transformer* rewrites the input KB $\mathcal{K}$ to a new KB $\mathcal{S}(\mathcal{K})$ while the module *Weak Transformer* changes the input query φ into a new query $\mathcal{W}(\varphi)$. Then the inner classical OWL reasoner is called.

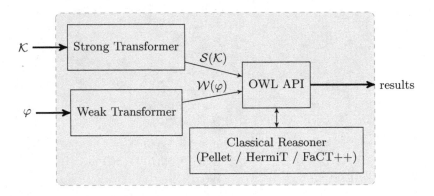

Fig. 1. PROSE architecture

In the initial version of PROSE, the only classical OWL reasoner Pellet is implemented, which is an open source Java based reasoner for OWL 2 DL developed by the Mind Swap group. PROSE is thus limited by some disadvantages of Pellet which are also limitations in reasoning on consistent ontologies.

OWL Reasoners Comparison

Comparsion of Pellet, HermiT and Fact++ in handling inconsistency with tansformation.

1.Select Ontologies

The first line are consistent ontologies and the second **inconsistent**.

✔ foaf.owl	✔ protege.owl	funding.owl	amino_acid.owl
✔ **bird.owl**	✔ **bad-food.owl**	**buggyPolicy.owl**	**family_inc.owl**

You have selected: foaf.owl bird.owl protege.owl bad-food.owl

2.Select Reasoners

Select one or more reasoners:

✔ Pellet	✔ HermiT	✔ Fact++

Turn the PROSE on? or self-decision system:

• Yes	○ No	○ Self-decision

3.Select Reasoning Tasks

Select one or more reasoning tasks:

✔ check consistency	☐ classification

Run	Reset

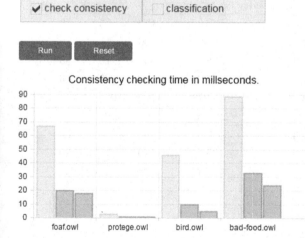

Consistency checking time in millseconds.

Fig. 2. Demo of using PROSE to evaluate Pellet, HermiT, and FaCT++

In the current version of PROSE, we have extended PROSE by including the other two classical OWL reasoners: HermiT and FaCT++. Both of them are widely used and have outstanding performance in specific reasoning tasks [5]. PROSE has good performance in reasoning over both consistent and inconsistent ontologies. It is evaluated empirically on various ontologies (e.g., containing lare-scale of classes or properties or individuals) in doing specific reasoning task and results are presented in the followling subsection. We give a critical estimation before using PROSE to perform paraconsistent reasoning. So, we can take full advantage of the highlights of different OWL reasoners through PROSE.

Based on the transformation algorithms, PROSE rewrites the input ontology to a new ontology which is then taken as input by the inner classical reasoner. The algorithms are implemented in Java 7 and are built on OWL-API 3.5.0 [26], which is a Java programming interface for manipulating OWL ontologies.The transformation algorithms traverse the ontologies, visit each component block and apply the corresponding rules which are introduced in the previous section.

The procedure is designed in the Visitor Pattern which is widely used in OWL-API. Since OWL-API is a general implementation for OWL ontologies and contains functions that do not have a tightly specified functionality, we use it to link PROSE with different classical reasoners. We have developed different programs for Pellet version 2.3.0, HermiT version 1.3.8, and FaCT++ version 1.6.2, but all of the programs are through specific interfaces in OWL-API 3.5.0. And these specific interfaces are supported by the three classical reasoners (see Fig. 2).

Table 3. The benchmark ontologies

Ontology	DL expressivity	#Axioms	N_C	N_P	N_I
100Class.owl	ALCHI	318	100	15	3
1000Class.owl	ALCHI	2732	1000	15	3
10000Class.owl	ALCHI	29269	10000	15	3
100Property.owl	ALCHI	479	100	105	3
1000Property.owl	ALCHI	2407	100	1005	3
10000Property.owl	ALCHI	21625	100	10005	3
100Individual.owl	ALCHI	473	100	15	100
1000Individual.owl	ALCHI	2283	100	15	1000
10000Individual.owl	ALCHI	20271	100	15	10000

5   Evaluation and Analysis

Evaluation The experiments were performed under Windows 7 64-bit on an Intel i5-4430S, 2.70 GHz CPU system with 8 GB memory. The program was written in Java 1.7 with maximum 4 GB heap space allocated for JVM. Nine tested ontologies were generated according to the increasing number of one component (e.g., classes) with the fixd number of others (e.g., properties and individuals). For instance, we built three ontologies consisting of 15 properties, 3 individuals, and 100, 1000, 10000 classes respectively. The experiments measured consistency checking time and classification time on each ontology. All stated runtimes are averaged over five independent runs. Note that all results are in millseconds. More results of an online demo of PROSE can be found in the website: http://prosev2.duapp.com/.

Results Analysis. Based on experimental results, we will find some phenomenon about the reasoning time of PROSE with calling the three classical reasoners (Pellet, HermiT, and FaCT++) under some different scenes:

– There are some differences between consistent ontologies and inconsistent ontologies or among ontologies with different sizes even when PROSE calls the same classical reasoner. In Figs. 3 and 4, let us consider the ontology "*100Individual.owl*", the ascending order of the reasoning time for the consistent ontology is as follows: FaCT++, HermiT, and Pellet while the ascending order of the reasoning time for the inconsistent ontology becomes the following: FaCT++, Pellet, and HermiT. Moreover, in Fig. 4, the ascending order of the reasoning time for the ontology "*100Property.owl*" which consists of 100 properties is as follows: FaCT++, HermiT, and Pellet while the ascending order of the reasoning time for the ontology *1000Property.owl* which consists of 1000 properties is as follows: HermiT, FaCT++, and Pellet. Addtionally, in Fig. 5, the ascending order of the reasoning time for the ontology "*100Individ-ual.owl*" which consists of 100 indivuduals is as follows: FaCT++, HermiT, and Pellet while the ascending order of the reasoning time for the ontology "*1000Individual.owl*" which consists of 1000 individuals is as follows: HermiT, Pellet, and FaCT++.

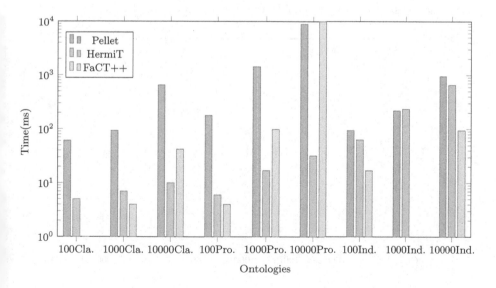

Fig. 3. Consistency checking time on consistent ontologies

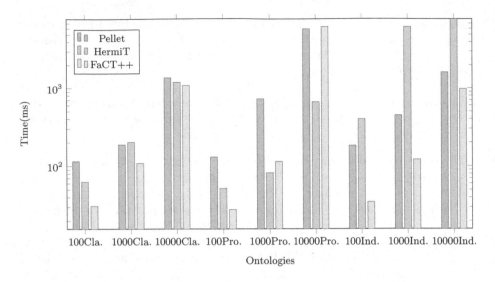

Fig. 4. QC-Consistency checking time on inconsistent ontologies

– There are also some differences among different reasoning problems or different dimensions even when PROSE calls the same classical reasoner on the same ontology. In Fig. 4 and by Fig. 5, let us consider the ontology "*1000Property.owl*", the ascending order of the reasoning time for the QC-consistency check on the ontology is as follows: HermiT, FaCT++, and Pellet while the ascending order of the reasoning time for QC-classification on the ontology is as follows: FaCT++, HermiT, and Pellet. Additionally, in Fig. 6, PROSE

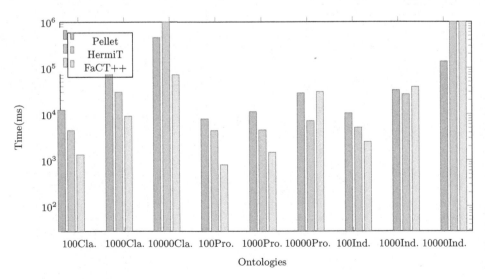

Fig. 5. QC-Classification time on inconsistent ontologies

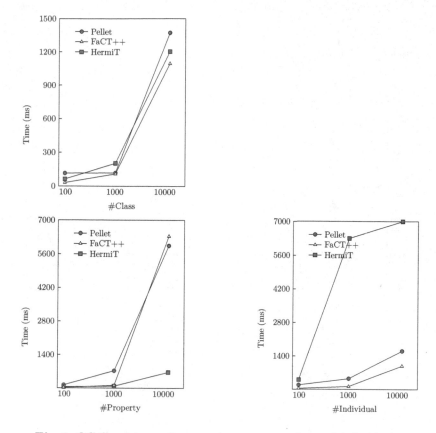

Fig. 6. QC-Consistency time on classes, properties, and individuals

with calling HermiT looks more efficient in handling properties than in handling individuals, and, in Fig. 7, PROSE with calling HermiT looks more efficient in handling properties than in handling classes.

In short, it is difficult to say which is the best one among the three classical OWL reasoners. Because of this, we provide users a multiple option in PROSE (see Fig. 2).

6   Conclusions

In this paper, we have implemented PROSE as a framework for paraconsistent reasoning for OWL ontologies and, within this framework, then we can select variable classical OWL reasoners to serve for inconsistent ontologies. Moreover, our experiment results show that classical OWL reasoners as the engine of PROSE have different performance in reasoning between consistent ontologies and inconsistent ontologies. In some sense, we provide a handy tool which gives more insight how different DL reasoners functions. In the future work, we present a revised weak transformer to characterize ParOWL [27] based on four-valued

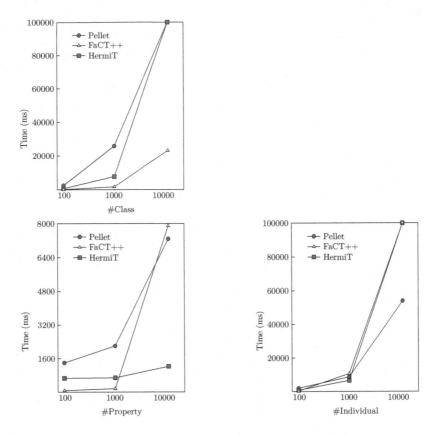

Fig. 7. QC-Classification time on classes, properties, and individuals

DL so that we can re-implement ParOWL in our framework PROSE. As another future work, it is also interesting to apply quasic-classical semantics to measure inconsistency of ontologies in DLs.

Acknowledgements. We would like to thank the anonymous reviewers for their comments which helped us to improve the paper. We gratefully acknowledge Zuoquan Lin, Kewen Wang, Guilin Qi, Yue Ma, and Guohui Xiao for discussions and their critical comments on our previous work of quasi-classical desription logics. This work is supported by the program of the National High-tech R&D Program of China (863 Program) under 2013AA013204 and the National Natural Science Foundation of China (NSFC) under 61502336, 61572353, 61373035. Xiaowang Zhang is supported by the project-sponsored by School of Computer Science and Technology in Tianjin University.

References

1. Baader, F., Calvanese, D., McGuinness, D.L., Nardi, D., Patel-Schneider, P.F. (eds.): Theory, Implementation, and Applications. Cambridge University Press, Cambridge (2003)

2. Belnap, N.D.: A useful four-valued logic. In: Epstein, G., Dunn, J.M. (eds.) Modern Uses of Multiple-Valued Logics, pp. 7–73. Reidel Publishing Company, Boston (1977)
3. Berners-Lee, T., Hendler, J., Lassila, O.: The semantic web. Sci. Am. **5**, 29–37 (2001)
4. Bertossi, L., Hunter, A., Schaub, T. (eds.): Inconsistency Tolerance. LNCS, vol. 3300. Springer, Heidelberg (2005)
5. Dentler, K., Cornet, R., Annette, T.T., De Keizer, N.: Comparison of reasoners for large ontologies in the OWL 2 EL profile. Semant. Web **1**, 1–5 (2011)
6. Fang, J., Huang, Z., van Harmelen, F.: Contrastive reasoning with inconsistent ontologies. In: Proceedings of WI 2011, IEEE CS, pp. 191–194 (2011)
7. Flouris, G., Huang, Z., Pa,n J.Z., Plexousakis, D., Wache, H.: Inconsistencies, negations and changes in ontologies. In: Proceedings of AAAI 2006. AAAI Press (2006)
8. Gamma, E., et al.: Design Patterns, p. 175ff. Addison-Wesley Publishing Co, Inc., Reading (1995)
9. Gómez, S.A., Chesñevar, C.I., Simari, G.R.: Reasoning with inconsistent ontologies through argumentation. Appl. Artif. Intell. **24**(1&2), 102–148 (2010)
10. Haase, P., van Harmelen, F., Huang, Z., Stuckenschmidt, H., Sure, Y.: A framework for handling inconsistency in changing ontologies. In: Gil, Y., Motta, E., Benjamins, V.R., Musen, M.A. (eds.) ISWC 2005. LNCS, vol. 3729, pp. 353–367. Springer, Heidelberg (2005)
11. Horridge, M., Bechhofer, S.: The OWL API: a java API for OWL ontologies. Semant. Web **2**(1), 11–21 (2011)
12. Horridge, M., Parsia, B., Sattler, U.: Explaining inconsistencies in OWL ontologies. In: Godo, L., Pugliese, A. (eds.) SUM 2009. LNCS, vol. 5785, pp. 124–137. Springer, Heidelberg (2009)
13. Huang, Z., van Harmelen, F., ten Teije, A.: Reasoning with inconsistent ontologies. In: Proceedings of IJCAI 2005, Professional Book Center, pp. 454–459 (2005)
14. Hunter, A.: Reasoning with contradictory information using quasi-classical logic. J. Log. Comput. **10**(5), 677–703 (2000)
15. Kamide, N.: Paraconsistent description logics revisited. In: Proceedings of the 23rd International Workshop on Description Logics (DL 2010), CEUR Workshop Proceedings, vol. 573 (2010)
16. Lang, C.: Four-valued logics for paraconsistent reasoning. Diplomarbeit von Andreas Christian Lang, Technische Universität Dresden (2006)
17. Lembo, D., Lenzerini, M., Rosati, R., Ruzzi, M., Savo, D.F.: Query rewriting for inconsistent DL-Lite ontologies. In: Rudolph, S., Gutierrez, C. (eds.) RR 2011. LNCS, vol. 6902, pp. 155–169. Springer, Heidelberg (2011)
18. Ma, Y., Hitzler, P., Lin, Z.Q.: Algorithms for paraconsistent reasoning with OWL. In: Franconi, E., Kifer, M., May, W. (eds.) ESWC 2007. LNCS, vol. 4519, pp. 399–413. Springer, Heidelberg (2007)
19. Maier, F., Ma, Y., Hitzler, P.: Paraconsistent OWL and related logics. Semant. Web **4**(4), 395–427 (2013)
20. Marquis, P., Porquet, N.: Computational aspects of quasi-classical entailment. J. Appl. Non-Classical Logics **11**(3&4), 295–312 (2001)
21. McGuinness, D.L., van Harmelen, F.: OWL web ontology language overview. W3C Recommendation (2009). http://www.w3.org/TR/owl-features/
22. Motik, B.: KAON2 - scalable reasoning over ontologies with large data sets. ERCIM New. **72**, 19–20 (2002)

23. Nguyen, L.A., Szałas, A.: Three-valued paraconsistent reasoning for semantic web agents. In: Jędrzejowicz, P., Nguyen, N.T., Howlet, R.J., Jain, L.C. (eds.) KES-AMSTA 2010, Part I. LNCS, vol. 6070, pp. 152–162. Springer, Heidelberg (2010)
24. Mu, K.: Responsibility for inconsistency. Int. J. Approx. Reasoning **61**, 43–60 (2015)
25. Odintsov, S.P., Wansing, H.: Inconsistency-tolerant description logic. part II: A tableau algorithm for CACL^c. J. Appl. Logic **6**(3), 343–360 (2008)
26. Owl, A.P.I.: A Java API for working with OWL 2 ontologies. http://owlapi.sourceforge.net/
27. ParOWL: Paraconsistent reasoner with OWL. http://neon-toolkit.org/wiki/1.x/ParOWL
28. Parsia, B., Sirin, E., Kalyanpur, A.: Debugging OWL ontologies. In: Proceedings of WWW 2005, pp. 633–640. ACM (2005)
29. Protégé: The protégé ontology editor and knowledge acquisition system. Website http://protege.stanford.edu/
30. Qi, G., Du, J.: Model-based revision operators for terminologies in description logics. In: Proceedings of IJCAI 2009, pp. 891–897. ACM (2009)
31. Qi, G., Liu, W., Bell, D.A.: A revision-based approach to handling inconsistency in description logics. Artif. Intell. Rev. **26**(1&2), 115–128 (2006)
32. Ross Anderson, A., Belnap, N.: The logic of relevance and necessity, vol. 1. Princeton University Press, Princeton (1976)
33. Schlobach, S., Cornet, R.: Non-standard reasoning services for the debugging of description logic terminologies. In: Proceedings of IJCAI 2003, Morgan Kaufmann, pp. 355–362 (2003)
34. Shearer, R., Motik, B., Horrocks, I.: A highly-efficient OWL reasoner. In: OWLED 2008, vol. 432 of CEUR Workshop Proceedings (2008)
35. Sirin, E., Parsia, B., Cuenca Grau, B., Kalyanpur, A., Katz, Y.: Pellet: a practical OWL-DL reasoner. J. Web Sem. **5**(2), 51–53 (2007)
36. Straccia, U.: A sequent calculus for reasoning in four-valued description logics. In: Galmiche, D. (ed.) TABLEAUX 1997. LNCS, vol. 1227, pp. 343–357. Springer, Heidelberg (1997)
37. TONES: Ontology repository. University of Manchester (2008). http://owl.cs.manchester.ac.uk/repository/
38. Tsarkov, D., Horrocks, I.: FaCT++ description logic reasoner: system description. In: Furbach, U., Shankar, N. (eds.) IJCAR 2006. LNCS (LNAI), vol. 4130, pp. 292–297. Springer, Heidelberg (2006)
39. Zhang, X., Lin, Z.: An argumentation framework for description logic ontology reasoning and management. J. Intell. Inf. Syst. **40**(3), 375–403 (2013)
40. Zhang, X., Lin, Z.: Quasi-classical description logic. Multiple-Valued Logic Soft. Comput. **18**(3&4), 291–327 (2012)
41. Zhang, X., Lin, Z., Wang, K.: Towards a paradoxical description logic for the semantic web. In: Link, S., Prade, H. (eds.) FoIKS 2010. LNCS, vol. 5956, pp. 306–325. Springer, Heidelberg (2010)
42. Zhang, X., Xiao, G., Lin, Z., Van den Bussche, J.: Inconsistency-tolerant reasoning with OWL DL. Int. J. Approx. Reasoning **55**(2), 557–584 (2014)
43. Zhang, X., Wang, K., Wang, Z., Ma, Y., Qi, G.: A distance-based paraconsistent semantics for DL-Lite. In: Zhang, S., Wirsing, M., Zhang, Z. (eds.) KSEM 2015. LNCS (LNAI), vol. 9403, pp. 1–13. Springer, Heidelberg (2015)
44. Zhang, X., Wang, K., Wang, Z., Ma, Y., Qi, G., Feng, Z.: A distance-based framework for inconsistency-tolerant reasoning, inconsistency measurement in DL-Lite. Technical report: TR20150928 (2015)

Meta-Level Properties for Reasoning on Dynamic Data

Yuting Zhao[1](✉), Guido Vetere[1], Jeff Z. Pan[2], Alessandro Faraotti[1], Marco Monti[1], and Honghan Wu[2]

[1] IBM Italia S.p.A., Roma, Italy
yuting.zhao@it.ibm.com
[2] Department of Computer Science, University of Aberdeen, Aberdeen, UK

Abstract. Dynamic features are important for data processing when dealing with real applications. In this paper we introduce a methodology for validating the construction of ontological knowledge base and optimising the query answering with such ontologies. In this paper, we firstly introduce some meta-properties of dynamic for ontologies. These meta-properties impose several constraints on the taxonomic structure of an ontology. We then investigate how to build up a meta-ontology with the constrains on these meta-properties. The goal of our methodology is *not* to help on providing strict logical conditions for judging inconsistency, but rather to help as heuristics for validating ontologies. Furthermore, some results on how to improve the reasoning on dynamic data by using these properties are also introduced.

1 Introduction

Dynamic is an important nature of data. Some data are always changing, *e.g.*, *the outdoor temperature of Rome*; some do not change, *e.g.*, *Chinese people speaks Chinese in general*. Some data will be valid for some period, *e.g.*, *Mr. David Cameron is the "Prime Minister of the UK"*; some data will change from one category to another, *e.g.*, if he retired, Mr. David Cameron would become the *"Former Prime Minister of the UK"*. Specially in the Big Data time, understanding the dynamic feature of data becomes a key issue for improving the quality of data, and developing efficient services based on the data.

Dynamic properties or stream data processing [1,2,4–7,11,12,14] have been studied in research communities of Artificial Intelligence, Database, and Linked Data for many years. Specifically, the first line of relevant techniques is those for modelling dynamics in the data level. Roughly speaking, a data stream is an (infinite) ordered sequence of data items. One of the key questions is how to model the timestamps in the data. The simplest approach is no timestamp associated with the data items, which means that the data is simply an ordered list of data items. The second approach is to attach a timestamp to each data item, which annotates when the item occurs. The third one is one step further from the second to add one more timestamp that means when the item was removed (or invalidated). The second type of approaches is the extensions on

© Springer International Publishing Switzerland 2016
G. Qi et al. (Eds.): JIST 2015, LNCS 9544, pp. 271–279, 2016.
DOI: 10.1007/978-3-319-31676-5_19

query languages for processing data dynamics. For example, CQL [2] is a continuous query language model for querying data streams. The stream data, infinite unbounded bag of data, is converted into discrete snapshots each of which is a finite bag of relations. So, the key question to solve here is how to select the right snapshot(s). Inspired by CQL, researchers in Linked Data community also proposed similar ideas to processing RDF data streams [1,4,11].

Our approach is a light weight solution, and does not request the special timestamp on the data. Furthermore, we are targeting a meta level description of the dynamic properties of the data items. One of the most related interesting works is OntoClean [9,10]. Guarino and Welty defined formal and domain independent properties of concepts (*i.e.* meta-properties) that aid the knowledge engineer to make correct decisions when building ontologies and therefore to avoid common mistakes. OntoClean originally defines four meta-properties as follow:

- **Identity:** the criteria for defining 'sortal' classes; classes all of whose instances are identified in the same way.
- **Unity:** a property that only holds of individuals that are 'wholes'.
- **Rigidity:** a property that only holds of individuals that have it and cannot change, is 'essential'.
- **Dependence:** a property that only holds for a class if each instance of it implies the existence of another entity.

According the methodology those meta-properties are used to analyze concepts in order to check hierarchies correctness, for example a 'rigid' concept like *Person* cannot be subsumed by an 'not-rigid' concept like *Physician*. In this paper, we go one step further and focus on formal description of changes and dynamic characterizations of the entities.

In general, "changes" in a dynamic knowledge base would be categorised in terms of meta-properties such as *Temporary, Reversible, Non-Reversible, Static, Rigid,* and *Periodic,* etc. If we model a dynamic knowledge base as a stream of Knowledge Bases (KBs), *i.e.* an ordered set of KBs sharing the same TBox where $KB(t)$ means the stream KB at time t, we could come up with a complete analysis in terms of meta-properties such as rigidity, cyclicity, monotonicity, etc. Then, one of the tasks could be that of inferring such meta properties for the KB TBox, based on observations of the stream KB at different times. Once we have a complete dynamic meta-level characterization of the TBox, we can decide how to optimize the access of the dynamic KB in many use cases. When applied to unary concepts, *Static* and *Temporary* relate to Rigid and Anti-Rigid in Guarinos formal ontology [8].

In this paper we analysis different kind of changes in a dynamic knowledge base, and categorise changes into some meta-properties. We have also investigated how to build up a meta-ontology with the constrains on these meta-properties. We stress the similar opinion as OntoClean, that our methodology is *not* to help on providing strict logical conditions for judging inconsistency, but rather to help as heuristics for validating the ontological knowledge bases, and optimising reasonings on them.

Currently the methodology of OntoClean does not aim at characterising ontology elements from the dynamic point of view and we believe that refining it by adding meta-properties about dynamics can be useful, when evolving domains (and datasets) are involved, both to design a well formed ontology and also to support reasoning. As an illustrative example we may have an individual which acquires a 'status' that cannot be changed any more, such kind of property is not rigid but become rigid somehow once acquired; we can imagine the case of person which once vaccinated against poliomyelitis remains immunized throughout his life. We called such meta-property 'Reversible' property. In a stream reasoning context, knowing that a property is not reversible makes it possible to leverage previous results so to avoid re-computation. So the second contribution in this paper is, we show some interesting results on how to improve the reasoning on dynamic knowledge base by using these properties.

We note in this paper we do not restrict our approach on any specific logic languages, but we use the convenience of normal expressive capabilities of *Description Logics* (DLs) [3] to present some logic properties in our discussion. DLs is a family of *Knowledge Representation* (KR) formalisms that represent the knowledge structure of an application domain, in terms of *concepts, roles, individuals*, and their relationships. We assume most of the readers are familiar with the DLs. At the same time, we also use the *First Order Logic* (FOL) to describe some semantics in our study, and we assume most of the readers are familiar with the FOL.

The paper is organised as follows. In Sect. 2 we introduce a formalization of dynamic knowledge base and various kind of "changes". In Sect. 3 we study the dynamic meta-properties and show how to build the meta-ontology. In Sect. 4, we show some interesting results on how to optimize Query Answering in a dynamic knowledge base with the meta-properties. A Conclusion is given in Sect. 5.

2 Knowledge Stream Bases and Changes

In this section we firstly define a generic formalization of dynamic knowledge base, then we classify several typical kind of changes in a dynamic knowledge base.

2.1 A Generic Knowledge Stream Bases

Here we formalize a *dynamic Knowledge Base* as a steam of knowledge bases, and then we use *change* to capture the difference between KBs.

Definition 1 (Knowledge Stream Bases (KSB)). *A* Knowledge Stream Bases (KSB) *is a steam of knowledge base* K_1, K_2, ..., K_n, *a change* $C(i)$ *is the difference between* K_{i-1} *and* K_i, *i.e.,*

$$C(i) = (K_i - K_{i-1}) \cup (K_{i-1} - K_i). \tag{1}$$

in where $(K_i - K_{i-1})$ is the set of data ADDED in K_i, donated by $C_{add}(i)$; and $(K_{i-1} - K_i)$ is the set of data REMOVED from in K_{i-1}, donated by $C_{del}(i)$. So we have:

$$C_{add}(i) = \{k | k \in K_i \text{ and } k \notin K_{i-1}\}; \tag{2}$$

$$C_{del}(i) = \{k | k \in K_{i-1} \text{ and } k \notin K_i\}. \tag{3}$$

Obviously above definition of "change" reflects the difference between the current KB (*e.g.* K_i) and the former KB (*e.g.* K_{i-1}). In general it contains two parts of information: the "new data" added in the current KB, and the "outdated data" removed from the previous KB.

2.2   Category of Changes in Knowledge Stream Bases

Based on common sense knowledge, there could be various kinds of changes on the data in a Knowledge Stream Bases. Without losing generality, the behavioural properties of "changes" in Stream data could be categorised as:

1. **Come and Remain (Change.CM):** A fact is detected and will never change, *e.g.* a person was born: *hasBirthday*(Marie Curie, 1867-11-07). Obviously Change.CM belongs to the added data, but does not belong to the removed data, i.e., $Change.CM \subseteq C_{add}$ and $Change.CM \nsubseteq C_{del}$.
2. **Removed Forever (Change.RF):** A fact is removed from the KB forever, *e.g.* if a patient is dead, then all in treatment data will be removed from the current KB, and move to the other KB.
3. **Come and Leave (Change.CL):** A fact appeared for a while and then disappeared, *e.g.* a patient is pregnant: *Pregnant*(Mary). Obviously Change.CL belongs to both the added data and the removed data, i.e., $Change.CL \subseteq C_{add}$ and $Change.CL \subseteq C_{del}$.
4. **Periodical Changes (Change.PC):** A fact has a value which will change periodically with repeated values, *e.g.* the changes of 4 seasons.
5. **Monotonic Changing the Value (Change.MC):** A fact has a value which will change monotonically, *e.g.* the ages of a person: $hasAge^-.\{Mary\}$.
6. **Non-monotonic Changes (Change.NC):** A fact has a value which will change NON-monotonically, *e.g.* the body temperature of a patient: $hasTemperature^-.\{Mary\}$.

In the following Table 1 we give the semantics to different kinds of categories of changes.

3   Meta-Properties for Dynamic and Meta Ontology

In this section we introduce a method to describe static and dynamic properties in a dynamic knowledge base. In general, these "changes" in knowledge streams would be categorised in terms of meta-properties such as Periodic, Monotonicity, etc. At the same time, we also need meta-properties such as Static and Rigid, in order to handle the real practical applications.

Table 1. Semantics of changes

Change.CM = $\{c \mid \exists t$, so that $c \in C_{add}(t)$, in where $i < t$, $c \notin K_i$, and $j \geq t$, $c \in K_j \}$
Change.RF = $\{c \mid \exists t$, so that $c \in C_{del}(t)$, in where $i < t$, $c \in K_i$, and $j \geq t$, $c \notin K_j \}$
Change.CL = $\{c \mid \exists i < j$, so that $c \in C_{add}(i)$ and $c \in C_{del}(j)$, in where $i \leq t < j$, $c \in K_t \}$
Change.PC = $\{c \mid c \in$ Change.CL, and $\exists t$, let $k = [1, 2, \cdots]$, so that if $c \in K_i$ then $c \in K_{i+k*t}$, and if $c \notin K_i$ then $c \notin K_{i+k*t} \}$
Change.MC = $\{c \mid$ if $i < j < k$, then either $c_{K_i} \geq c_{K_j} \geq c_{K_k}$, or $c_{K_i} \leq c_{K_j} \leq c_{K_k} \}$
Change.NC = $\{c \mid$ if $c \notin$ Change.MC $\}$

- **Temporary:** Temporary property holds for entities in which individuals are always changing. For example, concept *Student* is temporary, because in general most of the individuals of *Student* would not be *Students* if they graduated. According to the semantic of changes in Table 1, *Change.CL* (come and leave) is a typical dynamic property.
- **Static:** Static property only holds for entities in which all the individuals will never be withdrew. For example, somebody is a *Person*, and he/she is always a person. In some special cases an individual may not have the membership of an entity at the beginning, but when it gained the membership, the membership always holds. For example, if *Peter* gained the *PhD* degree, he will always be a *PhD*. We call this kind of entities *"semi-static"*. Obviously based on the semantic of changes in Table 1, *Change.CM* (come and remain) and *Change.RF* (removed forever) are semi-static.
- **Reversible:** An entity is *"reversible"*, if it is the case that an individual could lose the membership of this entity, but could regain the membership again. For example, *Alice* was *Preganant* several years ago, and she is *Preganant* again. According to Table 1, *Change.PC* (periodical changes) is a typically reversible.
- **Non-Reversible:** A temporary entity is *"non-reversible"*, if for this kind of entity, an individual lose the membership, he/she will never gain the membership again. For example, if someone had finished his/her job as the President of the USA, then he/she would never take this job again. So *president-of-USA* is not reversible. Based on the semantic of changes in Table 1, we find *Change.CM* (come and remain) and *Change.RF* (removed forever) are non-reversible.
- **Periodic:** A temporary entity is *"Periodic"*, if the individual regains the membership of this entity periodically. For example, if *"today is Sunday"*, then after every seven days, *"today is Sunday"* again. Obviously from Table 1, *Change.PC* (periodical changes) is a typical *Periodic* property.
- **Rigid:** a property is *"rigid"* if it is essential to all its possible instance; that only holds for a concept that have it and cannot change.

From now on we restrict the Knowledge Base to be an *ontology*, which $K = (T, A)$ contains a pair: T (*TBox*) is a set of terminology of concepts and roles, and A (*ABox*) is a set of assertions based on T and a set of individuals.

Now we can specify a dynamic ontology stream which is capable to embody dynamic characters of its knowledge entities in a meta level, as shown in Fig. 1.

Definition 2 (Dynamic Ontology Stream (DOS)). *A* Dynamic Ontology Stream (DOS) *contains a meta layer M, and a knowledge stream base (Definition 1) K_1, K_2, ..., K_n, in which $K_i = (T_i, A_i)$ is a normal ontology. The meta layer M contains:*

(1) following meta-concepts: Static, Rigid, Temporary, Reversible, Non-Reversible, *and* Periodic, *and a meta-TBox as:*

Rigid $\sqsubseteq$ Static;
Periodic $\sqsubseteq$ Reversible;
Reversible $\sqsubseteq$ Temporary;
Non-Reversible $\sqsubseteq$ Temporary.

(2) and assertions about the meta-concepts and with concepts and roles in T as individuals.

Here we have the following constraints for these meta-properties. Obviously this proposition always hold. For example, $Person \sqsubseteq Teenager$ is unsatisfiable, because typically if someone is a person, he/she is always a person; but being a teenager is just a short period of his/her life. Here we say "typically" because we prefer to use these constrains as heuristic for validating the ontological knowledge bases, instead of strict logical conditions for judging inconsistency.

Proposition 1 (Constrain Between Static & Temporary). *Given an ontology and two entities (concepts or roles) C and D, if C is* static *and D is* temporary, *then $C \sqsubseteq D$ is typically unsatisfiable.*

4   Optimised Reasoning and Query Answering

In the following we will show how to optimise query answering with dynamic mete-properties.

Query Answering with Dynamic Profiles. In the dynamic ontology stream defined above, we could leverage query answering with meta-level dynamic characterizations. For example, given a boolean query $Q_1(\ ) = Person(Peter)$, and at time t we know $Perter$ is a person. So the answer of this boolean query is $True$ at time t. Since concept $Person$ is static, the answer should be fixed. So in any future time $k > t$, we know the answer of the boolean query should always be $True$, even we do not need to check the knowledge base. Similarly if we have boolean query $Q_2(\ )=BlueEyes(Peter), Student(Peter)$, since $BlueEyes$ is static but $Student$ is temporary, the whole query Q_2 is temporary. We have the following result for conjunctions in Table 2. In this table, P and Q are knowledge

Table 2. Dynamic properties of conjunctions

P	Q	P∧Q
Static	Static	Static
Static	Temporary	Temporary
Static	Periodic	Periodic
Temporary	Temporary	Temporary$_{min(f(P),f(Q))}$
Temporary	Static	Temporary$_{min(f(P),f(Q))}$
Periodic	Periodic	Periodic$_{min(f(P),f(Q))}$

Table 3. Dynamic profile of conjunctive queries

Query profile	Execution	Result Validity
Static	single	unlimited
Temporary$_f$	periodic	at most up to f
Periodic$_f$	periodic	at least up to f

entities (concepts or roles). We also use a function $f(P)$ to assigning a frequency value [13] to entity P.

Furthermore, for any ontology characterized by dynamic meta-properties, we can calculate the dynamic profile of conjunctive queries (*e.g.* SPARQL). This would allow optimizing query execution over knowledge streams (Table 3).

Query Answering with Semi-Static Properties. In traditional way it is OK to check "What is in the KB", but difficult to find "What was in the KB". For example in the stream (upper part of Fig. 1), given query $Q(X) - President(X)$,

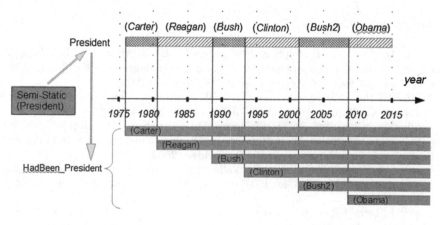

Fig. 1. Semi-static helps answering "What was in the KB".

the answer at time 1995 is *Clinton*, but at time 2010 is *Obama*. However it is difficult to answer "Who were the Presidents of the USA".

We note the dynamic nature of concept "Presidents of the USA" is *semi-static* and *non-reversible*. If someone was elected as the "Presidents of the USA", they will gain a property "had been the Presidents of the USA", and this property will never change in the future. For any interesting (being queried) semi-static concept P, an extra TBox axiom can be introduced (as shown in Fig. 1) in order to solve this kind of problem:

$$President \sqsubseteq HadBeing\_President \qquad (4)$$

5   Conclusion

Inspirited by OntoClean [9] from Guarino and Welty, in this paper we have investigated a methodology to analysis the dynamic properties in ontology streams. Guarino and Welty defined generic and domain independent properties of concepts (*i.e.* meta-properties) that aid the knowledge engineer to make correct decisions when building ontologies and therefore to avoid common mistakes. But in our approach, instead of attribute properties we mainly focus on the dynamic properties: *Static, Rigid, Temporary, Reversible, Non-Reversible,* and *Periodic*. We also investigate how to build up a meta-ontology with the constrains on these meta-properties. We have also shown some primary results on how to improve the reasoning on dynamic data by using these properties. Further enrich our methodology is main future work. Other extensions we are going to add are related to the cyclicality, trend (*e.g.* monotonic or non-monotonic) and variability (*i.e.* frequency of changes) of properties which allow to distinguish entities according their dynamics.

Acknowledgments. This work is partially supported by the FP7 K-Drive project (No. 286348) and the EPSRC WhatIf project (No. EP/J014354/1).

References

1. Anicic, D., Fodor, P., Rudolph, S., Stojanovic, N.: EP-SPARQL: a unified language for event processing and stream reasoning. In: Proceedings of the 20th International Conference on World Wide Web, pp. 635–644. ACM (2011)
2. Arasu, A., Babu, S., Widom, J.: The CQL continuous query language: semantic foundations and query execution. VLDB J. **15**(2), 121–142 (2006)
3. Baader, F., Calvanese, D., McGuinness, D.L., Nardi, D., Patel-Schneider, P.F. (eds.): The Description Logic Handbook: Theory, Implementation, and Applications. Cambridge University Press, Cambridge (2003)
4. Barbieri, D.F., Braga, D., Ceri, S., Valle, E.D., Grossniklaus, M.: C-SPARQL: a continuous query language for RDF data streams. Int. J. Semant. Comput. **4**(1), 3–25 (2010)
5. Botan, I., Derakhshan, R., Dindar, N., Haas, L., Miller, R.J., Tatbul, N.: Secret: a model for analysis of the execution semantics of stream processing systems. Proc. VLDB Endow. **3**(1–2), 232–243 (2010)

6. Calbimonte, J.-P., Jeung, H.Y., Corcho, O., Aberer, K.: Enabling query technologies for the semantic sensor web. Int. J. Semant. Web Inf. Syst. **8**(1), 43–63 (2012)
7. Flouris, G., Huang, Z., Pan, J.Z., Plexousakis, D., Wache, H.: Inconsistencies, negations and changes in ontologies. In: Proceedings of AAAI 2006, pp. 1295–1300 (2006)
8. Guarino, N.: Concepts, attributes, and arbitrary relations - some linguistic and ontological criteria for structuring knowledge bases. Data Knowl. Eng. **8**, 249–261 (1992)
9. Guarino, N., Welty, C.: An overview of ontoclean. In: Staab, S., Studer, R. (eds.) Handbook on Ontologies, pp. 151–159. Springer, Heidelberg (2004)
10. Guarino, N., Welty, C.A.: A formal ontology of properties. In: Dieng, R., Corby, O. (eds.) EKAW 2000. LNCS (LNAI), vol. 1937, pp. 97–112. Springer, Heidelberg (2000)
11. Le-Phuoc, D., Dao-Tran, M., Parreira, J.X., Hauswirth, M.: A native and adaptive approach for unified processing of linked streams and linked data. In: Aroyo, L., Welty, C., Alani, H., Taylor, J., Bernstein, A., Kagal, L., Noy, N., Blomqvist, E. (eds.) ISWC 2011, Part I. LNCS, vol. 7031, pp. 370–388. Springer, Heidelberg (2011)
12. Ren, Y., Pan, J.Z.: Optimising ontology stream reasoning with truth maintenance system. In: Proceedings of the ACM Conference on Information and Knowledge Management (CIKM) (2011)
13. Tamma, V.A.M., Capon, T.J.M.B.: Attribute meta-properties for formal ontological analysis. In: Gómez-Pérez, A., Benjamins, V.R. (eds.) EKAW 2002. LNCS (LNAI), vol. 2473, pp. 301–316. Springer, Heidelberg (2002)
14. Wang, S., Pan, J.Z., Zhao, Y., Li, W., Han, S., Han, D.: Belief base revision for datalog+/- ontologies. In: Kim, W., Ding, Y., Kim, H.-G. (eds.) JIST 2013. LNCS, vol. 8388, pp. 175–186. Springer, Heidelberg (2014)

Distance-Based Ranking of Negative Answers

Jianfeng Du$^{(\boxtimes)}$, Can Lin, and Kunxun Qi

Guangdong University of Foreign Studies, Guangzhou 510006, China
jfdu@gdufs.edu.cn

Abstract. Suggesting negative answers to users is a solution to the problem of insufficiently many answers in conjunctive query answering over description logic (DL) ontologies. Negative answers are complementary to certain answers and have explanations in the given ontology. Before being suggested to users, negative answers need to be ranked. However, there is no method for ranking negative answers by now. To fill this gap, this paper studies ranking methods that require only information on the given query and the given ontology. Three distance-based methods are proposed. Experimental results demonstrate that these methods can effectively rank negative answers to those conjunctive queries that have certain answers in the given DL ontology.

1 Introduction

Description logics (DLs) [1], as an important formalism for knowledge representation, underpin the standard Web Ontology Language (OWL). In DLs, an ontology is often expressed as a knowledge base consisting of both terminological knowledge in the TBox and assertional knowledge in the ABox. People have seen more and more DL ontology-based applications in the past decade. A notable application is the development of ontology-based data access (OBDA) systems in which a DL ontology is used as a unified view of information in variant data sources. Conjunctive query answering (CQA) is an important service provided in all OBDA systems. Traditional techniques for CQA focus on computing the set of certain answers to a conjunctive query. When an OBDA system is unable to return sufficiently many certain answers to a given query, it is desirable to suggest related answers in the descending order of their relevance degrees.

Negative answers [5,7] are well-defined related answers to conjunctive queries. These answers are complementary to certain answers and have *explanations* in the given consistent DL ontology, where an explanation for a negative answer is a set of assertions whose appending to the given ontology renders the negative answer into a certain answer while keeping the ontology consistent. Besides computing negative answers, ranking these answers by their relevance degrees to the given query is also important in making a user more satisfied with the CQA service. While there exist methods for computing negative answers [7], there is no method for ranking negative answers by now. To fill this gap, this paper proposes three distance-based methods for ranking negative answers.

© Springer International Publishing Switzerland 2016
G. Qi et al. (Eds.): JIST 2015, LNCS 9544, pp. 280–288, 2016.
DOI: 10.1007/978-3-319-31676-5_20

The first method ranks negative answers by k nearest neighbors, where k is determined by leave-one-out cross-validation on certain answers to the given query. The second method ranks negative answers by both *positive examples* and *negative examples*, where the set of positive examples is the set of certain answers to the given query and the set of negative answers is the set of certain answers to the negation of the given query. The last method ranks negative answers by positive examples only. Among all the above methods, the second method is relatively complicated since it requires computing all certain answers to the negation of a conjunctive query. This paper proposes a first method for this computation, which is applied to first-order rewritable ontologies. The expressivity of first-order rewritable ontologies is sufficient for many ontology-based applications such as the development of OBDA systems.

We generate test ontologies with different incompleteness degrees by randomly removing assertions from benchmark first-order rewritable ontologies. Experimental results show that all proposed ranking methods have significantly higher precisions in retrieving *true answers* (i.e. certain answers in the original ontology) than random ranking when test queries have certain answers. All proofs and more details about our experiments are available at http://www.dataminingcenter.net/jfdu/JIST15-full.pdf.

2   Preliminaries

We briefly introduce description logics (DLs) and refer readers to a handbook [1] for more details. A DL ontology consists of a TBox and an ABox, where the TBox is a finite set of axioms on relations between concepts and roles, and the ABox is a finite set of axioms (also called *assertions*) declaring instances of concepts and roles. We treat a DL ontology (a TBox or an ABox) as a set of axioms. We only consider *normalized* ABoxes that have only *basic assertions*, namely concept assertions of the form $A(a)$ and role assertions of the form $r(a, b)$, where A is a concept name and r is a role name. Other concept assertions and role assertions can be normalized to basic ones in a standard way. A DL ontology $\mathcal{O}$ is said to be *consistent* if it has at least one model, otherwise *inconsistent*.

A *conjunctive query* (*CQ*) $Q(\boldsymbol{x})$ is a formula of the form $\exists \boldsymbol{y}\, \phi(\boldsymbol{x}, \boldsymbol{y})$, where $\phi(\boldsymbol{x}, \boldsymbol{y})$ is a conjunction of atoms over concept or role names in the considering ontology or (in)equational atoms, $\boldsymbol{x}$ are *answer variables*, and $\boldsymbol{y}$ are *quantified variables*. A CQ without answer variables is called a *Boolean conjunctive query* (*BCQ*). In this paper a BCQ is written and treated as a set of atoms; e.g., the BCQ $\exists x\, A(x) \wedge B(x)$ is written as $\{A(x), B(x)\}$. A *disjunction* of CQs is a formula of the form $Q_1(\boldsymbol{x}) \vee \ldots \vee Q_n(\boldsymbol{x})$ where $n \geq 1$ and $Q_1(\boldsymbol{x})$, ..., $Q_n(\boldsymbol{x})$ are CQs. A disjunction of BCQs Q_D is said to be *entailed by* $\mathcal{O}$, denoted by $\mathcal{O} \models Q_D$, if Q_D is satisfied by all models of $\mathcal{O}$. Given a DL ontology $\mathcal{O}$ and a disjunction of CQs $Q_D(\boldsymbol{x})$, a tuple $\boldsymbol{t}$ of individuals with the same arity as $\boldsymbol{x}$ is called a *certain answer* to $Q_D(\boldsymbol{x})$ in $\mathcal{O}$ if $\mathcal{O} \models Q_D(\boldsymbol{t})$, where $Q_D(\boldsymbol{t})$ is a disjunction of BCQs obtained from $Q(\boldsymbol{x})$ by replacing variables in $\boldsymbol{x}$ with corresponding individuals in $\boldsymbol{t}$. The set of certain answers to $Q_D(\boldsymbol{x})$ in $\mathcal{O}$ is denoted by $\mathsf{ans}(\mathcal{O}, Q_D(\boldsymbol{x}))$.

Datalog$^\pm$ [3] extends datalog with *existential rules* R of the form $\forall \boldsymbol{x} \forall \boldsymbol{y}\, \phi(\boldsymbol{x}, \boldsymbol{y}) \to \exists \boldsymbol{z}\, \varphi(\boldsymbol{x}, \boldsymbol{z})$, where $\phi(\boldsymbol{x}, \boldsymbol{y})$ and $\varphi(\boldsymbol{x}, \boldsymbol{z})$ are conjunctions of atoms and $\boldsymbol{x}$, $\boldsymbol{y}$ and $\boldsymbol{z}$ are pairwise disjoint sets of variables. The parts of R at left-hand side and right-hand side of $\to$ are called the *body* and the *head* of R, respectively. An existential rule is called an *equality generating dependency* (*EGD*) if its head is of the form $x_1 = x_2$ where x_1 and x_2 are different variables appearing in the body of it; called a *constraint* if its head is empty; otherwise, called a *tuple generating dependency* (*TGD*). A TGD is said to be *linear* if its body contains a single atom; *multi-linear* if all atoms in its body have the same variables. A linear TGD is also a multi-linear TGD.

A datalog$^\pm$ program is a finite set of existential rules. Since existential rules are formulae in first-order logic with equality, a TBox expressed in some DLs can be translated to a datalog$^\pm$ program. It follows that an ontology expressed in some DLs can be translated to the union of a datalog$^\pm$ program and a normalized ABox. We call such an ontology *datalog$^\pm$-translatable*. For a datalog$^\pm$-translatable ontology with TBox $\mathcal{T}$, by $\mathcal{S}_{\mathcal{T}}^D$ we denote the set of TGDs translated from $\mathcal{T}$, by $\mathcal{S}_{\mathcal{T}}^C$ we denote the set of constraints translated from $\mathcal{T}$ and by $\mathcal{S}_{\mathcal{T}}^E$ we denote the set of EGDs translated from $\mathcal{T}$. Since the *unique name assumption* is adopted in datalog$^\pm$ programs, it is also adopted in datalog$^\pm$-translatable ontologies that are considered in this paper. This means that all individuals appearing in an ontology are interpreted as different in any model of the ontology.

A set $\mathcal{S}^D$ of TGDs is said to be *first-order rewritable* if, for every CQ $Q(\boldsymbol{x})$, there exists a finite disjunction of CQs $Q_D(\boldsymbol{x})$ such that $\mathsf{ans}(\mathcal{S}^D \cup \mathcal{A}, Q(\boldsymbol{x})) = \mathsf{ans}(\mathcal{A}, Q_D(\boldsymbol{x}))$ for all ABoxes $\mathcal{A}$. It has been shown [3] that a set $\mathcal{S}^D$ of TGDs is first-order rewritable if all TGDs in $\mathcal{S}^D$ are multi-linear. A set $\mathcal{S}^E$ of EGDs is said to be *separable* from a set $\mathcal{S}^D$ of TGDs if the following holds for every ABox $\mathcal{A}$: if there exists an EGD $\forall \boldsymbol{x}\, \phi(\boldsymbol{x}) \to x_1 = x_2$ in $\mathcal{S}^E$ and a ground substitution σ for $\boldsymbol{x}$ such that $\mathcal{S}^D \cup \mathcal{A} \models \phi(\boldsymbol{x}\sigma)$ and $x_1 \sigma \neq x_2 \sigma$, then there is a ground substitution θ for $\boldsymbol{x}$ such that $\mathcal{A} \models \phi(\boldsymbol{x}\theta)$ and $x_1 \theta \neq x_2 \theta$; otherwise, $\mathcal{S}^D \cup \mathcal{S}^E \cup \mathcal{A} \models Q$ if and only if $\mathcal{S}^D \cup \mathcal{A} \models Q$ for all BCQs Q. A datalog$^\pm$-translatable ontology with TBox $\mathcal{T}$ is called a *first-order rewritable* ontology if $\mathcal{S}_{\mathcal{T}}^D$ is first-order rewritable and $\mathcal{S}_{\mathcal{T}}^E$ is separable from $\mathcal{S}_{\mathcal{T}}^D$ [7].

3   Computing Negative Answers

When a CQ does not have sufficiently many certain answers in a DL ontology, users of CQA would probably like to see more related answers in the descending order of their relevance degrees to the CQ. In addition, users may also want to know why a related answer cannot be a certain answer. This requirement has been emphasized for achieving a high user satisfaction in OBDA systems [2]. To characterize related answers, a notion of negative answers was introduced in [5,7], where a *negative answer* t to a CQ $Q(\boldsymbol{x})$ in a DL ontology $\mathcal{O}$ is defined as a tuple of individuals such that t and $\boldsymbol{x}$ have the same arity and $\mathcal{O} \not\models Q(t)$. This notion of negative answers does not reflect relevancy to the given CQ. Hence we alter this notion below by considering explanations that reflect relevancy.

Definition 1 (Negative Answer). *Given a DL ontology $\mathcal{O}$ and a CQ $Q(x)$, a tuple t of individuals with the same arity as x is called a* negative answer *to $Q(x)$ in $\mathcal{O}$ if $\mathcal{O} \not\models Q(t)$ and there is a set $\mathcal{E}$ of assertions such that $\mathcal{O} \cup \mathcal{E}$ is consistent and $\mathcal{O} \cup \mathcal{E} \models Q(t)$, where $\mathcal{E}$ is called an* explanation *for $Q(t)$ in $\mathcal{O}$.*

Simply speaking, a negative answer is not a certain answer but can become one after a consistent set of assertions is appended to the given ontology. Its relevancy to the given CQ can explicitly be shown by explanations. Regarding the computational aspect, a method has been proposed in [7] to compute a certain portion of negative answers in a consistent first-order rewritable ontology. In the following, we adapt this method to computing all negative answers.

We assume that $\mathcal{O}$ is a consistent first-order rewritable ontology with TBox $\mathcal{T}$ and ABox $\mathcal{A}$, and that $\mathcal{S}_Q(x)$ is a set of CQs rewritten from the CQ $Q(x)$ such that $\mathsf{ans}(\mathcal{S}_{\mathcal{T}}^{\mathcal{D}} \cup \mathcal{A}', Q(x)) = \mathsf{ans}(\mathcal{A}', \bigvee \mathcal{S}_Q(x))$ for all ABoxes $\mathcal{A}'$. The negative answers to $Q(x)$ in $\mathcal{O}$ can be retrieved from ground substitutions for x by instantiating subsets of $Q'(x)$ into subsets of $\mathcal{A}$ for all CQs $Q'(x)$ in $\mathcal{S}_Q(x)$. Formally, we define a *bipartition* of a CQ $Q(x)$ as a pair of CQs $(Q_1(x_1), Q_2(x_2))$ such that $Q_1(x_1)$ and $Q_2(x_2)$ have no common atoms while the conjunction of $Q_1(x_1)$ and $Q_2(x_2)$ yields $Q(x)$, where x_1 and x_2 are answer variables in $Q_1(x_1)$ and $Q_2(x_2)$, respectively. By $\mathsf{bipart}(Q(x))$ we denote the set of bipartitions of $Q(x)$, by $\mathsf{ind}(S)$ we denote the set of individuals in S, and by $\Psi(\mathcal{T}, \mathcal{A}, Q(x))$ we denote $\{x\theta\sigma \mid Q'(x) \in \mathcal{S}_Q(x),\ (Q_1(x_1), Q_2(x_2)) \in \mathsf{bipart}(Q'(x)),\ \theta$ is a ground substitution for $Q_2(x_2)$ such that $Q_2(x_2)\,\theta \subseteq \mathcal{A}$, σ is a ground substitution for $x\theta$ such that $\mathsf{ind}(x\theta\sigma) \subseteq \mathsf{ind}(\mathcal{A})$ and $\mathcal{T} \cup \mathcal{A} \cup Q_1(x_1\theta\sigma)$ is consistent$\}$. Then $\Psi(\mathcal{T}, \mathcal{A}, Q(x)) \setminus \mathsf{ans}(\mathcal{O}, Q(x))$ is the set of negative answers to $Q(x)$ in $\mathcal{O}$. The following theorem shows the time complexity and correctness of this method.

Theorem 1. $\Psi(\mathcal{T}, \mathcal{A}, Q(x))$ *is computable in* PTIME *in data complexity, while $\Psi(\mathcal{T}, \mathcal{A}, Q(x)) \setminus \mathsf{ans}(\mathcal{O}, Q(x))$ is the set of negative answers to $Q(x)$ in $\mathcal{O}$.*

4   Ranking Negative Answers

Suggesting negative answers to users is a solution to the problem of insufficiently many answers to a CQ. To make this solution more satisfactory, the negative answers should be ranked before being suggested to users. In this section we propose three ranking methods based on a distance measure.

4.1   Distance Between Two Tuples of Individuals

We introduce a distance measure between two tuples of individuals that have the same arity. This measure is extended from the distance measure between two individuals [6]. Given a DL ontology $\mathcal{O}$, an exponent p and a feature set $\mathbf{F} = \{F_1, \ldots, F_m\}$ which is a set of DL concepts coming with weights w_i ($1 \leq i \leq m$), the distance between two individuals a and b is defined as

$$d_p^{\mathbf{F}}(a, b) = \left(\sum_{i=1}^{m} w_i \delta_i(a, b)^p \right)^{\frac{1}{p}},$$

where $p > 0$ and for all $i \in \{1, \ldots, m\}$, the dissimilarity function δ_i is defined as

$$\delta_i(a, b) = \begin{cases} 0 & (\mathcal{O} \models F_i(a) \text{ and } \mathcal{O} \models F_i(b)) \text{ or } (\mathcal{O} \models \neg F_i(a) \text{ and } \mathcal{O} \models \neg F_i(b)), \\ 1 & (\mathcal{O} \models F_i(a) \text{ and } \mathcal{O} \models \neg F_i(b)) \text{ or } (\mathcal{O} \models \neg F_i(a) \text{ and } \mathcal{O} \models F_i(b)), \\ \frac{1}{2} & \text{otherwise.} \end{cases}$$

We extend $d_p^{\mathbf{F}}(a, b)$ to a distance measure between two tuples of individuals by treating the feature set of a tuple as multiple copies of the feature set of an individual. Formally, for two tuples of individuals $t_1 = \langle a_1, \ldots, a_n \rangle$ and $t_2 = \langle b_1, \ldots, b_n \rangle$, the distance between t_1 and t_2 is defined as

$$d_p^{\mathbf{F}}(t_1, t_2) = \left(\sum_{i=1}^{m} \sum_{j=1}^{n} w_i \delta_i(a_j, b_j)^p \right)^{\frac{1}{p}}.$$

In our experiments, we define the feature set $\mathbf{F}$ as the set of concept names in $\mathcal{O}$ and the exponent p as 2. We also adopt the entropy measure [6] to estimate the weights for all features. In more details, let N_I denote the set of individuals in $\mathcal{O}$ and $|S|$ the cardinality of a set S, then the weight w_i for F_i is defined as

$$w_i = -p_{F_i} \log_2(p_{F_i}) - p_{\neg F_i} \log_2(p_{\neg F_i}) - p_{U_i} \log_2(p_{U_i}),$$

where $p_{F_i} = |\{a \in N_I \mid \mathcal{O} \models F_i(a)\}|/|N_I|$, $p_{\neg F_i} = |\{a \in N_I \mid \mathcal{O} \models \neg F_i(a)\}|/|N_I|$, and $p_{U_i} = 1 - (p_{F_i} + p_{\neg F_i})$. Intuitively, under an assumption that the set of instances of a feature or its negation is much smaller than the set of individuals in $\mathcal{O}$, this measure gives a higher weight to a feature F such that the number of instances of F and the number of instances of $\neg F$ are larger and closer.

4.2   KNN: Ranking by k Nearest Neighbors

The first ranking method (written KNN) estimates the relevance degree of a negative answer by its k nearest neighbors, where the aforementioned distance measure is used to find nearest neighbors. More precisely, let $\mathrm{NN}_k(t)$ denote the set of k nearest neighbors of the tuple t of individuals in the space made up of arbitrary tuples with the same arity, then the relevance degree of a negative answer t to a CQ $Q(x)$ in a DL ontology $\mathcal{O}$ is defined as

$$rd_1(t, Q(x), \mathcal{O}, k) = \frac{\sum_{t' \in \mathrm{NN}_k(t) \cap \mathrm{ans}(\mathcal{O}, Q(x))} \frac{1}{1 + d_p^{\mathbf{F}}(t', t)}}{\sum_{t' \in \mathrm{NN}_k(t)} \frac{1}{1 + d_p^{\mathbf{F}}(t', t)}}.$$

The definition of $rd_1(t, Q(x), \mathcal{O}, k)$ is slightly modified from the definition of the likelihood for t being a certain answer [6] by using $1/(1 + d_p^{\mathbf{F}}(t', t))$ instead of $1/d_p^{\mathbf{F}}(t', t)$ as the weight of a neighbor t' of t. This modification is needed since the distance between two tuples can be zero.

An appropriate value of k in $rd_1(t, Q(x), \mathcal{O}, k)$ can be determined by leave-one-out cross-validation on certain answers to $Q(x)$ in $\mathcal{O}$. More precisely, in our experiments the value of k is set as the minimum integer in a fairly wide range $[1, 100]$ such that $\sum_{t \in \mathrm{ans}(\mathcal{O}, Q(x))} rd_1(t, Q(x), \mathcal{O}, k)$ is maximized.

4.3   PosNeg: Ranking by Positive Examples and Negative Examples

The performance of KNN depends on the parameter k. In practice it is hard to set an accurate value for k to achieve the best performance. Thus, we may avoid this setting by considering all neighbors that are either certain answers to the given CQ or certain answers to the negation of the given CQ. We call certain answers to the given CQ $Q(x)$ *positive examples* for $Q(x)$ and certain answers to the negation of $Q(x)$ (written $\neg Q(x)$) *negative examples* for $Q(x)$. In the second ranking method (written PosNeg) we define the relevance degree of a negative answer t to a CQ $Q(x)$ in $\mathcal{O}$ as the sum of weights of all positive and negative examples for $Q(x)$, where the weight of a positive (resp. negative) example is defined in the same way as the weight (resp. opposite weight) of a neighbor. More precisely, we define the relevance degree as

$$rd_2(t, Q(x), \mathcal{O}) = \sum_{t' \in \mathsf{ans}(\mathcal{O}, Q(x))} \frac{1}{1 + d_p^{\mathbf{F}}(t', t)} - \sum_{t'' \in \mathsf{ans}(\mathcal{O}, \neg Q(x))} \frac{1}{1 + d_p^{\mathbf{F}}(t'', t)}.$$

As far as we know, there is no method designed for computing certain answers to the negation of a general CQ i.e. for computing $\mathsf{ans}(\mathcal{O}, \neg Q(x))$ for a general CQ $Q(x)$. In [8] an efficient method is proposed to compute $\mathsf{ans}(\mathcal{O}, \neg A(x))$ in a consistent first-order rewritable ontology, where A is a concept name and x is the unique answer variable. In the following we show that this method can be extended to handle general CQs in consistent first-order rewritable ontologies. Restricting the given ontology to a first-order rewritable ontology does not impair the applicability much, because in many ontology-based applications such as the development of OBDA systems, the back-end ontology is expressed in DLs in the DL-Lite family [4], while most of these DLs belong to the first-order rewritable class [3], including DL-Lite$_{\mathcal{R}}$ which underpins the QL profile of OWL 2.

We assume that $\mathcal{O}$ is a consistent first-order rewritable ontology with TBox $\mathcal{T}$ and ABox $\mathcal{A}$. Since $\mathcal{S}_{\mathcal{T}}^D$ is first-order rewritable, for every BCQ Q there must be a set $\mathcal{S}_Q$ of BCQs such that $\mathcal{S} \cup \mathcal{A}' \models Q$ if and only if $\mathcal{A}' \models \bigvee \mathcal{S}_Q$ for all ABoxes $\mathcal{A}'$; hence we denote the set $\mathcal{S}_Q$ by $\gamma(Q, \mathcal{S}_{\mathcal{T}}^D)$, considering that it is rewritten from Q through $\mathcal{S}_{\mathcal{T}}^D$. By $\rho(R)$ we denote the BCQ $\exists x\, \phi(x)$ if R is a constraint $\forall x\, \phi(x) \rightarrow$, or the BCQ $\exists x\, \phi(x) \wedge x_1 \neq x_2$ if R is an EGD $\forall x\, \phi(x) \rightarrow x_1 = x_2$.

The set of certain answers to $\neg Q(x)$ in $\mathcal{O}$ can be computed by rewriting $\neg Q(x)$ into a disjunction of CQs and by answering it over $\mathcal{A}$. In more details, given a BCQ Q' that has no variables appearing in $Q(x)$, we call a substitution θ for Q' a $Q(x)$-*restricted substitution* for Q' if $Q(x)$ is a subset of $Q'\theta$, where $Q(x)$ is treated as a set of atoms, and all answer variables and quantified variables in $Q(x)$ or Q' are treated as free variables. For example, suppose $Q(x) = A(x) \wedge r(x, x)$, then the substitution $\theta = \{y \mapsto x, z \mapsto x\}$ for $Q' = \{A(y), r(y, z), B(z)\}$ is a $Q(x)$-restricted substitution for Q', since $\{A(x), B(x)\}$ which is converted from $Q(x)$ is a subset of $Q'\theta$. Without loss of generality, we assume that all BCQs in $\{\gamma(\rho(R), \mathcal{S}_{\mathcal{T}}^D) \mid R \in \mathcal{S}_{\mathcal{T}}^C\} \cup \{\rho(R) \mid R \in \mathcal{S}_{\mathcal{T}}^E\}$ have no variables in $Q(x)$. By $\Xi(\mathcal{T}, Q(x))$ we denote the disjunction of CQs that is obtained from $\bigvee\{Q'\theta \setminus Q(x) \mid Q' \in \{\gamma(\rho(R), \mathcal{S}_{\mathcal{T}}^D) \mid R \in \mathcal{S}_{\mathcal{T}}^C\} \cup \{\rho(R) \mid R \in \mathcal{S}_{\mathcal{T}}^E\}, \theta$ is a $Q(x)$-restricted substitution for $Q'\}$ by treating $Q'\theta \setminus Q(x)$ as a CQ having x as answer variables and other variables as quantified variables. Then answering $\Xi(\mathcal{T}, Q(\boldsymbol{x}))$ over $\mathcal{A}$ gets all certain answers to $\neg Q(\boldsymbol{x})$ in $\mathcal{O}$, as shown in the following theorem.

Theorem 2. *For an arbitrary CQ $Q(\boldsymbol{x})$, $\mathsf{ans}(\mathcal{A}, \Xi(\mathcal{T}, Q(\boldsymbol{x}))) = \mathsf{ans}(\mathcal{O}, \neg Q(\boldsymbol{x}))$.*

4.4   PosOnly: Ranking by Positive Examples only

PosNeg requires computing negative examples, but it is still unknown whether negative examples can be practically computed in a consistent DL ontology that is not first-order rewritable. To achieve a more practical ranking method that can be applied to an arbitrary DL ontology, we neglect negative examples for a given CQ and define the relevance degree of a negative answer by only positive examples for the given CQ. In the third ranking method (written PosOnly) we define the relevance degree of a negative answer t to a CQ $Q(\boldsymbol{x})$ in $\mathcal{O}$ as

$$rd_3(\boldsymbol{t}, Q(\boldsymbol{x}), \mathcal{O}) = \sum_{t' \in \mathsf{ans}(\mathcal{O}, Q(\boldsymbol{x}))} \frac{1}{1 + d_p^{\mathrm{F}}(\boldsymbol{t}', \boldsymbol{t})}.$$

5   Summary of Evaluation Results

In this section we summarize experimental results on evaluating the effectiveness of the three ranking methods. More details are available at http://www.dataminingcenter.net/jfdu/JIST15-full.pdf. To develop a goal standard that tells if a negative answer is a true answer in reality, we generated incomplete DL ontologies with different incompleteness degrees by randomly removing assertions from benchmark first-order rewritable ontologies and treated a negative answer to a CQ in an incomplete DL ontology as a true answer if it is a certain answer to the same CQ in the original DL ontology. A generated ontology whose incompleteness degree is $\mu\%$ is a consistent first-order rewritable ontology modified from a benchmark ontology by removing $\mu\%$ of assertions from the ABox for every concept name and every role name. By Semintec$_{\mu\%}$, LUBM1$_{\mu\%}$ and LUBM10$_{\mu\%}$ we denote the ontology with incompleteness degree $\mu\%$ generated from Semintec[1], LUBM1[2] and LUBM10[2], respectively, where $\mu = 10, 20, 30, 40, 50$. We evaluated the methods on computing and ranking negative answers to all those CQs that are originally used to verify the completeness of a DL reasoner [9]. By using the SyGENiA tool[3] [9] we generated the same 54 CQs for all Semintec ontologies and the same 79 CQs for all LUBM ontologies.

Experimental results show that, when ranking negative answers to those test CQs that have certain answers, on the precision measure at top 5 negative answers all proposed methods have similar precisions as random ranking for Semintec$_{10\%}$, Semintec$_{20\%}$ and Semintec$_{30\%}$, and have significantly higher precisions than random ranking for other test ontologies. These results remain

---

[1] http://www.cs.put.poznan.pl/alawrynowicz/semintec.htm.
[2] http://swat.cse.lehigh.edu/projects/lubm/.
[3] https://code.google.com/p/sygenia/.

almost the same on the precision measure at top 10 and top 20 negative answers. The comparison results between the three proposed ranking methods show that PosOnly has the highest precision in almost all the test cases although it is relatively simple. By further comparing PosOnly with KNN and PosNeg in terms of the normalized discounted cumulative gain (NDCG) measure, it can be seen that PosOnly is still the best among all the proposed ranking methods.

6   Conclusion and Future Work

Ranking negative answers is as important as computing negative answers in a satisfactory solution to the problem of insufficiently many answers in CQA over DL ontologies. In this paper we have conducted a pioneer study on ranking negative answers and made the following contributions. First of all, we proposed three distance-based methods (KNN, PosNeg and PosOnly) for ranking negative answers and empirically showed that all these methods generally have higher precisions than random ranking for test CQs that have certain answers. Secondly, we proposed a PTIME (data complexity) method for computing all certain answers to the negation of a general CQ in a consistent first-order rewritable ontology. Finally, we proposed a PTIME (data complexity) method for computing all negative answers to a CQ in a consistent first-order rewritable ontology.

All the proposed methods for ranking negative answers have a limitation that they are not applicable to a failed CQ that has not any certain answer. This is because the considering k nearest neighbors in KNN must include some certain answers to the given CQ while the positive examples used in PosNeg and PosOnly are certain answers to the given CQ. To rank negative answers to a failed CQ, we need to develop ranking methods that do not depend on certain answers to the given CQ. In the future work we plan to exploit explanations for negative answers to estimate the relevance degrees of negative answers.

Acknowledgements. This work is partly supported by NSFC (61375056) and Guangdong Natural Science Foundation (S2013010012928).

References

1. Baader, F., Calvanese, D., McGuinness, D.L., Nardi, D., Patel-Schneider, P.F. (eds.): The Description Logic Handbook: Theory, Implementation, and Applications. Cambridge University Press, Cambridge (2003)
2. Borgida, A., Calvanese, D., Rodriguez-Muro, M.: Explanation in the $DL - Lite$ family of description logics. In: Tari, Z., Meersman, R. (eds.) OTM 2008, Part II. LNCS, vol. 5332, pp. 1440–1457. Springer, Heidelberg (2008)
3. Calì, A., Gottlob, G., Lukasiewicz, T.: A general datalog-based framework for tractable query answering over ontologies. J. Web Sem. **14**, 57–83 (2012)
4. Calvanese, D., Giacomo, G., Lembo, D., Lenzerini, M., Rosati, R.: Tractable reasoning and efficient query answering in description logics: The DL-Lite family. J. Autom. Reasoning **39**(3), 385–429 (2007)

5. Calvanese, D., Ortiz, M., Simkus, M., Stefanoni, G.: Reasoning about explanations for negative query answers in DL-Lite. J. Artif. Intell. Res. **48**, 635–669 (2013)
6. d'Amato, C., Fanizzi, N., Esposito, F.: Query answering and ontology population: an inductive approach. In: Bechhofer, S., Hauswirth, M., Hoffmann, J., Koubarakis, M. (eds.) ESWC 2008. LNCS, vol. 5021, pp. 288–302. Springer, Heidelberg (2008)
7. Du, J., Liu, D., Lin, C., Zhou, X., Liang, J.: A system for tractable computation of negative answers to conjunctive queries. In: Zhao, D., Du, J., Wang, H., Wang, P., Ji, D., Pan, J.Z. (eds.) CSWS 2014. CCIS, vol. 480, pp. 56–66. Springer, Heidelberg (2014)
8. Du, J., Pan, J.Z.: A system for tractable computation of negative answers to conjunctive queries. In: ISWC, pp. 339–355 (2015)
9. Grau, B.C., Motik, B., Stoilos, G., Horrocks, I.: Completeness guarantees for incomplete ontology reasoners: Theory and practice. J. Artif. Intell. Res. **43**, 419–476 (2012)

Contrasting RDF Stream Processing Semantics

Minh Dao-Tran[✉], Harald Beck, and Thomas Eiter

Institute of Information Systems, Vienna University of Technology,
Favoritenstraße 9-11, 1040 Vienna, Austria
{dao,beck,eiter}@kr.tuwien.ac.at

Abstract. The increasing popularity of RDF Stream Processing (RSP) has led to developments of data models and processing engines which diverge in several aspects, ranging from the representation of RDF streams to semantics. Benchmarking systems such as LSBench, SRBench, CSRBench, and YABench were introduced as attempts to compare different approaches, focusing mainly on the operational aspects. The recent logic-based LARS framework provides a theoretical underpinning to analyze stream processing/reasoning semantics. In this work, we use LARS to compare the semantics of two typical RSP engines, namely C-SPARQL and CQELS, identify conditions when they agree on the output, and discuss situations where they disagree. The findings give insights that might prove to be useful for the RSP community in developing a common core for RSP.

1 Introduction

In interconnected information technologies such as Internet of Things and Cyber-Physical Systems it is crucial to have simple access to data irrespective of their sources. The Semantic Web's RDF data model was designed to integrate such distributed and heterogeneous data. Recently, RDF Stream Processing (RSP) has been emerging to tackle novel problems arising from streaming data: to integrate querying and processing of static and dynamic data, e.g., information continuously arriving from sensors.

This has led to the development of data models, query languages and processing engines, which diverge in several aspects, ranging from the representation of RDF streams, execution modes [17], to semantics [2,4,5,12,16]. To deal with this heterogeneity, the RSP community[1] was formed to establish a standard towards a W3C recommendation.

A standardization must start from seeing the differences between existing approaches and thus comparing RSP engines is an important topic. Initial empirical comparisons were carried out in two benchmarking systems, namely SRBench [19] and LSBench [17]. The former defined functional tests to verify the query languages features by the engines, while the latter measured mismatch

---

This research has been supported by the Austrian Science Fund (FWF) projects P24090, P26471, and W1255-N23.

[1] https://www.w3.org/community/rsp/.

© Springer International Publishing Switzerland 2016
G. Qi et al. (Eds.): JIST 2015, LNCS 9544, pp. 289–298, 2016.
DOI: 10.1007/978-3-319-31676-5_21

between the output of different engines, assuming they are sound, i.e., all output produced by them are correct. Later on, CSRBench [8] introduced an oracle that pregenerates the correct answers wrt. each engine's semantics, which are then used to check the output returned by the engine. YABench [14] follows the approach by CSRBench with the main purpose of facilitating joint evaluation of functional, correctness, and performance testing. However, this approach allows only partial comparison between engines by referring to their ideal counterparts.

Due to the lack of a common language to express divergent RSP approaches, the three works above could just look at the output of the engines and did not have further means to explain beyond the output what caused the difference semantically.

Recently, [9] proposed a unifying query model to explain the heterogeneity of RSP systems. It shows the difference between two approaches as represented by representative engines in the RSP community, namely C-SPARQL [2], $SPARQL_{Stream}$ [5] and CQELS [16]. This work identified types of datasets that C-SPARQL/$SPARQL_{Stream}$ can handle while CQELS cannot, and vice versa. However, it does not point out systematically when and how the engines agree on the output.

In the stream processing community, SECRET [10] was proposed to characterize and analyze the behavior of stream processing engines, but at the operational level.

Latterly, a **L**ogic-based framework for **A**nalyzing **R**easoning over **S**tream (LARS) was introduced [3]. LARS can be used as a unifying language which stream processing/reasoning languages can be translated to. It may serve as a formal host language to express semantics and thus allows a deeper comparison that goes beyond mere looking at the output of the respective engines. Furthermore, the model-based semantics of LARS is a means to formalize the intuition of agreement between not only RSP engines but also engines from other fields, and to identify conditions where this holds.

In this paper, we exploit the capability of LARS to analyze the difference between the semantics of C-SPARQL and CQELS by: (a) providing translations that capture the push- and pull- execution modes for general LARS programs, (b) providing translations from C-SPARQL, CQELS to LARS, (c) introducing a notion of *push-pull-agreement* between LARS programs, and (d) identifying *conditions* where C-SPARQL and CQELS agree on their output, by checking whether the translated LARS programs push-pull-agree.

Due to space reasons, (a) and (b) will briefly mentioned while (c) and (d) will be presented in detail. For an extended version of this paper, we refer the reader to [7].

Our findings show that C-SPARQL and CQELS agree on a very limited setting, and give insights on their difference. This result might prove to be useful for the RSP community in developing a common core for RSP.

For the purpose of a theoretical comparison, we adopt (as in [9]) the assumption that execution time of RSP engines is neglectable compared to the rate of the input streams.

2   Preliminaries

From RDF to RDF Stream Processing. RDF [6] is a W3C recommendation for data interchange on the Web. RDF models data as directed labeled graphs whose nodes are resources and edges represent relations among them. SPARQL [13], a W3C recommendation for querying RDF graphs, is essentially a graph-matching query language. As the underlying RDF graphs are static, SPARQL's one-shot queries are not able to give answers under dynamic input. RSP was thus introduced to express queries on streaming RDF data.

Two representative RSP engines are C-SPARQL and CQELS. Their query languages and semantics are inspired by the Continuous Query Language (CQL) [1] in which queries are composed of three classes of operators, namely stream-to-relation (S2R), relation-to-relation (R2R), and relation-to-stream (R2S) operators (RStream, IStream, and DStream). The main adaptation for RSP is that R2R operators are SPARQL operators. However, C-SPARQL and CQELS diverge in two crucial aspects as presented in Table 1. This makes it non-trivial to compare the two engines semantically.

To make the comparison possible, we propose a unified model of RSP queries that covers all of the differences above. Extending the work in [18] for SPARQL queries, we model an RSP query as a quadruple $Q = (V, P, \mathcal{D}, \mathcal{S})$, where V is a result form, P is a graph pattern, $\mathcal{D}$ is a dataset, and $\mathcal{S}$ is a set of input stream patterns. Notably, $\mathcal{S}$ is a set of tuples of the form $(\mathbf{s}, \omega, g)$, where $\mathbf{s}$ is a stream identifier, ω is a window expression (e.g., [RANGE 10] for time-based windows and [COUNT 5] for count-based windows), and g is a basic RDF graph pattern. An RSP query Q in this model corresponds to a C-SPARQL query, denoted by $cs(Q)$ and to a set $cq(Q)$ of $2^{|\mathcal{S}|}$ CQELS queries.

We now informally introduce LARS, which will serve as theoretical underpinning for both languages.

Table 1. C-SPARQL vs. CQELS

	C-SPARQL	CQELS
Execution mode	Pull-based	Push-based
Snapshot creation	Merge patterns on input streams into the default graph	Apply patterns on input streams

LARS. The Logic-based Framework for Analyzing Reasoning over Streams [3] extends Answer Set Programming [11] for streaming data. Crucial extensions are the following:

– Streams are represented as tuples $S = (T, v)$ where T is a timeline and v is an evaluation function mapping each time point t to a set of facts. We call S a *data stream*, if it contains only extensional atoms. An *interpretation stream* expresses inferred information by additional intensional atoms.

The *projection* of a stream S to a predicate p is defined as $S|_p = (T, v|_p)$, where $v|_p(t) = \{p(\boldsymbol{c}) \mid p(\boldsymbol{c}) \in v(t)\}$. By $Ats(S) = \bigcup_{t \in T} v(t)$, we denote the set of all atoms appearing in S.

- Window functions are generically defined to capture all types of windows in practice. The function $w_\iota(S, t, \boldsymbol{x})$ of type ι takes as input a stream S, a time point t, and a vector of parameters $\boldsymbol{x}$ (which depend on type ι) and returns a substream S' of S.
- Window operators of the form $\boxplus_\iota^{\boldsymbol{x}}$ are the means to connect LARS formulas to the corresponding window functions $w_\iota(S, t, \boldsymbol{x})$.
- Three temporal operators allow fine-grained control for temporal reference. Given a formula φ, $\Diamond\varphi$ (resp. $\Box\varphi$) holds, if φ holds at some (resp. all) time point(s) in the selected window, and $@_t\varphi$ holds if φ holds exactly at time point t in the window.
- The output of a LARS program P wrt. to a data stream D at a time point t is represented as a set $\mathcal{AS}(P, D, t)$ of *answer streams*, which are minimal interpretation streams I satisfying the reduct of P under I at t (see [3] for a formal definition). Intuitively, an answer stream I adds to D inferred information according to P, and different possibilities exist in general. As RSP queries return just a single answer at a time point, we consider in this paper LARS programs that have a unique answer stream. By $\mathcal{AS}(P, D, t)$, we directly refer to the single element in $\mathcal{AS}(P, D, t)$.

The following examples illustrate how these concepts are combined:

- $\boxplus^7\Diamond signal$ holds, if the atom *signal* appeared (at least once) in the last 7 time units. (The implicit window type is *time-based*, see [3] for others.)
- $@_{T+3}alert \leftarrow \boxplus^5@_T signal$. This rule says: If in the last 5 time units, a signal occurred at time T, then 3 time units later, *alert* has to hold. Given a program P consisting only of this rule, the unique answer stream for $D = ([0, 10], \{6 \mapsto \{signal\}\})$ at $t = 8$ additionally contains the mapping $9 \mapsto \{alert\}$.

We next show how C-SPARQL and CQELS can be translated into LARS.

3   Translating RSP Queries into LARS

We propose the following translations (see details in [7]):

(1) Given a window expression ω in C-SPARQL or CQELS, $\tau(\omega)$ translates it to a corresponding window operator of LARS, for example, $\tau([\texttt{RANGE 10}]) = \boxplus^{10}$.
(2) Given a LARS program P and a pulling period $U > 0$, the translations $\triangleright(P)$ and $\triangleleft(P, U)$ encode the push- and pull- modes by LARS rules, respectively.
(3) Given an RSP query Q, translation τ_1 is applied on $cs(Q)$ and translation τ_2 is applied on $Q' \in cq(Q)$ and return LARS programs. Both share the core from translation τ from SPARQL to Datalog in [18] and differ due to two approaches by C-SPARQL and CQELS that deal with streaming input. As an extension of τ, static RDF triples are represented as a 4-ary predicate

of the form $\mathtt{triple}(S, P, O, G)$ while triples arriving at a stream $\mathbf{s}$ at time t contributes to the evaluation function v at t under a predicate $\mathtt{striple}$, that is, $\mathtt{striple}(s, p, o, \mathbf{s}) \in v(t)$. Auto-generated predicates of the form $\mathtt{ans}_i$ are used to hold answers of intermediate translations, and $\mathtt{ans}_1$ holds the answers of the queries.

So far, it has not been clear under which conditions the two engines will return the same output. Tackling this question now becomes possible at a formal level using LARS.

Given an RDF triple (s, p, o) and a basic graph pattern g, we say (s, p, o) *sub-matches* g, denoted by $sm(s, p, o, g)$, iff there exists a triple pattern $(S, P, O) \in g$ s.t. $[\![(S, P, O)]\!]_{\{(s,p,o)\}} \neq \emptyset$, where the notion of subgraph matching $[\![.]\!]$ is defined in [15]. Given a graph pattern P, let $trp(P)$ be the set of triple patterns appearing in P.

The following result identifies a class of RSP queries where the answer streams of the translated LARS programs by τ_1 and τ_2 coincide on the output predicate $\mathtt{ans}_1$.

Theorem 1. *Let $Q = (V, P, \mathcal{D}, \mathcal{S})$ be an RSP query where $\mathcal{D} = (G, G_n)$ contains a default graph G and a set G_n of named graphs, and $\mathcal{S} = \{(\mathbf{s}_1, \omega_1, g_1), \ldots, (\mathbf{s}_m, \omega_m, g_m)\}$. Let $P_1 = \tau_1(cs(Q))$, $P_2 = \tau_2(Q')$, for any $Q' \in cq(Q)$, D be a data stream, and t be a time point. If*

$$\forall g \neq g' \in \{\{trp(P) \setminus \bigcup g_i\}\} \cup \{g_1, \ldots, g_m\} \colon g \cap g' = \emptyset, \qquad (\star)$$

$$\forall \mathtt{striple}(s, p, o, \mathbf{s}_i) \in Ats(D) \colon sm(s, p, o, g_i) \, and$$

$$\forall g_j \neq g_i \in \mathcal{S} \colon \neg sm(s, p, o, g_j) \, and \, \neg sm(s, p, o, trp(P) \setminus \bigcup g_j) \qquad (\star\star)$$

then $AS(P_1, D, t)|_{\mathtt{ans}_1} = AS(P_2, D, t)|_{\mathtt{ans}_1}$.

Condition $(\star)$ requires that the graph patterns wrt. the static dataset and the input streams do not share triple patterns while $(\star\star)$ makes sure that triples arrived at stream $\mathbf{s}_i$ are not allowed to enter any other stream or to stay in the static dataset. Combining these two conditions intuitively means that all input streams and static dataset have disjoint input. Then, the two approaches in building snapshots correspond as the distinction of input due to stream graph patterns in CQELS also happens for C-SPARQL. Thus, the answer streams produced by two translated LARS programs coincide on the output predicate.

4   RSP Semantics Analysis Based on LARS

In the previous section, (3) presents translations from RSP queries on either C-SPARQL or CQELS branches into LARS programs. Under condition $(\star)$ and $(\star\star)$ in Theorem 1, the two translated LARS programs from a C-SPARQL and a CQELS queries, rooted from the same RSP query, produce the same output predicate $\mathtt{ans}_1$ (thus on $\mathtt{RStream}$ operator) when they are evaluated at the same time point.

However, C-SPARQL and CQELS are based on two different execution modes: push-based and pull-based, which are captured in (2) for general LARS programs. In order to theoretically analyze and compare the semantics of C-SPARQL and CQELS, we need to combine the above two results, together with taking into account the difference between `IStream` and `RStream` operators. But first of all, we must clarify what we mean by saying "C-SPARQL and CQELS agree on the output."

4.1   Agreement Between C-SPARQL and CQELS

We propose a characterization of agreement between C-SPARQL and CQELS using LARS. For the core notion, we concentrate on the agreement on the resulted mappings after non-aggregate SPARQL operators such as AND, UNION, etc. Extending to aggregate will be discussed in Sect. 5.

Intuitively, the two semantics are considered to agree on a timeline T with a pulling period U, if (1) they both start at the same time point 0, and (2) for every interval $(i \cdot U, (i + 1) \cdot U] \in T$, where $i \geq 0$, the union of outputs produced by CQELS in the interval coincides with the output produced by C-SPARQL at the right-end of the interval. To formalize the conditions for agreement, we need the notion of trigger time points and incremental output presented next.

Trigger Time Points. Let $t_1 < t_2$ be two time points. The set of *trigger time points* in a data stream D in the interval $(t_1, t_2]$ is defined as $ttp(t_1, t_2, D) = \{t \in (t_1, t_2] \mid v_D(t) \neq \emptyset\}$. For a time point $t \in T_D$ such that $t > 0$, the *previous trigger point* of t with respect to D is $prev(t, D) = \max(ttp(0, t - 1, D))$ if $ttp(0, t - 1, D) \neq \emptyset$ and is 0 otherwise.

Incremental Output. Next, we capture the incremental output strategy, i.e., the `IStream` operator by means of the difference between answer streams of two consecutive trigger time points. Let $I_t = AS(P, D, t)$. Then, the *incremental output* $inc(P, t)$ at a trigger time point t (i.e., $v_D(t) \neq \emptyset$) is $Ats(I_t \setminus I_{prev(t)})$ if $t > 0$, and $inc(P, 0) = Ats(I_0)$. Here, the difference between two streams $S_1 = (T, v_1)$ and $S_2 = (T, v_2)$ is defined as $S' = S_1 \setminus S_2 = (T, v')$ s.t. for all $t' \in T$, we have that $v'(t') = v_1(t') \setminus v_2(t')$.

Based on this, we define when two LARS programs, executed on push- and pull- modes, *agree* on an interval of time.

Definition 1. *Given two LARS programs P_1, P_2, a data stream $D = (T_D, v_D)$, and two time points $t_1 < t_2$ of T_D, let $A_1 = \bigcup_{t \in ttp(t_1, t_2, D)} inc(P_1, t) \cup \bigcap_{t \in ttp(t_1, t_2, D) \cup \{t_2\}} Ats(I_t)$ and $A_2 = Ats(AS(P_2, D, t_2))$. Let $R = \{p_1, \ldots, p_n\}$ be a set of predicates. We say P_1 and P_2 push-pull-agree on D and R*

(i) *during the interval $(t_1, t_2]$, denoted by $P_1 \equiv_{t_1, t_2}^{D, R} P_2$, iff $A_1|_R = A_2|_R$;*

(ii) *with pulling period U, denoted by $P_1 \equiv_U^{D, R} P_2$, iff $P_1 \equiv_{t_1, t_2}^{D, R} P_2$, where $t_1, t_2 \in T_D$ such that there exists some $i \in \mathbb{N}$, where $t_1 = i \cdot U$ and $t_2 = (i + 1) \cdot U$.*

Intuitively, push-pull-agreement during $(t_1, t_2]$ is established by comparing the answer stream evaluated at t_2 with the union of incremental answer computed at all trigger time points in the interval. The term $\bigcap_{t \in ttp(t_1,t_2,D) \cup \{t_2\}} Ats(I_t)$ ensures that for programs that always produce some output $p(\mathbf{c})$ at every time point, this output is also counted in comparing the incremental result and the result at t_2.

4.2 Agreement Conditions

Given an RSP query $Q = (V, P, \mathcal{D}, \mathcal{S})$, let $Q_1 = cs(Q)$ and $Q_2 \in cq(Q)$. We want to identify conditions guaranteeing that the LARS programs $\tau_1(Q_1)$ and $\tau_2(Q_2)$ agree on the output predicate $\mathbf{ans}_1$, that is $\tau_2(Q_2) \equiv_{t_1,t_2}^{D,\mathbf{ans}_1} \tau_1(Q_1)$.

Let $D = (T_D, \upsilon_D)$ be a data stream. The projection of D on an input stream identified by an IRI $\mathbf{s}$ is $D|_\mathbf{s} = (T_D, \upsilon_D|_\mathbf{s})$, where for all $t \in T_D$, we have that $\upsilon_D|_\mathbf{s}(t) = \{\mathtt{striple}(S, P, O, \mathbf{s}) \in \upsilon_D(t)\}$. That is, we keep only facts with $\mathbf{s}$ as the stream identifier. The *snapshot* of D with respect to $\mathcal{S}$ at time point t is defined as:

$$sn(D, \mathcal{S}, t) = \bigcup\nolimits_{(s,\omega,g) \in \mathcal{S}} w_{\tau(\omega)}(D|_\mathbf{s}, t).$$

Intuitively, for each input stream pattern $(\mathbf{s}, \omega, g) \in \mathcal{S}$, we apply the window function $w_{\tau(\omega)} = w_{\boxplus}$ to the projection of D on $\mathbf{s}$. Note that $\tau(\omega)$ translates the window expression ω to a window operator $\boxplus$, and $w_{\boxplus}$ is the window function behind $\boxplus$. The union of all substreams extracted by the window functions give us the snapshot. The following result identifies sufficient conditions where C-SPARQL and CQELS agree.

Theorem 2. *Let $Q = (V, P, \mathcal{D}, \mathcal{S})$ be an RSP query, where P contains neither MINUS nor FILTER NOT EXISTS, $\mathcal{D} = (G, G_n)$ contains a default graph G and a set G_n of named graphs, and $\mathcal{S} = \{(\mathbf{s}_1, \omega_1, g_1), \ldots, (\mathbf{s}_m, \omega_m, g_m)\}$ contains only time-based windows of the form [RANGE L]. Let $Q_1 = cs(Q)$, $Q_2 \in cq(Q)$, and $t_1 < t_2$. If $(\star)$ and $(\star\star)$ hold, and additionally*

$$\bigcup_{t \in ttp(t_1,t_2,D)} sn(D, \mathcal{S}, t) = sn(D, \mathcal{S}, t_2) \qquad\qquad (\star\star\star)$$

then

$$\tau_2(Q_2) \equiv_{t_1,t_2}^{D,\mathbf{ans}_1} \tau_1(Q_1).$$

This result can be straightforwardly extended to check whether $\tau_1(Q_1) \equiv_U^{D,\mathbf{ans}_1} \tau_2(Q_2)$, but is omitted due to space reason. The theorem shows that having agreement between C-SPARQL and CQELS is not easy to achieve, as discussed in the next section.

5   Discussion and Conclusion

Theorem 2 identifies sufficient conditions on which C-SPARQL and CQELS agree on their output, including (i) no MINUS or FILTER NOT EXISTS operator, (ii) only time-based windows with sliding size 1, (iii) only "disjoint" patterns and data in the static datasets and the input streams, and (iv) having the same snapshot collected in the interval as at the right end of the interval.

While (i)–(iii) correspond to useful fragments of queries for practical purposes, (iv) cannot be guaranteed in case of high throughput. The reason is that with dense input streams, the snapshots taken at time points near the left end of an interval will have high chances to collect more triples than the snapshot at the right end of the interval. Thus, having C-SPARQL and CQELS agreeing in practice is very unlikely, due to the strong semantic implications of push/pull-based querying. Consequently, data independent agreement conditions are unlikely to be found for queries that go beyond pure SPARQL.

One can easily find a counter example for the agreement when relaxing any of (i)–(iii) and keeping other conditions unchanged. For example, when FILTER NOT EXISTS or MINUS is allowed, the translated LARS programs are not positive. This takes away the monotonic property, i.e., having more input one any side (push-or pull-based) might lead to shrinking the output facts and introducing disagreement on the output.

For sliding windows with sliding size greater than 1, C-SPARQL can produce output that CQELS cannot, even when $(\star \star \star)$ is satisfied. Intuitively, this is because the window only slides after a certain amount of time and might miss some most recent input. In this case, we think that pull-based is preferable over push-based execution.

Finally, if the static datasets and the input streams share patterns by which triples are matched for R2R operators, the result of C-SPARQL and CQELS will be different. For instance, if these datasets share the same pattern and the static dataset contains some triples matching this pattern, then C-SPARQL can produce output even when no input triple arrives at the stream, as it cannot distinguish where a triple comes from. Besides, the stream graph pattern of CQELS has no mapping due to empty input, and thus produces no output. However, this situation should not happen often in practice as merged input streams will usually be distinguishable by an implicit schema.

The core notion of agreement does not consider aggregates. When considering aggregates, we observe that only certain types of aggregates allow for tracing agreement between pull- and push-based executions. For example, for COUNT, we can say that CQELS agrees with C-SPARQL in an interval $(t_1, t_2]$ iff the sum of output values produced by the former during the interval equal to the output value returned by the latter at t_2. Similar extension can be done for MAX, MIN. However, with MEDIAN or AVG, one cannot reproduce the result from CQELS' output to match that from C-SPARQL. In general, we can only give agreement notion for aggregates that can be recursively defined.

Conclusion and Outlook. This paper utilizes LARS to give insights on the contrast between two RSP semantics implemented in two representative engines, namely C-SPARQL and CQELS. In addition to [9], we introduced a notion of agreement between the engines and identified conditions when it holds.

The theoretical result is based on the assumption that engine execution time is neglectable to the input rate. For further practical comparison, we envision future work where this condition is dropped. Implementing the proposed translations is also on our agenda. In another direction, we are investigating equivalence for general LARS programs. Once this result is available, one can have an automatic equivalence checker which takes any two translated LARS programs of two continuous queries from any two stream processing languages, tell whether the two original queries are equivalent, and possibly even enumerate their different outputs due to our model-based approach.

References

1. Arasu, A., Babu, S., Widom, J.: The CQL continuous query language: semantic foundations and query execution. VLDB J. **15**(2), 121–142 (2006)
2. Barbieri, D.F., Braga, D., Ceri, S., Valle, E.D., Grossniklaus, M.: C-SPARQL: A continuous query language for RDF data streams. Int. J. Semant. Comput. **4**(1), 3–25 (2010)
3. Beck, H., Dao-Tran, M., Eiter, T., Fink, M.: LARS: A logic-based framework for analyzing reasoning over streams. In: AAAI (2015)
4. Bolles, A., Grawunder, M., Jacobi, J.: Streaming SPARQL - Extending SPARQL to process data streams. In: Bechhofer, S., Hauswirth, M., Hoffmann, J., Koubarakis, M. (eds.) ESWC 2008. LNCS, vol. 5021, pp. 448–462. Springer, Heidelberg (2008)
5. Calbimonte, J.-P., Corcho, O., Gray, A.J.G.: Enabling ontology-based access to streaming data sources. In: Patel-Schneider, P.F., Pan, Y., Hitzler, P., Mika, P., Zhang, L., Pan, J.Z., Horrocks, I., Glimm, B. (eds.) ISWC 2010, Part I. LNCS, vol. 6496, pp. 96–111. Springer, Heidelberg (2010)
6. Cyganiak, R., Wood, D., Lanthaler, M.: RDF 1.1 Concepts and Abstract Syntax (2014). http://www.w3.org/TR/rdf11-concepts/
7. Dao-Tran, M., Beck, H., Eiter, T.: Contrasting RDF stream processing semantics. Technical report, Institut für Informationssysteme, TU Wien (2015)
8. Dell'Aglio, D., Calbimonte, J.-P., Balduini, M., Corcho, O., Valle, E.D.: On correctness in RDF stream processor benchmarking. In: Alani, H., Kagal, L., Fokoue, A., Groth, P., Biemann, C., Parreira, J.X., Aroyo, L., Noy, N., Welty, C., Janowicz, K. (eds.) ISWC 2013, Part II. LNCS, vol. 8219, pp. 326–342. Springer, Heidelberg (2013)
9. Dell'Aglio, D., Valle, E.D., Calbimonte, J.-P., Corcho, O.: RSP-QL semantics: A unifying query model to explain heterogeneity of RDF stream processing systems. IJSWIS 10(4) (2015)
10. Dindar, N., Tatbul, N., Miller, R.J., Haas, L.M., Botan, I.: Modeling the execution semantics of stream processing engines with SECRET. VLDB J. **22**(4), 421–446 (2013)
11. Cabalar, P.: Answer set programming? In: Balduccini, M., Son, T.C. (eds.) Logic Programming, Knowledge Representation, and Nonmonotonic Reasoning. LNCS, vol. 6565, pp. 334–343. Springer, Heidelberg (2011)

12. Groppe, S.: Data Management and Query Processing in Semantic Web Databases. Springer, Heidelberg (2011)
13. Harris, S., Seaborne, A.: SPARQL 1.1 Query Language (2013). http://www.w3.org/TR/sparql11-query/
14. Kolchin, M., Wetz, P.: Demo: YABench - Yet another RDF stream processing benchmark. In: RSP Workshop (2015)
15. Pérez, J., Arenas, M., Gutierrez, C.: Semantics and complexity of SPARQL. ACM Trans. Database Syst. **34**, 16:1–16:45 (2009)
16. Le-Phuoc, D., Dao-Tran, M., Xavier Parreira, J., Hauswirth, M.: A native and adaptive approach for unified processing of linked streams and linked data. In: Aroyo, L., Welty, C., Alani, H., Taylor, J., Bernstein, A., Kagal, L., Noy, N., Blomqvist, E. (eds.) ISWC 2011, Part I. LNCS, vol. 7031, pp. 370–388. Springer, Heidelberg (2011)
17. Le-Phuoc, D., Dao-Tran, M., Pham, M.-D., Boncz, P., Eiter, T., Fink, M.: Linked stream data processing engines: Facts and figures. In: Cudré-Mauroux, P., et al. (eds.) ISWC 2012, Part II. LNCS, vol. 7650, pp. 300–312. Springer, Heidelberg (2012)
18. Polleres, A.: From SPARQL to rules (and back). In: WWW, pp. 787–796 (2007)
19. Zhang, Y., Duc, P.M., Corcho, O., Calbimonte, J.-P.: SRBench: A streaming RDF/SPARQL benchmark. In: Cudré-Mauroux, p, et al. (eds.) ISWC 2012, Part I. LNCS, vol. 7649, pp. 641–657. Springer, Heidelberg (2012)

In-Use

.

Towards an Enterprise Entity Hub: Integration of General and Enterprise Knowledge

Haklae Kim[✉], Jeongsoo Lee, and Jungyeon Yang

Software R&D Center, Samsung Electronics Co., Ltd., Samsung-ro 189,
Maetan-dong, Yeongtong-gu, Suwon, Gyeonggi, Korea
{scot.kim,jeong.s.lee,jy190.yang}@samsung.com

Abstract. Today's enterprises possess vast quantities of data generated
by their own applications and services across different organisational
systems. However, the data tends to be stored in many data-silos with
redundant, duplicated, and outdated information. This means that finding the right information and obtaining valuable insights is difficult. To
address this problem, we develop the enterprise entity hub that enables
users to search, analyse, share, and filter information across large-scale
data sources. One of its core components, the base knowledge is used to
organise and interlink the organisation's knowledge model, we then construct a base knowledge through the integration of several data sources,
and expand this knowledge by interlinking a set of enterprise content.

1 Introduction

For years, the key to success for enterprises has resided in data as their strategic
asset. Managing data strategically implies not only having full access to enterprise data but also having a comprehensive view of data that must be available
when required. However, one of the challenge that many companies face today
is managing vast quantities of data derived from their own applications and services to different organisational systems, such as enterprise resource planning,
customer relationship management, supply chain management, and others [5].
Those data tends to be stored in many data-silos with redundant, duplicated,
and out-dated information [11]. The data in a silo remains in inherent isolation
from the rest of the organisation, and it cannot exchange content with other
systems, such as grain in a farm silo is closed off from the outside elements.

With changes to the enterprise IT landscape, various approaches, techniques
and methods have been introduced in order to solve the data integration challenges [5]. Linked data, in contrast to traditional approaches, can continually
evolve to keep pace with the change of data in an organisation [12,13].

A major challenge is to bridge the gap between domain-specific knowledge
and general knowledge in an enterprise. In this sense, a background knowledge is
essential because companies must connect all silo'ed data into a knowledge base
that allows access to any information at any time on an as-needed basis. Several

© Springer International Publishing Switzerland 2016
G. Qi et al. (Eds.): JIST 2015, LNCS 9544, pp. 301–310, 2016.
DOI: 10.1007/978-3-319-31676-5_22

open knowledge bases (e.g. DBpedia[1], YAGO[2], Wikidata[3]) are candidates for having global and general knowledge, even though they do not consider specific needs in enterprise domains. Thus, to create an enterprise knowledge base, which integrates global knowledge into enterprise knowledge from business-specific data sources is necessary.

In this study, we introduce the overall process of building an enterprise entity hub. We develop an semi-automatic knowledge base framework for creating a base knowledge from various data sources. In addition, various domain knowledge can be expanded by interlinking this base knowledge. The remainder of this paper is organised as follows. In Sect. 2, we introduce the vision of the enterprise entity hub, which aims to build a single, interlinked database using linked data technologies. In Sect. 3, we describe the knowledge model, which governs the structure of the base knowledge, including general and enterprise knowledge. In Sect. 4, we provide some use cases of interlinking general knowledge and enterprise knowledge. Related studies are described in Sect. 5 and a conclusion is provided in Sect. 6.

2   Towards Enterprise Entity Hub

The hybrid enterprise entity hub aims to become a single, up-to-date canonical source of data as a crucial infrastructure meant to structure, integrate and manage knowledge in enterprises. It aims to organise and distribute enterprise data and enable its universally accessibility for data consumers. By means of linked Data technologies, silo'ed data in enterprises can be interlinked, and thus allow people to use data with ease and share data across heterogeneous sources.

To realise the enterprise entity hub, we must construct a base knowledge, one that provides a stable set of reusable concepts and entities with consistent identifiers to enterprise systems and applications. All content and data can be interlinked to this base knowledge, and data consumers can take advantage of richer knowledge for individual entities through interlinked knowledge. It comprises of both global knowledge for describing things in the world and enterprise knowledge that provides various domain-specific business data. A domain-specific knowledge is transformed into a graph-based knowledge base by interlinking the base knowledge. When entities in a set of domain knowledge are not present in the base knowledge, this information is appended to the base knowledge.

The base knowledge comprises a set of knowledge sources and enterprise taxonomies. In particular, two types of knowledge are contained therein: (1) global (or general) knowledge used to understand and uniquely identify entities in the world, (2) domain knowledge used to describe specific subjects of enterprise systems and applications. We develop a global knowledge by collecting and integrating open knowledge bases and datasets such as Wikidata, Freebase, or

---

[1] http://dbpedia.org.
[2] www.mpi-inf.mpg.de/yago/.
[3] www.wikidata.org/.

crawled data from the Web, whereas a set of domain datasets in the enterprise are transformed into graph based knowledge bases. Both the base and domain knowledge are connected through knowledge association, which refers to inter-linking and annotating data based on linked data technologies [1,6].

In the following sections, we will introduce the overall knowledge structures for describing our knowledge base as well as the approach for constructing the *hybrid* enterprise entity hub, which combines of both general-purpose and domain-specific enterprise knowledge through the lightweight, semantic integration of enterprise data.

3 Data Model of the Knowledge Base

3.1 A Formal Definition of Knowledge Base

Formally, the data model of a knowledge base is a directed, labelled graph [8]:

$$(V, E, N_L, E_L)$$

where V is a set of nodes, $E \subseteq V \times V$ is a multi-set of edges, N_L is a set of node labels and E_L is a set of edge labels. Each node $v \in V$ is assigned to a label $l(v) \in N_L$ and each edge $e \in E$ is assigned to a label $l(e) \in E_L$. Each node represents an entity as an instance. For example, a node v with label $l(v)$ = Barack Obama represents the president of United States. Likewise, each edge describes a relationship between two entities. For instance, if w is a node with label $l(w) = 1961$, the fact that Barack Obama was born in 1961 is represented by an edge $e = (v, w)$ with the edge label $l(e) =$ bornInYear. Thus, knowledge base facts can be represented by their label of nodes connected by the edge (i.e., simply stated, Barack Obama (president) bornInYear 1961).

3.2 Knowledge Model

A comprehensive and sophisticated ontology is essential for representing a wide variety of enterprise knowledge. First, we investigate existing vocabularies for both general-purpose and enterprise domain knowledge, and then evaluate ontology matching tools to ensure the feasibility of automatic ontology creation among these ontologies. We obtain a reference ontology model, and then extract and design classes with their hierarchies using this model. Finally we create our ontology model, called the knowledge model, by assigning a set of properties into those classes. The following explains each step in detail.

Reference Ontologies. To secure a highly qualified and reliable ontology model, we create the *knowledge model* to represent our knowledge using manual mapping. First, we use schema.org as the fundamental ontology, because it aims to identify things and describe their characteristics and relationships to other things on the Web. Although the scope of the current model is limited with respect to details concerning some of the domains, it can be an acceptable starting point to create an extensible ontology model. Other vocabularies

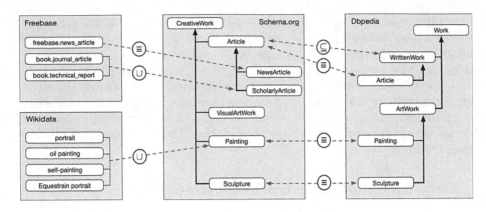

Fig. 1. The knowledge model via aligning existing vocabularies. ≡ describes an *equivalent* relationship among classes, and ∪ depicts *belong to*, meaning that those classes are merged as the Painting class. For example, freebase.new_article is equivalent with the NewsArticle class of schema.org, whereas book.journal_article and book.technical_report are mapped and merged to the ScholarlyArticle of schema.org. ⊆ refers to a subclass relationship. The Article class of schema.org is represented by a subclass of the WrittenWork class of DBpedia.

such as Wikidata, Freebase, and DBpedia are aligned to classes and properties of schema.org. We do not consider a set of candidates and extensions of schema.org for the sake of clarifying the scope of integration in the initial version.

Taxonomy of Classes. We extract an alignment among several ontologies using Falcon that provides the best scores compared to other systems. Note that Wikidata content is derived from Wikipedia, and the values of instance-of and subclass-of properties of Wikidata also depend on categories of Wikipedia. In general, categories of Wikipedia can be thought of as a simple subject of each article that is not a controlled vocabulary or well-defined information. Furthermore, values of two properties might be created by a set of Wikidata bots with no consideration for ontological semantics. This means that it is not useful to convert or align those values from two properties directly into classes of our model.

Aligning Freebase schema is extremely complicated: a topic concerning some aspect in the world has a multi-faceted nature, and this feature is represented by a type and domain. A type represents categorical collections of instances that are most commonly required to describe a particular aspect of information. A topic can have any number of types assigned to it. For example, the topic *Bob Dylan* in Freebase has several types such as book author, company, and music artist. A domain refers to a collection of types which share a namespace. Commons and Bases are special types of domains. Just as properties are grouped into types, types themselves are grouped into domains. In addition, Freebase introduces a compound value type that is used to represent two or more properties of a single topic. For example, in order to represent the population of a city that changes over time, Freebase is largely used to a reified statement. Because this concept is only for representing Freebase instances, when it is transformed, changing the semantics of original vocabularies including domain and range values is not required.

Figure 1 illustrates a simplified process of ontology alignment. Classes are shown in rectangles with rounded corners, e.g., in schema.org, Painting is a specialization (subclass) of CreativeWork. The first step in integrating ontologies is matching, which identifies correspondences, namely, the candidate entities to be merged or to have subsumption relationships under an integrated ontology. Once the correspondences between ontologies have been determined, a candidate is determined: for instance, the classes with labels portrait, oil painting, self-painting, and equestrain portrait in Wikidata are the candidates to be merged as Painting in schema.org, whereas the class with label WrittenWork in DBpedia subsumes the class Article in schema.org. As a result, we extracted 1,363 classes from 10, 500 items from instance-of properties: an item is eliminated if the item does not exist as an entity on Wikidata, even though the value of this property is assigned. Extracted classes are then mapped into 359 classes of schema.org with its hierarchies. A mapping between Wikidata and schema.org is straightforward: each class of Wikidata is aligned to a subclass of schema.org if a certain class exists in both vocabularies. We also extracted 91 domains and 2,071 types from Freebase, and in total 2,010 total classes are aligned to the knowledge model. In addition, a domain of each type is converted into through the k:hasDomain property.

Properties. After designing hierarchies of extracted classes, a set of properties are assigned to those classes. All properties of schema.org are assigned to original classes by default with domain and range values. We then examine the possibility of using the Wikidata properties. Because Wikidata properties have no domain and range values, assigning proper values of domain and range for each property is necessary. Although Wikidata has approximately 1,000 properties, 516 properties are obtained.

In order to secure expressivity for describing general knowledge, a considerable number of properties in Freebase are also integrated. Multiple classes of Freebase can be mapped into a class of the knowledge model because of a different class expression. For example, muisic.musician and music.artist of Freebase is mapped to the class MusicArtist of the knowledge model. In these cases, we create new properties instead of using original properties, because constraints (domain and range) of each property can be vague.

We introduce an extension mechanism that allows developers to extend our proposed schema when they must describe the content of new domain. Anyone can create new schema that are connected to the vocabularies on the current ontology model, and they should create new ones. We use the '/' character after the base URI to create extensions of existing vocabulary specifications. For example, The <http://k.samsung.com/product> refers to a namespace for the product domain.

4   Bridge Between Enterprise Knowledge and the Base Knowledge

As we discussed in Sect. 2, the enterprise entity hub plays a central role in connecting many important categories of enterprise knowledge. The base knowledge of the enterprise entity hub is interlinked to various enterprise data as reference knowledge. For example, products of consumer electronics are transformed into a knowledge base: each product has its own specification that comprises a set of technical features, images, and reviews. Some features can be applied to multiple products, and they present a shared instance. For example, specifications in UHD TV and Galaxy S6 have the same HDMI feature, and further similar cases exist throughout all products of consumer electronics. In this case, defining this feature and its detailed descriptions as an entity in the base knowledge is more effective than describing it with respect to its individual categories (or domains). Hence, if this entity already exists, this entity is interlinked into the entity URI in the base knowledge, whereas it is added onto the base knowledge if the entity is not defined.

As of March 2015, the Entity Hub has approximately 4 billion entities and 20 billion facts for general-purpose and domain-specific knowledge. Some domain-specific knowledge is already interlinked into the enterprise entity hub, including 1.2 million entities for products and specifications, 24 million entities for entertainment contents, five million entities for restaurants and dining, 46 million entities for multimedia contents, and 24 million geo-spatial entities.

The entity hub is allowed to be accessed by using a set of open application programming interfaces (APIs) and SPARQL Protocol and RDF Query Language (SPARQL) endpoints, and some federated queries are also supported. Figure 2 shows an example of the manner in which base and domain-specific knowledge can be utilised in conventional enterprise applications. This application provides real-time news for various categories. The Knowledge Tag offers a set of semantic entities from a certain article through the named entity recognition API, which is based on the enterprise entity hub. When an entity is selected, this panel shows the details of the entity. An entity consists of several subjects. For example, Fig. 2-A shows a set of technical features from the article as entities in the Knowledge Tag. Each entity has detailed information and links to other knowledge sources. In this example, the entity Wi-Fi is connected to several products of mobile devices. If a user selects a product, the entity panel offers its specification in detail, and provides a set of real-time reviews, etc. Furthermore, this panel also provides contextual information about the selected entity through lightweight reasoning. For example, if an entity is a type of Place, this panel provides information concerning weather and temperature for or nearby that place by combining other open APIs, as illustrated in Fig. 2-B. This can be expanded to various application scenarios based on entity types such as (Company: stock and location), (Person: event, organisation, and social network). As illustrated in Fig. 2, information on the entity panel can be a starting point to discover other entities beyond subjects using the base and domain knowledge. People may choose a whatever link they want to see in detail, and the entity panel present that entity on the right side.

5 Related Work

As we presented in this paper, a general-purpose knowledge base would be a cornerstone to construct knowledge of a wide variety of enterprise content. Wikidata [14] is a free knowledge base about the world that can be read and edit by humans and machines alike. Things described in the Wikidata knowledge base are called items and can have labels, descriptions and aliases in all languages, which does not aim at offering a single truth about things, but providing statements given in a particular context. DBpedia [9] is a project aiming to extract structured, multilingual knowledge from Wikipedia and makes it freely available on the Web using Semantic Web and Linked Data technologies. Since it covers a wide variety of topics and sets of RDF links pointing to various external data sources, it has developed into the central interlinking hub in the Web of Linked Data. One of the projects that pursue similar goals to DBpedia is YAGO [7,10]. YAGO combines the information from the Wikipedia in multiple languages. It fuses multilingual knowledge with English WordNet to build a coherent knowledge base. The main differences between YAGO and DBpedia include: one is that the DBpedia ontology is manually maintained, while the YAGO ontology is backed by WordNet [4] and Wikipedia leaf categories. The other is that the integration of attributes and objects in infoboxes is done via crowd-sourced mappings in DBpedia, while the YAGO implements it by expert-designed declarative rules. Freebase is a large collaborative knowledge base consist of metadata composed mainly by its community members [2].

Recently, Google announced Knowledge Vault [3], which is a Web-scale probabilistic knowledge base that combines extractions from Web content with prior knowledge derived from existing knowledge repositories. It employs supervised machine learning methods for fusing these distinct information sources and features a probabilistic inference system that computes calibrated probabilities of fact correctness.

6 Conclusion and Future Work

In this study, we introduced the enterprise entity hub that is to realise interlinked enterprise knowledge bases through integration of both general-purpose and domain-specific knowledge. We presented the knowledge model for representing terminological knowledge of a wide variety of enterprise domains. In particular, this model is created by matching and integrating existing vocabularies and also provides an extension mechanism to allow various domain knowledge to be extended.

Furthermore, we presented some use cases in which enterprise knowledge are interlinked to the base knowledge, and introduced a real mobile application for

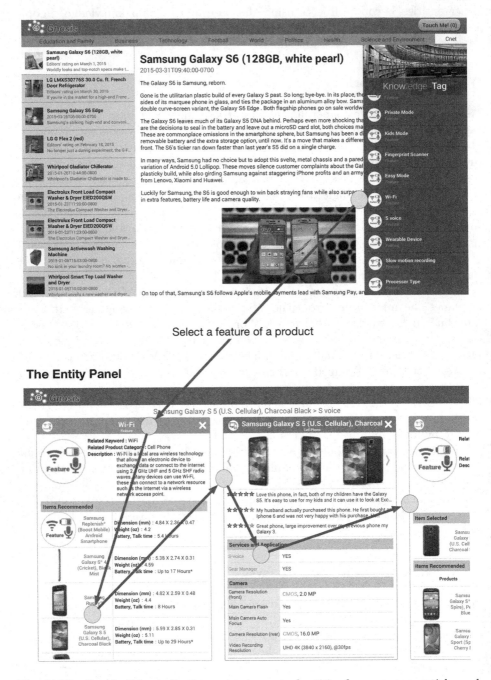

Fig. 2. The Gnosis: this application extracts a set of entities from an news article, and then provides details information of the chosen entity on the knowledge tag. A set of entities on the entity panel are discoverable to look at further information.

demonstrating how the knowledge base is used in enterprise. We conclude that an enterprise knowledge base based on linked data technologies has great potential and can result in considerable benefits.

However, we are aware that additional challenges than the aforementioned must be addressed when trying to create a sophisticated enterprise knowledge base and its infrastructure. We consider for future research the investigation of ontology matching and instance enrichment as a subproblem of the data integration challenges in an enterprise.

References

1. Bizer, C., Heath, T., Berners-Lee, T.: Linked data - the story so far. Int. J. Semantic Web Inf. Syst. **5**(3), 1–22 (2009)
2. Bollacker, K., Evans, C., Paritosh, P., Sturge, T., Taylor, J.: Freebase: a collaboratively created graph database for structuring human knowledge. In: Proceedings of the ACM SIGMOD International Conference on Management of Data, SIGMOD 2008, pp. 1247–1250. ACM, New York (2008)
3. Dong, X., Gabrilovich, E., Heitz, G., Horn, W., Lao, N., Murphy, K., Strohmann, T., Sun, S., Zhang, W.: Knowledge vault: a web-scale approach to probabilistic knowledge fusion. In: The 20th ACM SIGKDD International Conference on Knowledge Discovery and Data Mining, KDD 2014, pp. 601–610, New York, 24–27 August 2014
4. Fellbaum, C. (ed.): WordNet: An Electronic Lexical Database. MIT Press, Cambridge (1998)
5. Frischmuth, P., Auer, S., Tramp, S., Unbehauen, J., Holzweißig, K., Marquardt, C.-M.: Towards linked data based enterprise information integration. In: Coppens, S., Hammar, K., Knuth, M., Neumann, M., Ritze, D., Sack, H., Sande, M.V. (eds.) Proceedings of the Workshop on Semantic Web Enterprise Adoption and Best Practice Co-located with 12th International Semantic Web Conference (ISWC), Sydney, Australia, 22 October 2013, vol. 1106, CEUR Workshop Proceedings (2013). CEUR-WS.org
6. Heath, T., Bizer, C.: Linked Data: Evolving the Web into a Global Data Space. Synthesis Lectures on the Semantic Web. Morgan & Claypool Publishers, San Rafael (2011)
7. Hoffart, J., Suchanek, F.M., Berberich, K., Lewis-Kelham, E., de Melo, G., Weikum, G.: YAGO2: exploring and querying world knowledge in time, space, context, and many languages. In: Srinivasan, S., Ramamritham, K., Kumar, M.A., Ravindra, P., Bertino, E., Kumar, R. (eds.) Proceedings of the 20th International Conference on World Wide Web, WWW, Hyderabad, India, 28 March – 1 April 2011 (Companion Volume), pp. 229–232. ACM (2011)
8. Kasneci, G., Suchanek, F.M., Ifrim, G., Ramanath, M., Weikum, G.: NAGA: searching and ranking knowledge. In: Alonso, G., Blakeley, J.A., Chen, A.L.P. (eds.) Proceedings of the 24th International Conference on Data Engineering, ICDE, Cancún, México, pp. 953–962. IEEE, 7–12 April 2008
9. Lehmann, J., Isele, R., Jakob, M., Jentzsch, A., Kontokostas, D., Mendes, P.N., Hellmann, S., Morsey, M., van Kleef, P., Auer, S., Bizer, C.: DBpedia - a large-scale, multilingual knowledge base extracted from wikipedia. Semant. Web J. (2014)

10. Mahdisoltani, F., Biega, J., Suchanek, F.M.: YAGO3: a knowledge base from multilingual wikipedias. In: CIDR, Seventh Biennial Conference on Innovative Data Systems Research, Asilomar, CA, USA, 4–7 January 2015, Online Proceedings (2015). www.cidrdb.org

11. Masuch, L.: Enterprise knowledge graph - one graph to connect them all. In: Proceedings of the Unlimited Human Potential M-Prize, M-Prize Proceedings. MPrize (2014)

12. Mihindukulasooriya, N., Garcia-Castro, R., Gutiérrez, M.E.: Linked data platform as a novel approach for enterprise application integration. In: Hartig, O., Sequeda, J., Hogan, A., Matsutsuka, T. (eds.) Proceedings of the Fourth International Workshop on Consuming Linked Data, COLD, CEUR Workshop Proceedings, vol. 1034, Sydney, Australia, 22 October 2013. CEUR-WS.org

13. Servant, F.-P.: Linking enterprise data. In: Bizer, C., Heath, T., Idehen, K., Berners-Lee, T. (eds.) Proceedings of the WWW Workshop on Linked Data on the Web, LDOW 2008, CEUR Workshop Proceedings, vol. 369, Beijing, China, 22 April 2008. CEUR-WS.org

14. Vrandečić, D., Krötzsch, M.: Wikidata: a free collaborative knowledgebase. Commun. ACM 57(10), 78–85 (2014)

Ontology Development for Interoperable Database to Share Data in Service Fields

Towards Evaluation of Robotic Devices for Nursing Care

Satoshi Nishimura[✉], Ken Fukuda, Kentaro Watanabe,
Hiroyasu Miwa, and Takuichi Nishimura

Human Informatics Research Institute,
National Institute for Advanced Industrial Science and Technology,
2-3-26 Aomi, Koto-ku, Tokyo, 135-0064, Japan
{satoshi.nishimura,ken.fukuda,kentaro.watanabe
h.miwa,takuichi.nishimura}@aist.go.jp

Abstract. Sharing data from various systems such as sensor networks, time and motion study data, and text data is important. To process and manage the collected data, the authors have proposed a database framework called the COTO database. As described in this paper, the authors propose a COTO-ontology as the basis of the COTO database. The COTO-ontology was developed based on the data collected by DANCE, which is an evaluation tool of quality of care, from the real settings. It will be applied as a platform for the evaluation of robotic devices used for nursing care.

Keywords: Service engineering · Elderly nursing care · Ontology development

1 Introduction

According to recent population aging in Japan, costs related to nursing insurance were 75 billion US dollars in 2013 [1]. The low quality of nursing care and the high burden on nursing care givers are social problems in Japan.

Under these circumstances, the Ministry of Economy, Trade and Industry (METI) launched the "Project to Promote the Development and Introduction of Robotic Devices for Nursing Care," which is expected to contribute to the independence of elderly people and to reduce the burden on caregivers [2].

For development of brand-new robotic devices for nursing care, it is important to evaluate the efficiency of these devices in the dimensions of safety, functionality, and mechanics. In addition, the efficiency of life of elderly people and the efficiency for work of care workers and managers must be evaluated. Such efficiencies should be assessed in actual settings while using the robotic devices. Especially, the latter efficiency is defined by some different axis such as sensitiveness of care workers, role of the nursing care home, and variety of elderly people.

Ontology is used as an interoperable framework for semantics. It is useful to develop an ontology for sharing subjective information among many real settings.

© Springer International Publishing Switzerland 2016
G. Qi et al. (Eds.): JIST 2015, LNCS 9544, pp. 311–320, 2016.
DOI: 10.1007/978-3-319-31676-5_23

However, the nursing elderly care domain is still immature and lack standards to develop an ontology in a conventional way. For example, some nursing procedures are not common among different nursing homes because of the variety of elderly people they serve. Furthermore, the workers are unfamiliar with information technology. Therefore, conventional methodologies for ontology development do not suit this situation.

However, some experience exists in such settings in system development domain [3, 4]. The methodology, called "Participatory design," is reported to be effective under these circumstances. We have developed three evaluation tools: DANCE [5], Quality Study [6], and DRAW [7]. They can capture subjective information to evaluate robotic devices used for nursing care. To store and analyze such collected information, we have developed 'COTO' database (COTO DB) which is RDF triple store [8]. In this paper, we define COTO DB as a database to store "experience" data of the stakeholders in the service fields. The "experience" data, such as time, thing, theme, tone, talk, technique and any other aspects to represent work practices, situations and environments, are also necessary for evaluation of robotic devices.

In this paper, we propose an ontology named "COTO-ontology" for sharing data in the elderly nursing care domain. The feature of COTO-ontology is to be developed based on the data collected by evaluation tools that are used in elderly nursing care facilities. This methodology includes requirement analysis and structuring the concepts. The result of the requirement analysis reflects the circumstances in various nursing facilities. This feature suggests that the development process of the COTO-ontology includes know-how in real settings.

This paper is structured as described below. Section 2 describes related work. Section 3 presents required specifications for the ontology to share the knowledge. Section 4 presents a proposed explanation of the COTO-ontology with some examples from the evaluation tool called DANCE. Section 5 presents discussion of future work and concludes this paper in the following section.

2   Related Work

Several groups have proposed methodologies for ontology development, such as TOVE methodology [9], Activity-First Method [10], the methodology by Uschold and King [11], Methontology [12] and a guide to develop ontology [13]. For TOVE methodology, five steps are developed to produce an ontology and one particular feature is the first step: "motivating scenarios" [9]. The motivated scenarios clarify the specification of the ontology. In Activity-First Method, an ontology is separated to task and domain ontologies [10]. Then, the task ontology is specifically examined to identify the necessary concepts. The methodology does not contribute to the creation of useless concepts out of the emphasized domain. Enterprise Ontology was developed for deep and common understanding enterprise models. Uschold et al. proposed the methodology for the development of Enterprise Ontology [11]. The feature of this methodology is to identify key concepts by informal techniques. They emphasize that the informal techniques provide 'semi-informal' ontology consisting of carefully defined terms expressed in a restricted natural language." López et al. proposed Methontology, which is derived from their experience of developing chemical ontology [12]. The feature of the methodology is development according to a software life cycle such as the "water fall model."

The present study was conducted to develop an interoperable ontology for the entire elderly nursing care service domain. However, the motivated scenarios are dependent on nursing care facilities, which are impossible to describe. Moreover, the robotic devices intended for use in nursing care have been developed one after another. Therefore, it is difficult to identify the activity around them. In this study, we strive to develop an ontology for the elderly nursing care domain given such circumstances.

3   Requirement Specification of COTO-Ontology

As discussed in Sect. 1, it is necessary for the elderly nursing care domain to evaluate the quality of care with comparison among different facilities. In addition, it is important to collect subjective information of care workers and participants related with care such as elderly people and their family. This section presents clarification of the requirement specifications for COTO-ontology that is fundamental for COTO DB.

3.1   Evaluation Tools for Quality of Care

In cooperation with care workers, we have developed three tools to evaluate the quality of care. These tools collect subjective information such as awareness of their surroundings. The accumulated information helps care workers to explicate the idea and to discuss it collectively [25]. Such information and activities are important for evaluating the efficiency of robotic care devices in real settings. Especially, enhancement of such information and know-how of the communication activity is effective to promote the introduction of the robotic devices.

The first tool, Dynamic Action and kNowledge assistant for Collaborative sErvice fields (DANCE), supports the arrangement and sharing of information effectively to improve the quality of care. DANCE has been introduced and operated in an elderly care facility called "wakoen" from Febrary 2014 and some other facilities will introduce the system in this year. DANCE manages information of two types. The first is constructed of an attribute and an attribute value (e.g. "Number of bed guards" is an attribute; "two" is an attribute value.). We designate this information as a "face sheet" because of the necessity of daily work for each care receiver. The second type of information is represented by text or multimedia, such as a figure or sound. We designate this information as "handover information" because the information is more important than the "face sheet" as a point of information sharing or "handover." COTO-ontology must make such two types of information interoperable.

The second tool is Quality Study, which specifically examines the time necessary to perform tasks, similarly to conventional time and motion study. It also examines the subjective information of care workers from the perspective of quality of care. It is difficult to evaluate the quality of care adequately because the care is dependent on many elements such as care workers, care receivers, devices for nursing care, and the environment.

Therefore, the subjective information of care-experts is helpful to evaluate the quality of care. To support evaluation of the quality of care, Miwa et al. classified the nursing care tasks in a care facility [14]. They investigate which elements are necessary for evaluating the quality of care in each task.

COTO-ontology is necessary to represent metadata of tasks, such as the kind of task, other tasks performed with the task simultaneously, and the time necessary to perform the task. In addition to such metadata of the tasks, information related to the quality of care is denoted by the evaluator.

The third tool is a Design Representation tool for Autonomous Work systems (DRAW), which supports care workers in recalling the daily work events. Then the care workers can mutually discuss the recalled events. In this discussion, the care workers can refer to the information in the COTO DB using the DRAW system. Because of this usage of DRAW, the information that DRAW should manage is much broader than that of DANCE or Quality Study. Ideally COTO-ontology is necessary to cover all the information. However, in this study, we start to develop COTO-ontology to cover the information that DANCE manages.

3.2 Requirement for Information Extraction from COTO DB

The following three requirements occurred from the perspective of information extraction.

- Data model for information extraction
- Simple SPARQL query for information extraction
- Short response time for information extraction

For system users, the COTO DB provides information collected via COTO-ontology by the evaluation tools. Therefore, the COTO-ontology is necessary for the data model for information extraction.

COTO DB uses the Resource Description Framework (RDF) [15] as a data model. RDF has been developed as a semantic web technology. When system users extract information from the RDF database, they use SPARQL Protocol and RDF Query Language (SPARQL) [16], which is a kind of query language for information extraction from the semantic web. However, system users who are unfamiliar with semantic web technology have difficulty using it. Some research has been undertaken to tackle the problem, such as development of a data model for information extraction and sharing of SPARQL queries [17, 18], which is one challenge to the development of a data model to extract information in COTO-ontology.

COTO DB is intended for use in discussion related to designing new work processes by care workers. Therefore, it is desirable for the care workers to get information from COTO DB in short time.

3.3 Requirements for Sustainable Use of COTO DB

In the future, COTO DB is expected to manage the information gathered not only from those three tools but also from other tools, such as various devices for nursing care and sensors. However, information in COTO DB is maintained for consistency, even with increasing kinds of devices and information. COTO-ontology is expected to provide a perspective of the consistency.

4 COTO-Ontology

This section presents an explanation of the COTO-ontology which is developed according to the above requirements. First, we provide an overall picture of COTO-ontology. Second, we explain details of COTO-ontology with the data of DANCE.

4.1 Overall Picture of COTO-Ontology

Figure 1 presents an is-a hierarchy of COTO-ontology. Is-a hierarchy is a so-called generalized-specialized, super-sub hierarchy constructed with concepts and is-a relations, which denote the relation between generalized concept and specialized concept such as a mammal and a human. The upper level of the ontology is developed with reference to other conventional top-level ontologies [19–21]. By virtue of that, the COTO-ontology is presumed to incorporate inter-domain knowledge. To obtain concepts which are specific in nursing elderly care domain, we analyzed the collected

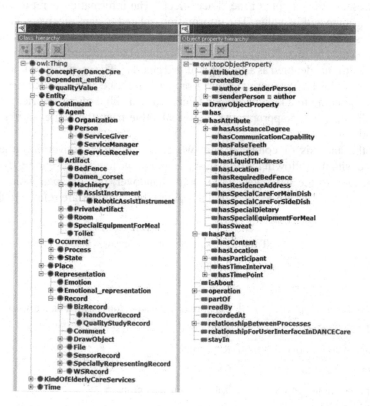

Fig. 1. Is-a hierarchy of class and object property in COTO-ontology.

information using the tools. Then we clarify the concepts which construct the information and construct the is-a hierarchy under the upper level of the ontology. In the COTO-ontology, the concepts are treated as classes and the collected data are treated as instances. The COTO-ontology is described using Web Ontology Language (OWL) [22] and Protégé as an ontology editor [23]. We use OWL Lite as a fragment of OWL because the COTO-ontology is only used for schema of COTO DB for interoperable data access. In this project, it is important for sharing data among the various facilities in elderly nursing care domain but the facilities do not want the automatic analysis discovery of knowledge by ontology reasoning. The statistical information of COTO-ontology is as followed, the number of classes is 214, the number of object properties is 61, and the number of data properties is 72.

4.2 Details of COTO-Ontology

Information Collected by DANCE. As described above, DANCE collects information of two types.

- Face sheet
- Handover information

Table 1 presents an example of the "Face sheet." The information is represented by a set of an attribute and its value. The information, related to care receivers, is classified to two types from the perspective of quality value. The first type includes attributes with value represented by text. For example, "two" is a kind of attribute value corresponding to a kind of attribute denoted as a "number of bed guards." The second type is an attribute for which the value is a categorical value. For example, "rice gruel" is a kind of attribute value corresponding to a kind of attribute denoted as a "kind of cooked rice." The user of DANCE chooses the appropriate categorical value from the value set that is fixed beforehand.

After the analysis of concrete data, we clarify that "Handover information" is constructed with the following information: sender of the handover, sending date, a care receiver related to the handover, topic of the handover, photographs, voice memos, and a text message. A care receiver and a topic of the handover are related to the "face

Table 1. Examples of face sheet

Attribute	Attributevalue
Number of bed guards	Two
Transferring support	Supportedby three persons (with a bath towel)
Main care giver	Awife of the first-born son
Kindof cooked rice	Rice gruel
Stateof side dish	Chopped
Special equipmentfor mea	Table spoon and apron
The others	Three bottles of milk drink in a refrigerator

sheet." The photographs, voice memos, and text representation can be reused at a conference for review of the daily work. DRAW can refer and reuse the information if such information is identified uniquely.

Definition of Handover Record in COTO-Ontology. Figure 2 presents a definition of a handover record in COTO-ontology. "HandoverRecord" is a subclass of "BizRecord," which denotes a record for business. "BizRecord" is a subclass of "Record" and "Representation". The "Representation" is the same as "representation" in YAMATO [21]. "HandoverRecord" is related to a service receiver and occurrent as contents of the record. The "ServiceReceiver" denotes an elderly person who received care services. The "Occurrent" denotes an event such as falling down during transfer from a bed to a wheelchair. "Face sheet" is defined as a set of attribute of a service receiver. "HandoverRecord" represents the information as content of the record. Similarly, "handover information" is defined as occurrent, which is represented in text form as a "HandoverRecord." "SendDateTime" is defined as a metadata of the handover record. The distinction between content and metadata are helpful when a system user analyzes the collected data. System can identify that the content of the handover record in DANCE and the content of the record in Quality Study are the same.

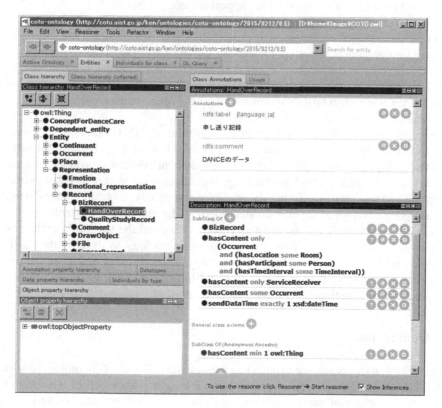

Fig. 2. Definition of handover record in COTO-ontology.

4.3 Use-Case Scenario

We provide one use-case scenario [7]. Care workers take the handover records by DANCE tool in daily work. One day, a care worker realizes there is a risk which makes an elderly person fall in a room. The care worker gathers other care workers pertain to the elderly and then they discuss risk management with referring to DANCE records. The result of the discussion is described by DRAW tool. Finally, the realized risk is extinguished because of rearrangement of the room. The data of DRAW are related to the record of DANCE by COTO-ontology, so another care worker who did not attend the discussion can access the DANCE record from DRAW data. The relation of data expresses the reason why the care worker rearranged the room.

5   Future Work

The 1st future challenge is to support the querying process. The user of COTO DB will be a care worker in the elderly nursing home. A user will use the COTO DB at a conference for reviewing and improving the daily work. However, such users are not familiar with the SPARQL query. For appropriate querying, they should know the structure of the COTO-ontology. It is important to support the query process.

The 2nd challenge is to decrease the time for querying. The care workers cannot wait for a long time. Therefore, the querying process should be done in short time. Decreasing the time for querying is important for smooth use of COTO DB at the conference. One possible approach is the ontology based data access (OBDA) which is a system which translates SPARQL queries to SQL queries over the original data source on the fly. Especially, some virtual OBDA systems, such as the open source project Ontop [26] or the commercial system Stardog [27], might be useful.

The 3rd future challenge is to identify the coverage of COTO-ontology and to evaluate it. It is difficult to identify the coverage of COTO-ontology because the aim of COTO-ontology is to interoperate the information about events and processes occurred in every service domain. The third challenge is to develop the methodology of ontology extension and management. To use the ontology sustainably, it is important to identify the necessary coverage at the domain and to extend the ontology consistently.

The 4th challenge is to reuse existing resources such as the top-level ontology and domain ontology. We already use some top-level ontology to develop the COTO-ontology. It is difficult to identify necessary parts of the ontology and reuse the parts. There are some useful domain ontologies and linked data such as the International Classification of Functioning, Disability and Health [24].

6   Conclusion

As described in this paper, we explain COTO-ontology. The COTO-ontology is a base of COTO DB that enables data interoperability among nursing facilities. Thanks to the ontology, semantics of various systems in each facility will be standardized. In the future, collected data based on the ontology will facilitate the evaluation of robotic devices for nursing in a real setting.

The COTO-ontology will also contribute in service science domain which is growing related to real settings [28]. Moreover, services are attracting attentions of some researchers in ontology engineering [29, 30]. Our proposed paper contributes to both domains.

Acknowledgements. A part of this study was supported by AMED Project to Promote the Development and Introduction of Robotic Devices for Nursing Care. We appreciate Wakoen and momo-chan's earnest support for this study.

References

1. Ministry of Health, Labour and Welfare. Annual Health, Labour and Welfare Report 2012–2013 (2013). http://www.mhlw.go.jp/english/wp/wp-hw7/dl/10e.pdf
2. The Japan Agency for Medical Research and Development (AMED). Robotic Care Devices Portal. http://robotcare.jp/?lang=en
3. Watanabe, K., Fukuda, K., Nishimura, T.: A technology-assisted design methodology for employee-driven innovation in services. In: Technology Innovation Management Review, pp. 6–14 (2015)
4. Buranarach, M., Thein, Y.M., Supnithi, T.: A community-driven approach to development of an ontology-based application management framework. In: Takeda, H., Qu, Y., Mizoguchi, R., Kitamura, Y. (eds.) JIST 2012. LNCS, vol. 7774, pp. 306–312. Springer, Heidelberg (2013)
5. Nishimura, T., Fukuhara, T., Yamada, K., Hamazaki, M., Nakajima, M., Miwa, H., Watanabe, K., Fukuda, K., Motomura, Y.: Proposal of handover system for care-workers using community intelligence. In: Mochimaru, M., Ueda, K., and Takenaka, T. (Eds), Serviceology for Services - Selected Papers of the 1st International Conference of Serviceology Part V, pp. 135–142 (2013)
6. Miwa, H., Fukuhara, T., Nishimura, T.: Service process visualization in nursing-care service using state transition model. In: Freund, L.E. (ed.), Advances in the Human Side of Service Engineering, pp. 3–12 (2012)
7. Watanabe, K., Nishimura, T.: Employee-driven design activities for services: A case study in elderly-care. In: AHFE International Conference (2015)
8. Watanabe, K., Fukuda, K., Nishimura, T., Motomura, Y., Mochimaru, M.: A strategic approach to implement service design activities and technologies in service industries. In: Proceedings of the 2nd International Conference on Serviceology (ICServ 2014) (2014)
9. Grüninger, M., Fox, M.S.: Methodology for the desig and evaluation of ontologies. In: Workshop on Basic Ontological Issues in Knowledge Sharing, IJCAI 1995 (1995)
10. Mizoguchi, R., Ikeda, M., Seta, K., Vanwelkenhuysen, J.: Ontology for modeling the world from problem solving perspectives. In: Workshop on Basic Ontological Issues in Knowledge Sharing, IJCAI 1995 (1995)
11. Uschold, M., Gruinger, M.: Ontologies: Principles, methods and applications. Knowl. Eng. Rev. **11**(2), 93–136 (1996)
12. López, M., Gómez-Pérez, A., Sierra, P.: Building a chemical ontology using methontology and the ontology design environment. IEEE Intell. Syst. **14**(1), 37–46 (1999)
13. Noy, N.F., McGuinnesss, D.L.: Ontology Development 101: A Guide to Creating Your First Ontology (2001). http://protege.stanford.edu/publications/ontology_development/ontology 101.pdf

14. Miwa, H., Fukuhara, T., Nishimura, T.: Service process visualization in nursing-care service using state transition model. Advances in the Human Side of Service Engineering, pp. 3–12. CRC Press, Boca Raton (2012)
15. W3C Working Group: RDF 1.1 Primer. http://www.w3.org/TR/2014/NOTE-rdf11-primer-20140624/
16. W3C Working Group: SPARQL 1.1 Overview. http://www.w3.org/TR/2013/REC-sparql11-overview-20130321/
17. Kozaki, K., Yamagata, Y., Kou, H., Imai, T., Ohe, K., Mizoguchi, R.: Publishing linked open data from a disease ontology toward a knowledge infrastructure. Trans. Japan. Soc. Artif. Intell. **29**(4), 396–405 (2014). (in Japanese)
18. Hamasaki, M.: Proposal of SPARQL query sharing system for linked open data. In: Proceedings of The 29th Annual Conference of the Japanese Society for Artificial Intelligence, 1G3-OS-08b-2 (2015). (in Japanese)
19. BFO Basic Formal Ontology. http://ifomis.uni-saarland.de/bfo/
20. DOLCE a Descriptive Ontology for Linguistic and Cognitive Engineering. http://www.loa.istc.cnr.it/old/DOLCE.html
21. Mizoguchi, R.: YAMATO: Yet another more advanced top-level ontology. In: Proceedings of the Sixth Australasian Ontology Workshop, pp. 1–16 (2010)
22. W3C OWL Working Group: OWL 2 Web Ontology Language Document Overview (Second Edition). http://www.w3.org/TR/2012/REC-owl2-overview-20121211/
23. Stanford University: protégé. http://protege.stanford.edu/
24. World Health Organization: International Classification of Functioning, Disability and Health (ICF). http://www.who.int/classifications/icf/en/
25. Fukuda, K., Watanabe, K., Fukuhara, T., Hamasaki, M., Fujii, R., Horita, M., Nishimura, T.: Text-mining of hand-over notes for care-workers in real operation. In: Meiselwitz, G. (ed.) SCSM 2015. LNCS, vol. 9182, pp. 30–38. Springer, Heidelberg (2015)
26. Ontop. http://ontop.inf.unibz.it/
27. Stardog. http://stardog.com/
28. Spohrer, J., Maglio, P.P.: The emergence of service science: Toward systematic service innovations to accelerate co-creation of value. Prod. Oper. Manage. **17**(3), 238–246 (2008)
29. Nardi, J.C., Falbo, R.A., Almeida, J.P.A., Guizzardi, G., Pires, L.F., Sinderen, M.J., Guarino, N., Fonseca, C.M.: A commitment-based reference ontology for services. J. Inf. Syst. **54**, 263–288 (2015)
30. Sumita, K., Kitamura, Y., Sasajima, M., Mizoguchi, R.: Are services functions? In: Snene, M. (ed.) IESS 2012. LNBIP, vol. 103, pp. 58–72. Springer, Heidelberg (2012)

Efficiently Finding Paths Between Classes to Build a SPARQL Query for Life-Science Databases

Atsuko Yamaguchi[1]($\boxtimes$), Kouji Kozaki[2], Kai Lenz[3], Hongyan Wu[1],
Yasunori Yamamoto[1], and Norio Kobayashi[3]

[1] Database Center for Life Science (DBCLS),
Research Organization of Information and Systems,
178-4-4 Wakashiba, Kashiwa, Chiba 277-0871, Japan
{atsuko,wu,yy}@dbcls.rois.ac.jp
[2] The Institute of Scientific and Industrial Research (ISIR),
Osaka University, 8-1 Mihogaoka, Osaka, Ibaraki 567-0047, Japan
kozaki@ei.sanken.osaka-u.ac.jp
[3] Advanced Center for Computing and Communication (ACCC),
RIKEN, 2-1 Hirosawa, Wako, Saitama 51-0198, Japan
{kai.lenz,norio.kobayashi}@riken.jp

Abstract. Many databases in life science are provided in Resource
Description Framework (RDF) model with SPARQL Protocol and RDF
Query Language (SPARQL) endpoints. However, it may be difficult for
users who are not familiar with Semantic Web technologies to write a
SPARQL query. Therefore, assisting users to build SPARQL queries is
important task to expand the range of users of RDF databases. We developed a web application called SPARQL Builder (http://sparqlbuilder.
org/) that enables users to access life-science RDF datasets by assisting
them in writing SPARQL queries. One of the key technologies used in
SPARQL Builder is to extract possible relationships in an RDF dataset
between two classes of input and output data. We express such relationships by paths on a labeled graph called class graph representing
class–predicate–class relations in a dataset. In addition, we present an
efficient algorithm to compute all the possible paths between two classes
on a class graph. To show the performance of the proposed algorithm,
we compared our algorithm with a naive method using RDF datasets of
various class sizes and confirmed that our algorithm runs much faster
when the numbers of classes and relations are relatively large.

Keywords: Semantic web · SPARQL · Intelligent query generation ·
Database integration · Life-science databases

1 Introduction

Semantic processing of a vast number of data published on web, such as semantic
search, semantic integration of databases, intelligent analysis of semantic data,

© Springer International Publishing Switzerland 2016
G. Qi et al. (Eds.): JIST 2015, LNCS 9544, pp. 321–330, 2016.
DOI: 10.1007/978-3-319-31676-5_24

is an important technique for efficient use of various data and related knowledge. Semantic web technologies give foundations for semantic processing. Linked Data is particularly a well-designed best practice for publishing open data on the web in machine readable ways based on semantic web technologies [1]. The Linked Open Data cloud diagram shows that many kinds of datasets are published on the web as Linked Open Data in many domains. These datasets are linked each other through semantic relations (named properties). As the result, these datasets could form a global database on the web.

In Linked Data, datasets are technically represented in graph structures based on the Resource Description Framework (RDF). That is, they are represented by resources and properties which correspond to nodes and links respectively in graphs. Currently, many important life-science databases also provide their data in RDF model. For example, UniProt [2], the largest protein database, has employed an RDF data model since 2008 to handle the many interlinks to various existing databases. Around the same time, the Bio2RDF project [3] published Linked Data originally generated from numerous major biological databases in RDF. In October 2013, an RDF platform [4] was made publicly available by the European Bioinformatics Institute (EBI). Through this platform, users can access the RDF data of six EBI databases.

These databases can be retrieved by SPARQL Protocol and RDF Query Language (SPARQL) according to graph patterns which represent semantic meanings of data. However, it is intractable to construct a SPARQL query for biologists who are unfamiliar with programming languages and RDF data schema; such difficulty is involved not only a simple difficulty to write a program but also a user need to know the graph structure of the datasets to be searched and construct suitable graph patterns as queries. This is why the authors developed an intelligent tool named SPARQL Builder that assists biological researchers in building SPARQL queries [5] without giving them such difficulties.

A key technology of SPARQL Builder is to find appropriate paths between classes because they are necessary for building SPARQL queries to obtain sets of data with related data through semantic relationships. Such queries are used to gather related data according to meanings of their relations. It is an important foundation for semantic processing using Linked Data. In practice, since the number of classes in these databases is very large especially in life science domains, it is important to find paths between classes efficiently. This paper discusses how to find these paths in the SPARQL Builder with some evaluation experiments. We suppose that the proposed method can be applicable to other domains and related semantic technologies as a fundamental technique for semantic processing using RDF and SPARQL.

The rest of this paper is organized as follows: Sect. 2 shows related work of this paper. Section 3 introduce SPARQL Builder developed for assisting users in writing SPARQL queries. In Sect. 4, we explain a class graph used in the path finder module of SPARQL Builder. We propose an efficient algorithm for finding possible paths of a class graph in Sect. 5 and show computational experimental result for our algorithm in Sect. 6.

2   Related Work

Many SPARQL based search tools and methods that allow users to access their interesting RDF resources have been proposed to date. The most popular method for instance searching is faceted search. For example, Ferré *et al.* proposed query-based faceted search (QFS) as a navigational support tool for faceted searching by logical information system query language (LISQL) [6]. QFS employs a SPARQL endpoint, enabling searching of large datasets [7]. Ferré *et al.* also developed a web based tool named Sparklis, which supports complex queries and exploratory searching for SPARQL endpoints. The tool presents users with lists of classes in its target endpoints and allows users to make queries through faceted based graphical user interfaces (GUIs). Although the tool presents queries in a logical language format, interactive GUIs for building SPARQL queries are provided by other systems such as NITELIGHT [8], iSPARQL[1] and RDF-GL [9].

Popov proposed an exploratory search called Multi-Pivot [10]. This search method extracts concepts and relationships from the ontologies, visualizes them for semantic searches among instances (data) associated with ontology terms. Kozaki *et al.* [11] also proposed a user-guided divergent ontology exploration tool. Multi-Pivot and Kozaki *et al.*'s tool are good examples of semantic searching for instances based on ontologies as conceptual structures. Following Popov's approach, our proposed SPARQL Builder is designed for quick discovery of possible paths between instances of selected classes. Although some systems are designed to accelerate RDF data retrieval [12,13], the proposed system accelerates only the computation of possible paths among preprocessed metadata. Currently, users can build SPARQL queries for life-science databases using GUI-based support tools, such BioSPARQL [14], which supports paths among instances in selected classes while limiting the target to local data.

From the view point of finding possible paths in graph efficiently, it is one of classic graph problems that is very well studied [15]. However, as far as we know, there is no algorithm proposed for enumerating unsimple paths on a multiedge graph although it is necessary for us to consider such graphs and paths for our goal. As a multiedge graph representation of datasets, Zhang *et al.* presented a method converting an ontology into a multiedge graph [16]. As we wrote in the Sect. 4, class–class relationships we used in this paper is not exactly same as OWL ontology, we defined a new graph representation for the relationships.

3   SPARQL Builder

SPARQL Builder is an intelligent tool by which users with no knowledge of SPARQL can generate SPARQL queries and retrieve results satisfying their requirement. Because the target user is a biological researcher having no knowledge of RDF and SPARQL, a key factor in the design of SPARQL Builder should be an information browsing method for RDF data. It is very difficult for such a user to find the desired data by browsing through RDF data, which tend in

---

[1] http://oat.openlinksw.com/isparql/index.html.

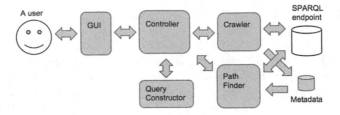

Fig. 1. System overview. The GUI module works on the user's web browser, and other modules work on the server side.

the life sciences to comprise a large and complex network because of their volume and heterogeneity. The user does not know how to find data efficiently by browsing. We accordingly placed limits on the forms of SPARQL queries for connecting data based on relations between two concepts (represented as classes in RDF), corresponding to the use of SPARQL queries mainly for omics data analysis.

There may be several relationships between two classes. For example, if one class contains genes and the other class proteins, there are many relationships between the two classes, such as coding of a gene for a protein, or DNA-protein interaction. In addition, a relationship may be represented by sequentially connected multi-step relations of subject-predicate-object (or object-predicate-subject) in RDF data. Thus, for a user to specify a desired relationship between two classes, the system should display a set of paths including candidate relationships between two classes.

Figure 1 shows an overview of the architecture of our system. It comprises five main modules. The GUI module includes lists specifying a SPARQL endpoint, two classes called a start and an end class, a panel for displaying a rooted tree generated by possible relationships from a start to an end class, and a text box for showing a SPARQL query generated by the system. The path finder module includes a function to compute possible relationships between a start and an end class. The query constructor module includes a SPARQL query generator to generate a SPARQL query using a relation specified by a user between a start and an end class. The crawler module enables the path finder module to be much faster by extracting metadata from SPARQL endpoints in advance.

A user can access SPARQL Builder at the URL http://sparqlbuilder.org/. As of September 2015, by using SPARQL Builder, a user can write a SPARQL query for 39 life-science RDF datasets.

4   Class Graph

To enumerate the relationships between input and output classes in the path finder module, we employed a specialized graph for our system whose nodes and edges correspond to classes and the class–class relations with predicates, respectively. We call the graph *class graph* formally defined as follows: Given an RDF dataset R, we denote by C the set of all classes in R. A class graph $G_R = (V, E, c, p)$ of R is a directed labeled multigraph defined as follows: V is a $|C|$-sized set of nodes and c is a one-to-one mapping from V to a set of URLs of C. E is a multiset of directed edges between the nodes of V, and p maps E to a set of URLs of predicates in R. To construct E and p from R, we add to E a directed edge e_{pred} from node n_d to n_r, where $c(n_d) = class_d$ and $c(n_r) = class_r$, and define $p(e_{pred}) = pred$ if $pred$ satisfies either of the following two conditions: (1) both the triples "$pred$ rdfs:domain $class_d$" and "$pred$ rdfs:range $class_r$" exist in R for some classes $class_d$ and $class_r$; (2) there exist three triples "sub $pred$ ob", "sub rdf:type $class_d$", "ob rdf:type $class_r$" in R, where sub and ob are resources and $class_d$ and $class_r$ are classes.

Note that class graph is constrcted from RDF dataset, not only from OWL ontology. For example, ideally, the property schema should be derived from rdfs:domain and rdfs:range. In practice, however, the property domains and ranges in RDF datasets may not be defined. Moreover, even when the domain d and range r are defined for a property p, erroneous triples "s p o" may exist for which the s and o classes written to the RDF dataset using rdf:type do not specify d and r.

Given a class graph G_R, we define a *class path* p from a start class *start* to an end class *end* by a sequence $(n_0, e_1, n_1, e_2, \ldots, n_k)$, where the nodes n_i and edges e_i of G_R satisfy the following conditions: (1) $c(n_0) = start$, $c(n_k) = end$, (2) $c(n_i) \neq end$ for any $i \neq k$, (3) e_i is a directed edge from n_{i-1} to n_i or from n_i to n_{i-1}. An edge e_i directed from n_{i-1} to n_i or from n_i to n_{i-1} is called *forward* or *reverse* directed, respectively. The length of a class path $(n_0, e_1, \ldots, n_k)$ is defined as k. To compute possible class paths between two classes in a practical time, the maximum length of class paths is given in advance and is currently set as five. Note that a class path corresponds to a SPARQL query to obtain the instances of an end class from the instances of a start class by relating a sequence of predicates $p(e_i)$. By searching the possible class paths from the start class to the end class, we can obtain the candidates of a SPARQL query that match the user's purpose.

We now explain how a SPARQL query is constructed from a class path $(n_0, e_1, n_1, e_2, \ldots, n_k)$. Basically, because a class path indicates a relationship between a start and an end classes, the WHERE clause of the SPARQL query should include "?s_i $p(e_i)$?o_i" or "?o_i $p(e_i)$?s_i" if the direction of e_i is forward or reverse, respectively. In addition, because s_i and o_i should be restricted to instances of classes $c(n_{i-1})$ and $c(n_i)$, the WHERE clause should also include two triples "?s_i rdf:type $c(n_{i-1})$" and "?o_i rdf:type $c(n_i)$" for any i. Therefore, for a class path $(n_0, e_1, n_1, e_2, \ldots, n_k)$, the following SPARQL query is constructed.

SELECT ?r_m WHERE {
$\quad$?r_0 $p(e_1)$?r_1. (or ?r_1 $p(e_1)$?r_0.)
$\quad$?r_1 $p(e_2)$?r_2. (or ?r_2 $p(e_2)$?r_1.)
$\quad$ ...

$\quad$?r_{k-1} $p(e_k)$?r_k. (or ?r_k $p(e_k)$?r_k.)
$\quad$?r_0 rdf:type $c(n_0)$.

$?r_1$ rdf:type $c(n_1)$.

...

$?r_k$ rdf:type $c(n_k)$

}

5   Method to Find Paths on Class Graph

Because SPARQL Builder is an interactive system using GUI, an efficient algorithm to enumerate all the possible class paths between two classes is required. Because a class graph may be a multigraph with high degree, if we use a breathfirst method on a class graph to compute all the possible class paths with length at most k, computational time explosively becomes large as k increases.

We here introduce an undirected class graph for a class graph. For a class graph $G_R = (V, E, c, p)$, a undirected class graph $G'_R = (V, E', c, p')$ is an undirected labeled graph defined as follows: E' is a set of undirected edges $\{n_i, n_j\}$ between nodes in V such that (n_i, n_j) or (n_j, n_i) exists in E. p' maps E' to a set of collections of pairs $(direction, pred)$, where $direction$ corresponds to a direction of an edge and $pred$ corresponds to a URL, constructed as follows: For each edge $e = (n_i, n_j)$ in E, a pair $((n_i, n_j), p(e))$ is constructed and added to a collection $p'(\{n_i, n_j\})$.

Using an undirected class graph, we present an algorithm to find all the paths between given two classes from a class graph G_R as shown in Algorithm 1. The algorithm first converts a class graph into the undirected class graph according to the definition of an undirected class graph. Then, it finds all the paths between two nodes on the undirected class graph. After that, for each path on the undirected class graph, a set of corresponding class paths is computed by using a set $p'(e)$ for each edge e in the undirected path. Although computational time for converting a class graph into an undirected class graph constantly needs for any k, it may be expected to take much shorter time on finding paths than a method finding paths on a class graph directly if the class graph is large and includes many high-degree multiedges. In the next section, we show the performance of our algorithm comparing to the method finding paths directly on a class graph for various sizes of RDF datasets.

6   Computational Experiment

To show the performance of our proposed algorithm, we compared it with naive algorithm of finding all the paths on G_R directly. We selected three RDF datasets Allie, Reactome and NCBI gene as of December 2014 with various sizes of classes and class–class relations from life-science RDF datasets. RDF dataset of Allie [17], that is a abbreviation-longform dictionary in life science, is available at Allie data portal (http://data.allie.dbcls.jp/). RDF dataset of pathway database Reactome is available at EBI RDF Platform (https://www.ebi.ac.uk/rdf/platform) [4]. RDF dataset of gene database NCBI gene is available at the site of Bio2RDF [3] Release 3 (http://download.bio2rdf.org/release/3/release.html). Detailed information of these datasets are shown in Table 1.

Algorithm 1. Algorithm for finding all the paths between two classes.

Input: A class graph $G_R = (V, E, c, p)$, two nodes n_{start} and n_{end} in V, maximum
    length k of paths.
Output: A set of all the possible class paths with length at most k between n_{start}
    and n_{end}.

Step1 Convert class graph G_R into undirected class graph G'_R:
    For each $e = (n_i, n_j) \in E$, $E' := E' \cup \{\{n_i, n_j\}\}$, $p'(\{n_i, n_j\}) := p'(\{n_i, n_j\}) \cup$
    $\{((n_i, n_j), p(e))\}$.
Step2 Find a set P' of all the paths on G'_R between n_{start} and n_{end} with length at
    most k:
    $P' := \emptyset$,
    $L_0 := \{(n_{start})\}$,
    for $i = 1$ to k,
        for each path $(n_{start}, ..., n_{i-1})$ in L_{i-1}
            for each adjacent node n_i of n_{i-1} on G'_R
                if $n_i = n_{end}$
                    $P' := P' \cup \{(n_{start}, ..., n_{i-1}, n_{end})\}$
                else
                    $L_i := L_i \cup \{(n_{start}, ..., n_{i-1}, n_i)\}$
                endif
            endfor
        endfor
    endfor
Step3 Convert P' into a set P of class paths by computing corresponding paths on
    G_R for each path in P': $P := \emptyset$,
    for each path $(n_0, ..., n_{m-1}, n_m)$ in P'
        for each combination $(e_1, ..., e_m)$ such that e_i corresponds to a pair in
    $p'(\{n_{i-1}, n_i\})$
            $P := P \cup \{(n_0, e_1, n_2, ..., n_{m-1}, e_m, n_m)\}$
        endfor
    endfor
Step4 Return P.

From the three datasets, the class graphs for them were constructed in
advance. We recorded running time to find possible paths between each two
classes using our algorithm and naive algorithm for maximum length $k = 1$
to 5. Algorithms were implemented using Java 1.7 and running times for them
were measured using the function System.currentTimeMillis(). The machine with
CentOS 6.6 and Intel® Xeon® CPU X5675 3.07 GHz was used for our experiment. As for memory, by using -Xmx and -Xms options, we set heap as 20GB.

Figures 2 are results for comparison between naive method and our algorithm.
Because Step 1 and Step 3 in Algorithm 1 are not used in the naive method
because the naive method searches paths on a class graph directly. Therefore,
for RDF datasets with small number of classes such as Allie, the naive method
runs a little faster than our algorithm. However, as the numbers of nodes and

Table 1. Information of datasets used for computational experiment. #cls, #e, and ave(#p) show the number of classes, the number of edges, and the average number of paths between two classes, respectively.

Name	#cls	#e	ave(#p)	SPARQL endpoint
Allie	10	32	902	http://data.allie.dbcls.jp/sparql
Reactome	46	330	8,273	http://www.ebi.ac.uk/rdf/services/reactome/sparql
NCBI-gene	58	1,024	701,478	http://cu.ncbigene.bio2rdf.org/sparql

edges increase, because ratio of the overheads of Step1 and Step3 become small comparing to whole running time, our algorithm runs faster than the naive method, especially for path length $k \geq 4$.

For practical use in the path finder module of SPARQL Builder, because computational time of finding paths for a large class graph is a problem for user interactive system, our proposed algorithm is suitable for the path finder module. Especially, since recent life-science RDF datasets become complicated with various types of bio-medical entities and the depth of RDF graphs becomes deeper, the result that our proposed algorithm enables to discover longer paths using a prevalent computer is practically important.

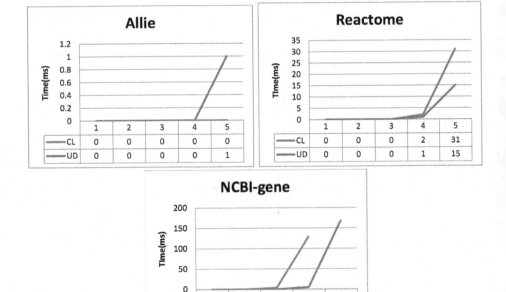

Fig. 2. Experimental result for Allie, Reactome, NCBi-gene. CL and UD indicate the naive method and our algorithm. x-axis and y-axis correspond to the maximum path length and the average of running time [ms] to find all the possible paths for one pair of classes. For NCBI gene, the program of the naive method was stopped with OutOfMemory error when $k = 5$.

7   Conclusion

We presented a key algorithm used in the path finder module of SPARQL Builder which intelligently generates a SPARQL query that searches relationships between two classes. By implementation of newly developed path finder module, SPARQL Builder can display more and longer paths to users in a practical time.

As one of important technologies used in SPARQL Builder which assists users to write a SPARQL query, we introduced the concept of class graph representing class–class relations in RDF datasets. We presented our efficient algorithm, that is implemented in the path finder module, to find possible paths between two nodes by converting a class graph into an undirected class graph, finding paths on the undirected class graph, and taking each undirected unlabeled path into labeled directed paths using a map of the undirected class graph. The performance of our proposed algorithm was shown by computational experiment comparing to a naive method of finding paths by breath-first way on a class graph directly. The experimental result showed that our algorithm is much faster than the naive method for a relatively large class graph, such as Reactome and NCBI gene, and for a relatively large length of paths, such as four and five. Althouhg the naive method runs a little faster than our algorithm only for a smallest dataset Allie, the delay of only 1 ms on average is not a problem for practical use.

Future work should include path ranking method because a very large number of paths may be found as shown in Table 1 because the path finder module can now compute a large number of paths owing to its efficiency. For example, although NCBI gene RDF dataset includes only 58 classes, the average number of class paths between two classes with maximum length five is 701,478. It is very difficult for a user to find desired relationship from these paths. Therefore, path ranking method may be important as the next step of the path finder module of SPARQL Builder.

Although we designed the proposed algorithm as the request of semiautomatic SPARQL query building for life-science RDF datasets, we believe that the idea to find class paths efficiently using a class graph and an undirected class graph can be helpful to other applications.

Acknowledgments. This work was supported by JSPS KAKENHI Grant Number 25280081, 24120002 and the National Bioscience Database Center (NBDC) of the Japan Science and Technology Agency (JST).

References

1. Linked Data. http://www.w3.org/DesignIssues/LinkedData.html
2. The UniProt Consortium: Reorganizing the protein space at the Universal ProteinResource (UniProt). Nucl. Acids Res. **40**(D1), D71–D75 (2012)

3. Belleau, F., Nolin, M.A., Tourigny, N., Rigault, P., Morissette, J.: Bio2RDF: towards a mashup to build bioinformatics knowledge systems. J. Biomed. Inform. **41**(5), 706–716 (2008)

4. Jupp, S., Malone, J., Bolleman, J., Brandizi, M., Davies, M., Garcia, L., Gaulton, A., Gehant, S., Laibe, C., Redaschi, N., Wimalaratne, S.M., Martin, M., Le Novére, N., Parkinson, H., Birney, E., Jenkinson, A.M.: The EBI RDF platform: linked open data for the life sciences. Bioinform. **30**(9), 1338–1339 (2014)

5. Yamaguchi, A., Kozaki, K., Lenz, K., Wu, H., Kobayashi, N.: An Intelligent SPARQL Query Builder for Exploration of Various Life-science Databases. In: CEUR Workshop Proceedings 1279 of the 3rd International Workshop on Intelligent Exploration of Semantic Data (IESD 2014), Riva del Garda, Italy.(1279)

6. Ferré, S., Hermann, A.: Reconciling faceted search and query languages for the semantic web. IJMSO **7**(1), 37–54 (2012)

7. Guyonvarch, J., Ferré, S.: Scalewelis: a scalable query-based faceted search elena work. Multilingual Question Answering over Linked Data (QALD-3), Valencia, Spain (2013)

8. Russell, A., Smart, P.R., Braines, D., Shadbolt, N.R.: NITELIGHT: a graphical tool for semantic query construction. In: Semantic Web User Interaction Workshop (SWUI 2008), Florence, Italy (2008)

9. Hogenboom, F., Milea, V., Frasincar, F., Kaymak, U.: RDF-GL: A SPARQL-based graphical query language for RDF. In: Chbeir, R., Badr, Y., Abraham, A., Hassanien, A.-E. (eds.) Emergent Web Intelligence: Advanced Information Retrieval, pp. 87–116. Springer, London (2010)

10. Popov, I.O., Schraefel, M.C., Hall, W., Shadbolt, N.: Connecting the dots: a multipivot approach to data exploration. In: Aroyo, L., Welty, C., Alani, H., Taylor, J., Bernstein, A., Kagal, L., Noy, N., Blomqvist, E. (eds.) ISWC 2011, Part I. LNCS, vol. 7031, pp. 553–568. Springer, Heidelberg (2011)

11. Kozaki, K., Hirota, T., Mizoguchi, R.: Understanding an ontology through divergent exploration. In: Grobelnik, M., Simperl, E., Parsia, B., Plexousakis, D., Leenheer, P., Pan, J., Antoniou, G. (eds.) ESWC 2011, Part I. LNCS, vol. 6643, pp. 305–320. Springer, Heidelberg (2011)

12. Li, F., Le, W., Duan, S., Kementsietsidis, A.: Scalable keyword search on large RDF data. IEEE Trans. Knowl. Data Eng. **26**, 2774–2788 (2014). doi:10.1109/TKDE.2014.2302294

13. Tran, T., Ladwig, G., Rudolph, S.: Managing structured and semistructured RDF data using structure indexes. IEEE Trans. Knowl. Data Eng. **25**(9), 2076–2089 (2013)

14. Kobayashi, N., Toyoda, T.: BioSPARQL: ontology-based smart building of SPARQL queries for biological linked open data. In: SWAT4LS, pp. 47–49, London, UK (2011)

15. Grossi, R.: Enumeration of paths, cycles, and spanning trees. In: Kao, M.Y. (ed.) Encyclopedia of Algorithms, pp. 1–7. Springer, New York (2015)

16. Zhang, H., Li, Y., Tan, H.B.K.: Measuring design complexity of semantic web ontologies. J. Syst. Softw. **83**, 803–814 (2010)

17. Yamamoto, Y., Yamaguchi, A., Bono, H., Takagi, T.: Allie: a database and a search service of abbreviations and long forms. Database (2011). doi:10.1093/database/bar013

Author Index

Printed in the United States
By Bookmasters